The Problem of Piracy in the Early Modern World

# The Problem of Piracy in the Early Modern World

## *Maritime Predation, Empire, and the Construction of Authority at Sea*

*Edited by*
*John Coakley,*
*C. Nathan Kwan and*
*David Wilson*

Routledge
Taylor & Francis Group
LONDON AND NEW YORK

First published in 2024 by Amsterdam University Press Ltd.

Published 2025 by Routledge
4 Park Square, Milton Park, Abingdon, Oxon OX14 4RN
605 Third Avenue, New York, NY 10158

*Routledge is an imprint of the Taylor & Francis Group, an informa business*



ISBN: 9789463720960 (hbk)
ISBN: 9781041188896 (pbk)
ISBN: 9781003707455 (ebk)
NUR 685

Cover illustration: "Attempt to rescue prisoners at Arosbay, Madura" from *Le second livre, journal ou comptoir, contenant le vray discours et narration historique, du voyage fait par les huit navires d'Amsterdam, au mois de mars l'an 1598, sous la conduite de l'admiral Jaques Cornille Nec, & du vice-admiral Wibrant de Warwic* (1609). Courtesy of the John Carter Brown Library.

Cover design: Coördesign, Leiden

DOI: 10.5117/9789463720960



For Product Safety Concerns and Information please contact our EU representative:
GPSR@taylorandfrancis.com
Taylor & Francis Verlag GmbH, Kaufingerstraße 24, 80331 München, Germany

# Table of Contents

## Section III Representations

# List of Abbreviations Commonly Used in Notes

## Printed Sources

| | |
|---|---|
| AFM | J. O'Donovan (ed.), *Annals of the Kingdom of Ireland by the Four Masters* |
| AC | M. Freeman (ed.), *Annála Connacht: The Annals of Connacht* |
| AMisc | S. Ó hInnse (ed.), *Miscellaneous Irish Annals, AD 1114–1437* |
| AU | W. Hennessey and B. MacCarthy (eds.), *Annála Uladh (Annals of Ulster): A Chronicle of Irish Affairs, AD 431 to AD 1540* |
| CDI | H. S. Sweetman (ed.), *Calendar of Documents Relating to Ireland, 1171–1307* |
| CPR | H. C. Maxwell Lyte (ed.), *Calendar of Patent Rolls: Henry IV, 1399–1413* |
| Cal. Carew MSS | J. S. Brewer (ed.), *Calendar of Carew Manuscripts Preserved in the Archiepiscopal Library at Lambeth,* 6 vols. |
| CSPC | *Calendar of State Papers, Colonial: America and the West Indies*, 45 vols. |
| ES | A. O. Anderson (ed.), *Early Sources of Scottish History, A.D. 500–1286* |

## Manuscript Sources

| | |
|---|---|
| A1895-78 | New York Colonial Council Minutes, New York State Archives, Albany, NY. |
| ADM/L | Navy Board, Lieutenant's Logs, Caird Library National Maritime Museum, UK |
| ARA | Algemeen Rijksarchief Brussel |
| BL | British Library |
| Mass Arch. | Massachusetts Archives Collection, 328 vols. |
| MHS | Massachusetts Historical Society |
| MSA | Massachusetts State Archives |

| | |
|---|---|
| IOR | India Office Records and Private Papers, British Library, London |
| NRAS | National Register of Archives for Scotland |
| NYPL | New York Public Library |
| RB | Rijksarchief Brugge |
| SAA | Stadsarchief Antwerpen |
| SCSJCF | Suffolk County (Massachusetts) Supreme Judicial Court Files |
| TNA | The National Archives, UK |
| CO | Colonial Office Papers, The National Archives, UK |
| ADM | Admiralty Office Papers, The National Archives, UK |
| HCA | High Court Admiralty Papers, The National Archives, UK |
| SP | State Papers, The National Archives UK |
| UBG | Ghent University Library |

# List of Illustrations

**Tables**



**Maps**

# Introduction

*John Coakley, C. Nathan Kwan, David Wilson*

Piracy and the historical study of it has brought many problems for states and scholars alike. In the early modern period, both legal and illegal maritime predation was a common occurrence in seas and oceans across the globe. Piracy in all its forms was, and still is, a worldwide phenomenon. As has been recently shown, too, it is persistent, ebbing and surging in response to political and economic pressures but never dying out completely.[1] Though piracy therefore reaches far beyond Europe historically and geographically, the expansion of European maritime empires in the early modern period exacerbated existing and created new problems of piracy, which states had to navigate and address. At times, European states addressed this problem in different ways according to their resources and interests. They might attempt to contain piracy to certain regions, co-opt maritime raiders for the state's benefit, deny maritime predation or their involvement in it, or suppress predation militarily and legally. As we have written elsewhere, contrary to the popularly held myth of pirates as the common enemies of all peoples, states only exercised power over pirates occasionally and for specific purposes.[2] This present volume extends that argument, deeply examining the relationship between European states and maritime predation, especially in Asian, Atlantic, and European waters between the fifteenth and eighteenth centuries.

In the early modern period, there were many different forms of maritime predation. Most raids were deemed lawful, conducted by either public or private agents of widely recognised states, but a significant minority was considered unlawful and therefore branded "piracy" by some or all of the states. Hans Hägerdal points to the "porous line between state-condoned warfare and sheer piracy," meaning that at times the

1 Amirell and Müller, *Persistent Piracy.*

2 Coakley, Kwan, Wilson, "Piracy and Occasional State Power."

Coakley, J., C. Nathan Kwan and D. Wilson, *The Problem of Piracy in the Early Modern World: Maritime Predation, Empire, and the Construction of Authority at Sea.*
Taylor & Francis Group, 2024
DOI 10.5117/9789463720960_INTRO

only difference between the two was one of perception or interpretation. States therefore eagerly sought to draw clear lines to distinguish between them in law and popular perception in order to justify their own predation and vilify that which diminished their power.[3] Definitions of piracy abounded in early modern law, especially in the nascent field of international law that some European jurists tried to create. But due to competing interests between states, making a universal definition of piracy was easier said than done. Since there were so many ways to be a pirate, and no one could agree at the time what piracy was, one of the first problems scholars come across is a definitional one.[4] This book addresses this fundamental problem of piracy by looking at specific instances of maritime predation, privileging local case studies. Here we hope to find meaning in the detail.

This book presents three connected sections, each dealing with a different facet of the study of maritime predation: first, states' attempts to exercise *jurisdiction* over seafarers and their actions, whether pirates or not; second, the multiple predatory marine *practices* that were at various historical moments considered "piracy"; and finally, the many *representations*—in speech, print, or other means—made about piracy by states or the seafarers themselves. Through the early-modern case studies presented in these sections, the book seeks the meanings and motivations behind piracy in this period. Across the entire book, several themes emerge as common throughlines that are represented in all sections: the relationship between pirates and states; the numerous and overlapping motivations for maritime predation; and, finally, the ways in which certain sea raiders were rhetorically made into pirates. We see that pirates and their practices share similarities with each other, but they defy the broad definitions that states attempted to make. The reasons for individual acts arise from local or regional realities—the multi-imperial political economy of the Caribbean, for example—mixed with wider transoceanic issues often stemming from the metropole, like declarations of war. States and other victims of these acts of predation sometimes crafted media campaigns against raiders, helping to shape popular perceptions of pirates, and inadvertently leaving historians popular but unreliable primary sources. This book reveals, therefore, that while European states attempted to fashion piracy into a global and homogenous phenomenon, it was largely a local and often idiosyncratic issue.

3 Hägerdal, "The Bugis-Makassar Seafarers," p. 119.

4 Coakley, "Little Privateers," pp. 6–26.

## Pirates and States

The question of the relationship between piracy and state power and the extent to which states could exercise jurisdiction over pirates dates back to antiquity. The Roman jurist Cicero famously considered pirates beyond lawful community and thus the "common enemy of all." No agreement or obligations were to be made towards pirates.[5] His beliefs were never widely applied in Europe, however, until states began expanding further overseas in the early-modern period, at which point they enacted and enforced piracy laws to counter hostile sea raiding. However, issues of practicality, as well as cultural and political differences between imperial territories and their metropoles made it difficult to enforce laws against piracy or even determine whether a particular mariner was a pirate. The Treaties of Tordesillas (1494) and Zaragoza (1529), mediated by the pope, divided the world into spheres of influence under Spain and Portugal. Spain considered any foreign ship navigating on its side of the line as piratical.[6] It was in defiance of Portuguese claims over the waters of Southeast Asia that Hugo Grotius made his famed defence of the 1603 Dutch capture of the *Santa Catarina* by arguing for the right to free navigation of the open oceans that all states held. All states then had an obligation to take action against those who infringed on this right, such as pirates.[7] Different European powers thus had divergent understandings of the law in force in the various sea spaces where they were active.

Though Grotius and others helped inaugurate a concept of international law, it was still in its infancy in the seventeenth century. In a misquoted paraphrase of Cicero, these international jurists gave pirates the label of *hostes humani generis*, "common enemies of all mankind." By the late-seventeenth century, the criteria of *hostes humani generis* and *animo furandi* ("private motives") became key components in defining forms of maritime depredation as piracy, but states and theorists alike still did not agree on a common definition of piracy.[8] Pirates were expected to be high seas robbers who did not act on behalf of any recognised sovereign power, but the lines between lawful and unlawful maritime predation remained blurred. Only at the turn of the eighteenth century did a "modern" vision of piracy begin

5 Heller-Roazen, *Enemy of All*, p. 16.

6 Benton, "Legal Spaces of Empire," p. 702. See also Wilson, *Suppressing Piracy*, p. 4.

7 It was in this context that Grotius devised the principle of *mare liberum* ("free seas"). Anand, *Origin*, pp. 77–79.

8 Rubin, *Law of Piracy*, pp. 11, 82–83.

to emerge when apolitically motivated Atlantic pirates began attacking shipping indiscriminately in a region that international law designated as open to navigation by all nations. In doing so, these pirates violated international order and law as defined by Europeans and could thus be considered enemies of all members of the family of nations. Yet, even this was a relatively short-lived phenomenon.[9] Nevertheless, the principle of universal jurisdiction, by which pirates as enemies of all could be tried in any competent tribunal, was popular with English legal theorists at the time, but as Lauren Benton points out, these cases were almost always punished by municipal criminal law and not under the less-reliable statutes of international law.[10] Therefore, despite attempts to universalise piracy in the early modern period, states continued to apply individual piracy laws on a case-by-case basis.

States dealt with piracy to such an extent because doing so aided their expansionist efforts. European imperial expansion extended the reach of European concepts of legality into new territories and their surrounding waters.[11] This gave states more opportunity to exert control over their own and foreign seafarers. As Peter Linebaugh and Marcus Rediker emphasise, the "hydrarchy" of mariners active in the world's oceans was both a potential engine of and a challenge to economic development and expansion in the late-seventeenth and early-eighteenth centuries. States made serious efforts to control the seafarers' hydrarchy to harness their potential as agents of capitalism and imperialism.[12] Suppressing piracy was therefore a key part of the effort, because it threatened the imperial trade and order states sought to impose in colonial waters. As Janice Thomson makes clear, in searching for a monopoly on violence, states needed sovereignty over maritime predation, either by claiming it as their own under the rubric of privateering, or by labelling it piracy and violently rooting it out.[13] They came to control not just European seafarers abroad, too. In extra-European contexts, colonial authorities imposed European understandings of piracy on indigenous and non-European maritime activities and sought to suppress them as such. This tactic was consistently used to undermine the legitimacy of non-European competitors in order to facilitate aggressive campaigns that aimed to extend imperial control

9 Benton, *Search for Sovereignty*, p. 131. See also Anand, *Origin*, pp. 77–89.

10 Rubin, *Law of Piracy*, p. 94; Benton, "New Legal History," p. 231. Mark Chadwick has a somewhat different interpretation, see Chadwick, *Universal Jurisdiction*.

11 Kempe, "Globalized Piracy," p. 354.

12 Linebaugh and Rediker, *Many-Headed Hydra*, pp. 144–45.

13 Thomson, *Mercenaries*, p. 54.

over certain coasts and seas.[14] In this volume, Martin Müller (chapter 2) discusses how the British state benefited by calling indigenous actors "pirates" in Southeast Asia in the early-nineteenth century while Anna Diamantouli (chapter 8) shows that the United States' involvement in North Africa in the eighteenth century relied on claims of piracy and otherness against Muslim corsairs.

It is in this context that the label of piracy became a useful one for dealing with internal and external threats to state-building in imperial spaces. Prosecuting pirates could thus be a means towards an imperial end and sailors sought to avoid stigmatisation as pirates. Simon Layton notes, however, that by the eighteenth century, particularly in Asian waters, accusations of piracy began to be levelled at polities and sovereigns rather than individual seafarers. Suppressing piracy could then be employed to advance an "imperialism of free seas," in which actions against alleged pirates could justify violence and appropriation of territory and maritime space.[15] In these ways, piracy was both a function of and a contributor to state expansion: expansion sparked predation, which resulted in increased piracy claims and suppression efforts, allowing states to take more. However, the long distances and limited capacities of states limited the authority that metropoles could impose on colonies and distant sea spaces, leaving room for much local initiative. Colonies hired sea raiders for defence and trade without state authorisation, individual raiders switched allegiances depending on the availability of raiding commissions, and local courts often failed to convict pirates because they continued to provide useful services the state could not.[16] Despite state ambitions, therefore, they could never prevent the stubbornly local persistent problem of piracy.

## Motivations for Piracy

State actions towards pirates comprises one major theme of this book, but the case studies do not approach the problem of piracy solely from the state perspective; most contributors opt instead to begin their analyses by looking at private seafarers and their actions. In the early modern period,

14 Wilson, "Indigenous Marine Dispossession." On European (mis)understandings of piracy in Asia, see Risso, "Cross-Cultural Perceptions."

15 Layton, "Hydras and Leviathans," pp. 224–25. On pirates' legal self-defence see Benton, *Search for Sovereignty*, p. 116.

16 For example, see John Coakley, "Jamaica's Private Seafarers." On local initiatives to suppress piracy see Wilson, *Suppressing Piracy*.

the word "piracy" was used to describe highly variable and diverse acts, which were intrinsically tied to and shaped by specific maritime contexts, traditions, and events. Practices could range from a single act committed by an individual crew, or several acts committed by several intermingling crews linked to the same communities. In contrast to modern definitions,[17] attacks described as piracy in the early modern period could occur not only on the high seas but also in nearshore waters and coastal expanses, crossing loosely defined and malleable jurisdictional zones. And piratical acts were not bound to the sea, but were borne by the sea, since pirates regularly left their vessels to commit attacks on land. As pirates cruised from one region to another or from one ocean to another, their practices changed and adapted to the maritime contexts in which they pursued their prey.[18] When committing these acts, pirates could be driven by diverse motivations ranging from opportunism to vengeance and from resistance to necessity. Such motivations were not mutually exclusive. As piracy was generally a collective act, there were also multiple overlapping and even contradictory motivations driving any one act.[19] Just as the motivations for one individual act of piracy could be multifaceted and contradictory, so too were the reasons driving regional surges in piracy, in which multiple crews committed several acts of piracy in a particular area.[20]

In their separate studies of the global history of piracy, John L. Anderson and David J. Starkey argue that the primary causal factors of piracy are episodic wars and fluctuating economies. Intermittent wars generate an unstable labour market that in turn causes the initial rise and sustenance of piracy through demand for black-market trading for provisions. These predatory societies are only suppressed when markets become stable, trade

17 The most prominent modern definition of piracy comes from the United Nations Convention on the Law of the Sea (UNCLOS), which defines piracy as any illegal acts of violence or detention, or any act of depredation, committed for private ends by the crew or the passengers of a private ship and directed either (i) on the high seas, against another ship, or against persons or property on board such ship or (ii) against a ship, persons or property in a place outside the jurisdiction of any State. United Nations Convention on the Law of the Sea, December 10, 1982, Part VII, Article 101.

18 See, for example, Andrews, *Trade, Plunder, and Settlement*; Bialuschewski, *Raiders and Natives*; Galvin, *Patterns of Pillage*; Hanna, *Pirate Nests*; Lane, *Pillaging the Empire*; Starkey and McCarthy, "A Persistent Phenomenon."

19 For examples on diverse motivations driving maritime predation see Bahar, "People of the Dawn"; Bromley, "Outlaws at Sea"; Cathcart, "Maritime Dimension"; Hahn, "Atlantic Odyssey"; Jowitt, "Shadow States"; Rediker, "Seaman as Pirate"; Rediker, *Deep Blue Sea*.

20 See Anderson, "Piracy and World History"; Hanna, *Pirate Nests*; Kempe, "Pirate Round"; McDonald, *Pirates, Merchants*; Starkey, "Pirates and Markets."

increases, and piracy becomes a hindrance rather than an advantage to maritime communities.[21] Here, it is the connection between markets, traders, and pirates that explains when, where, and why there are significant surges in piracy. Unemployed and displaced mariners turn to piracy so long as there is market demand for plundered goods, whether to sustain developing or declining markets. Pirates needed ready markets that they could access to sell plundered goods, which they could exchange for supplies that would enable them to continue their activities. Traders provided these supplies at inflated prices in exchange for plundered commodities that they otherwise had little or no access to. With episodic wars and volatile employment, and where ready markets for plunder could be found within local or regional economies, sustained acts of piracy continued.[22] In the early modern period, maritime conflict and trade cannot be separated from colonial expansion and mercantilism, which both created and exacerbated the conditions for maritime predation between competing powers.[23]

The rise and perpetuation of sustained episodes of piracy (or so-called piracy) cannot be separated from these tangled geopolitical and market conditions, as they played out in individual locales. While metropolitan issues, such as European declarations of war or sudden trade declines, created the conditions that could *encourage* a surge in piracy, this was not inevitable and did not lead to widespread piracy from all ports and places impacted by these issues as they reverberated across the oceans. As Benton remarks, "all piracy politics is local *and* regional."[24] Several of the case studies in this volume demonstrate these local motivations, showing that spikes in maritime predations occurred when broader geopolitical and market fluctuations mixed with highly localised contexts and conflicts. Simon Egan (chapter 3) reveals a constellation of regional issues combined with climate and political changes driving maritime predation in the late medieval Irish Sea. As Coakley shows in chapter 1, Henry Morgan's raids from Jamaica in the 1660s resulted from concerns over nearby Spanish depredations and the popularity of plundered goods in island markets. Similarly, Steven J. Pitt demonstrates in chapter 4 that Boston's turn toward piracy suppression was tied to the local logwood market and the actions of specific raiders. It was how metropolitan events played out in distinctive

21 Anderson, "Piracy"; Starkey, "Markets."

22 Bialuschewski, "Pirates, Markets and Imperial Authority"; Zahedieh, "Trade, Plunder, and Economic Development"; Zahedieh, "Merchants of Port Royal."

23 Benton, "Legal Spaces of Empire"; Benton, *Search for Sovereignty*; Hanna, *Pirate Nests*; Stein and Stein, *Silver, Trade, and War*; Steinberg, *Social Construction*, ch. 3.

24 Benton, "Pirate Passages," p. 276.

localities that stimulated the fluctuation in mariners committing acts of piracy in those colonial spaces, often as a reaction to broader factors like displacement, opportunity, or reprisal.

## Making Pirates

Local material conditions form a consistent throughline in this volume, but equally important are the more diffuse cultural issues surrounding the stories told about pirates. To their benefit or detriment, maritime predators could not escape their reputations; at times, some were lauded as heroes, but they could also find themselves rhetorically made into pirates, regardless of the legality of their actions. Just as piracy as a crime defies neat legal definitions and as piracy as a practice defies simple description, representations of "the pirate" as both an individual and a collective defies consistent categorisation. When an individual, group, or community is labelled as being piratical, this cannot be detached from the underlying motives of those who apply the label.[25] For example, beyond just legally prosecuting pirates, state actors sometimes crafted media campaigns against individuals, such as Henry Every or William Kidd. These stories competed in a saturated market with laudatory tales. The tone and content of narratives spread about them could be important to the raiders themselves; Henry Morgan successfully sued for libel against the publisher of a negative account of the sack of Panama.[26]

Untold numbers of seafarers have committed diverse acts of maritime predation and yet only certain individuals and groups dominate the popular imagination as pirates. Public campaigns and popular media have both deliberately and inadvertently constructed dominant piratical myths. Such representations blend fact with fiction by mixing historical events with false suppositions or by simplifying real life figures to fit contemporary desires. Such distorted representations of pirates and piracy also appear when reading the first-hand accounts of maritime predation across diverse historical records not intended for public consumption. In these documents, singular piratical acts are represented in strikingly different ways depending

25 See Bahar, *Indians and Empires*; Elliot, "Pirates, Polities and Companies"; Elliot, "The Politics of Capture"; Kwan, "Piracy and Occasional Interstate Power"; Kwan, "Barbarian Ships"; Layton, "Moghul's Admiral"; Lipman, *Saltwater Frontier*; Risso, "Cross-Cultural Perceptions"; Sicking, "Pirate and the Admiral."

26 Lincoln, "Henry Every"; Ritchie, *Captain Kidd*; Gibbs, "A Certain False."

on the author's perspective. Certain ideas then circulate in the contemporary media, depending on cultural desires for particular stories, often salacious, adventurous, heroic, or damning. Regardless of accuracy, these stories are then refashioned and repurposed across different popular forums over subsequent decades and centuries, coming to shape present-day popular images of pirates. Dominant representations of "piracy" and especially particular "pirates" are also determined by local and regional connections to distinctive individuals and groups and the local histories and myths that have emerged therein.[27] Across these representations, the ways that those dubbed pirates represent themselves is often lost as individual crewmembers rarely had the chance to shape the dominant narrative and, instead, were reduced to the nameless, heterogenous, and largely unheard followers of leading figures or communities.[28] In this volume, Wim de Winter (chapter 5) shows groups of Southern-Netherlandish sailors traversing the globe committing acts of piracy, yet always perceiving and portraying themselves as legitimate traders and privateers. James Rankine (chapter 7) also discusses raiders' self-perception, focusing on a single crew member forced to join a pirate vessel and never committed to the lifestyle.

Often relying on the same limited source material as popular representations of piracy, the historical scholarship has also been guilty of homogenising representations of pirates. This has further added to our simplification of who was a pirate and what a pirate was. This has led most predominantly to representations of pirates as groups existing outside of society, whether as a result of their isolation from landed society and the proactive pursuit of pirates by monolithic state forces or, alternatively, as a choice made by individual pirates to escape societal norms and create alternative social orders on board vessels or in island and coastal outposts.[29] While such perspectives have provided important insights into shipboard life and labour conditions in the maritime world more generally, the similarities between distinctive crews, crewmembers, and contexts are often exaggerated using limited and often problematic evidence while differences are either smoothed over cursorily or left largely unexplored. This has changed in the past two decades as historians have delved into the complexities of

27 See Burwick and Powell, *British Pirates*; Eastman, "Blood and Lust"; Jowitt, *Culture of Piracy*; Lincoln, "Henry Every"; Lincoln, *British Pirates*; Macfarlane, "Pirates and Publicity"; Rennie, *Treasure Neverland*.

28 Hanna, "Well Behaved Pirates."

29 Examples include Bromley "Outlaws"; Kinkor, "Black Men"; Rediker, "Hydrachy and Libertalia"; Rediker, "Seaman as Pirate"; Rediker, *Deep Blue Sea*; Rediker, "Under the Banner"; Rediker, *Villains of All Nations*.

piracy as a label as well as the sustained connections of pirates to landed societies.[30] In chapter 6, Rebecca James critically analyses one of the most popular primary sources, Charles Johnson's *General History of the Pyrates*; James surveys different versions of the text for the ways they discuss or elide the fellowship of pirate groups in the narrative.

## Chapter Summaries

This collection is divided into three thematic sections, each containing chapters covering different chronological and geographical contexts. In the first section of this collection, *Jurisdiction*, the authors discuss the limits of states in exercising jurisdiction over "piracy." Each case study suggests that colonial regimes often struggled to impose European understandings on and exercise control over the problems that maritime predation posed to state control in the seventeenth-century Caribbean and nineteenth-century insular Southeast Asia.

John Coakley (chapter 1) discusses the limits of metropolitan authority over the newly established English colony of Jamaica from the 1650s to 1670s. He argues that England's jurisdictional claims over its far-flung empire were tenuous and limited. London nominally administered Jamaica but lacked the capacity and resources to support the colonial project there. Instead, Coakley shows that the island's affairs were in fact largely in the hands of governors and local authorities who could often act independently of the metropole. Equipped with vaguely worded commissions, governors of Jamaica harnessed the island's seafaring power, employing it against Spanish targets. Mariners with such commissions became known as "privateers." While acting on geopolitical motives, the privateers were principally driven by local concerns. Thus, they continued to attack Spanish shipping and territory despite the negative impacts this might have on England's treaty negotiations with Spain. That maritime predation continued in defiance of restrictions from London revealed the limits of English control over Jamaica. As the example of Henry Morgan's expeditions reveal, maritime and military activity sometimes served imperial interests but were run locally and executed by private individuals. Even after the Treaty of Madrid of 1670 rendered maritime reprisals illegal, English authorities in Jamaica still relied on private seafarers to engage in illicit trade with Spanish territories

30 Appleby, *Women and English Piracy*; Hanna, *Pirate Nests*; Wilson, "Caribbean to Craignish"; Wilson, *Suppressing Piracy*.

in violation of the treaty. Transgressions against metropolitan decrees and international treaties during the first two decades of the English presence in Jamaica reveal the limits of England's control over the island and the extent to which local authorities held *de facto* jurisdiction over Jamaica's private seafarers and maritime affairs.

Moving beyond the Caribbean and the seventeenth century, Martin Müller (chapter 2) examines the piratical label that British and other colonial regimes applied to the activities of "sea nomads" in Southeast Asia in the first half of the nineteenth century and how this characterisation was used as a means of exerting control and jurisdiction over the literal floating population of the region. European observers considered the transient seaborne peoples of insular Southeast Asia to be less advanced and hence prone to activities that approximated piracy in European understandings. Through various, often problematic, explanations, piracy came to be seen as a characteristic of the sea nomads. While the sea nomads were considered piratical, their activities, which often took place close to shore, in rivers, on land, or otherwise outside of the realm of the high seas, did not necessarily fit the definition of piracy in international law. Suppressing piratical peoples, however, helped justify violence in areas outside of the jurisdiction of international maritime law. In the process, maritime groups were de-politicised and de-legitimised, their sphere of activity reterritorialised as imperially controlled space. Colonial regimes sought to assimilate them into sedentary society or to channel their activity away from predation and towards gathering marine products and thus integrating them into a capitalist economy, activities that would inform policies in the seventeenth-century Caribbean as well as nineteenth-century Southeast Asia.

In the different contexts discussed in this section, state authority dealt with piracy in a variety of ways: relying on local authorities and actors, and seeking to convert seafarers into economic agents in the capitalist system of global trade. In each case, exercising, or at least claiming, jurisdiction over seafarers, often by using the label "pirate" helped states extend control, but in all cases, such measures were only partly successful. One of the principal reasons why imposing legal control over pirates around the world was such a challenge was that the practices of piracy, discussed in the next section, were diverse and localised, not conforming to legal definitions and expectations.

In the second section of the volume, *Practices*, each chapter charts the complex and often deeply localised surges of piracy, providing multiple reasons why individuals turned to piracy, how they operated, and where they concentrated. They demonstrate that piracy regularly transcended the localities in which such practices initially emerged, both over the short- and long-term, whether by following the same prey to new localities, by adapting practices to new circumstances and waters, or by positioning their practices as lawful and legitimate compared to equivalent practices being committed by competitors.

Simon Egan (chapter 3) investigates this process in the late medieval Irish Sea, demonstrating that episodes of piracy (loosely defined) cannot be disentangled from the geopolitical and commercial competition surrounding Ireland and the resources of the Irish Sea. For example, increased herring stocks off Ireland's western coast, prompted by lowered ocean temperatures during the Little Ice Age, led to greater fishing and commercial traffic in Irish waters. Increased fishing opportunities and the resulting commercialisation then enabled and encouraged Irish lords to grow their own fleets to protect their control of these resources while also undercutting the economic base of their Irish and non-Irish rivals. These localised projections of power also paired with intermittent archipelagic and European warfare, in which maritime predation was intrinsically connected to the conflicts between the Irish, Scottish, Welsh, Manx, and English as well as their European rivals, particularly the Norse, French, and Spanish. Multiple groups committed various acts of piracy for varying complex reasons throughout the late medieval period, and such motivations cannot be easily disentangled from the backdrop of warfare, colonisation, and commercialisation over the long term.

Focusing on a shorter fifteen-year period, Steven J. Pitt (chapter 4) examines the role of Boston and the logwood trade in the rise and decline of piracy in the early eighteenth century. Pitt demonstrates the centrality of competition and conflict over the logwood trade in the surge in piracy that occurred in the western Atlantic following the Peace of Utrecht in 1713. This surge is traditionally considered to have resulted from the significant unemployment of mariners following the end of the War of the Spanish Succession, but Pitt offers a more nuanced evaluation of the process through which mariners were driven to piracy instead of other alternatives. Increased British participation in the logwood trade, particularly from Boston, centred on Laguna de Terminos following the declaration of peace, provided alternative employment for seafarers as both sailors and logwood cutters. The sheer volume of participation quickly left the market oversaturated and led to a severe drop in prices alongside famine amongst logwood cutters. The logwood cutters then turned to piracy, prompting retaliation by the Spanish who were already frustrated at the presence of traders and cutters in a region that they claimed fell under their possession. A Spanish assault on Laguna de Terminos ended the British trade and displaced the logwood cutters who turned to piracy once again and provided a significant boost to a population of pirates already gathering at the Bahamas. As Pitt shows, this was not the end of the intrinsic relationship between piracy, logwood, and Boston, as the decline of piracy was also shaped by the re-establishment of a logwood trade centred on the Bay of Honduras, which coincided with renewed employment opportunities for mariners and also to greater policing of these waters by British Royal Navy vessels in the mid-1720s at a time when pirates were operating there in a much reduced capacity.

Wim de Winter (chapter 5), meanwhile, focuses his examination of the practices of piracy on the shipboard perspectives of Southern-Netherlandish Sailors who traversed the North Sea, Indian Ocean, and Pacific Ocean in the eighteenth century. In his examination, De Winter demonstrates that Ostend sailors often engaged in maritime predation but did not consider themselves as pirates or these acts as piracy. Such terms were instead reserved for European and non-European competitors who were portrayed as "heathens," "moors," or "outlaws." In this way those who committed piracy were positioned as "others" whose actions were unlawful and unjustified compared to the actions of Ostenders, who legitimised their own practices as lawful privateering on dubious grounds. Crucially, De Winter discusses that the tactics that Ostenders used were similar to those whom they deemed pirates, especially as Ostenders adapted their practices to regional contexts and geographies. This included operating from bases on islands, small creeks, sand banks, or river deltas and employing a system of passes to justify predation. Likewise, the same local and regional issues that encouraged surges in predation of those groups and communities they deemed pirates, such as famine and market fluctuations, also encouraged Ostenders to turn to piracy as the impact of such issues led to declining opportunities in regional trade. Such similarities in practices did not diminish shipboard perspectives of the legitimacy of their actions compared to other groups who were deemed as illegitimate. Similar to the legal posturing of pirates and the legal ambiguities surrounding piracy, mariners could perceive and portray the same actions concentrated on the same regions using the same tactics as both piratical and not piratical practices.

By focusing on specific case studies covering different seas and groups, the three chapters in this section align with sentiments that geopolitical and market fluctuations shaped piratical practices while emphasising the importance of specific regional and local realities. At the same time, each of these chapters demonstrates the problems of representing predatory practices on the sea as outright piracy. This depended on whose perspectives were emphasised and whose activities were questioned.

It is to these issues that we turn in the third and final section of the book, *Representations*. In spite of the more nuanced representations of piracy that have emerged across the historiography, the representations of pirates as outlaws operating beyond society continues to influence both popular and academic perceptions of piracy. The three chapters in this section each tackle these issues from different perspectives, collectively demonstrating the problems of accepting and advancing the one-dimensional representations of pirates that have been constructed through testimonies, reports, newspapers, literature, plays, film, and historical scholarship.

Focusing on *A General History of the Pyrates* by the pseudonymous Captain Charles Johnson, Rebecca James (chapter 6) provides a closer reading of this infamous text than the countless others who have mined its pages for insights into a pirate world that was often more fantastical than factual. James compares five of the twelve versions of the text produced between 1724 and 1734 to observe the representations of pirates across different editions of the same text. James investigates how descriptions of fellowship amongst pirates alters from edition to edition with varying emphasis on the communality of pirates and individual protagonists. James argues that there is not a stable representation of piracy across the different editions but instead varied portrayals of piracy that were nonetheless simplified and narrativised to meet the needs and expectations of intended audiences. The text takes the complex motivations and feelings of pirates and distils these into erratic depictions of fellowship, community, and treachery that removes most individuals while also elevating certain individuals as archetypes. By focusing on different editions of this famous text, James offers a nuanced and critical reading of a text that has significantly influenced popular representations of piracy over the past three centuries.

Where James focuses on variable representations within different versions of the same text, James Rankine (chapter 7) instead investigates the variable experiences of crewmembers on board the same pirate vessel. Through a close examination of the experiences of Henry Glasby, a member of Bartholomew Roberts' crew between 1720 and 1722, Rankine argues against the representations of individual crewmembers as "monolithic extensions of their commanders" that Johnson and others have constructed. Rankine demonstrates that, in order to survive, Glasby had to assume different roles on board the pirate vessel on which he had been forcibly recruited. After being forced to sign the crew's articles following violent intimidation, Glasby worked as a pirate while actively resisting his reduction to compliant crewmember through attempts to escape and, later as ship's master, by exerting his influence over collective decisions. By focusing on Glasby's experience and the interpersonal networks that he navigated, Rankine convincingly demonstrates that the crew held highly variable attitudes and levels of involvement in piracy, in which the vast majority of the crew were those who were "neither fully committed to lives as criminals nor completely innocent." At the same time, there were a core of hardened veterans committed to piracy who had to carefully manage the internal tensions of a larger crew through violence, intimidation, and surveillance but also by granting privileges, security, and influence. Rankine argues that the representations of piracy that are provided through the rare accounts of rank-and-file crewmembers provides a fuller understanding of the historical realities of piracy than the myths and legends of figures who have dominated the narrative ever since they emerged in print.

Rather than concentrating on a specific text or crew, Anna Diamantouli (chapter 8) instead examines the representation of entire states as "piratical" to reveal how such descriptions were employed in politically and culturally charged attacks that have endured over the long term. Focusing on the "Barbary crisis" (1784–1797), Diamantouli charts the process by which diplomatic correspondence and newspaper articles in the United States began depicting North African corsairing sponsored from Algiers, Tunis, Tripoli, and Morocco as piracy being conducted on behalf of piratical states. This occurred within the context of the larger anxieties felt within the newly established United States, where diplomatic decision-makers struggled to communicate and negotiate effectively with the North African Ottoman states following the loss of protection via British treaties. The inability of the United States to effectively repel corsairing led to attempts to delegitimise the North African states in print, perpetuating a racially charged binary "between evil and good, Muslim and Christian, African and European." Having demonstrated the context in which this binary emerged, Diamantouli argues that this is a narrative that continues today, in which a line is drawn between the actions of "Barbary pirates" and modern-day fundamentalist groups, in order to continue the narrative of a conflict between the United States and a monolithic "Islamic World" as imagined in print. The purposeful and persistent reduction of North African corsairing to "Barbary piracy" demonstrates the intent of the US to deligitimise those Muslim states, and reveals a long history of racial and religious othering.

By interrogating representations of piracy across popular print, newspapers, and shipboard views, these chapters collectively emphasise that the designations of "pirate," "piracy," and "piratical" speaks more profoundly of the perspectives of those who deploy the terms than it does of the actions of those accused or characterised as such. The dominant representations that each of the chapters argue against not only indicate the ways that "pirates" have been simplified and romanticised but also demonstrate that the designation of "piracy" has been employed in targeted campaigns to deride, delegitimise, and propagandise. It is only by closely examining the sources and contexts of these representations—such as on board pirate vessels, during diplomatic crises, or in dominant literary texts—that we can begin to understand why particular representations of piracy emerged and recognise the complex power relations that fed into their construction. Uncritical engagement with such representations will only continue to undermine our understanding of pirates and piracy in the past. Perhaps more importantly, this will also continue to impact our perceptions of and approaches towards those who are deemed "pirates" and "piratical" in the present and future.

## Conclusion

Early modern states did not solve the multiple and intersecting problems of piracy that they largely created. Modern scholars may never fully untangle them either, but the authors collected here address the problems of piracy by carefully analysing the local contexts behind individual cases of maritime predation. They reveal that piracy was a label as much as anything else. States were most concerned with using this label against certain maritime practices that threatened state expansion or complicated their jurisdictional claims to territory and sea space. Legal manoeuvers to make pirates the "enemies of all"—and the crafting of international law in this period—had to do with states seeking control not just of seafarers but of the labels used to describe them. Military campaigns to suppress piracy furthered state's expansionist efforts, and media campaigns against certain individuals were assertions of primacy in a losing battle. Despite claims to the contrary, states did not win these many wars against the pirates, but rather they exerted their power occasionally in attempts to co-opt some seafarers and root out others. The motivations and practices of maritime predation proved too diverse to control entirely. As such, states perhaps never achieved a complete monopoly on violence, but they made significant strides towards a monopoly on *representing* violence by strategically utilising the pirate label. In doing so, persistent popular images of pirates emerged alongside dominant and flawed legal constructions of piracy, giving rise to cultural fascination, ongoing legal questions, and an entire field of study.

## Bibliography


Amirell, Stefan Eklöf and Leos Müller, eds. *Persistent Piracy: Maritime Violence and State-Formation in Global Historical Perspective.* Basingstoke: Palgrave Macmillan, 2014.

Amirell, Stefan Eklöf, Bruce Buchan, and Hans Hägerdal, eds. *Piracy in World History.* Maritime Humanities, 1400–1800. Amsterdam: Amsterdam University Press, 2021.

Anand, Ram P., *Origin and Development of the Law of the Sea.* The Hague: Martinus Nijhoff, 1983.

Andrews, Kenneth R. *Trade, Plunder, and Settlement: Maritime Enterprise and the Genesis of the British Empire, 1480–1630.* Cambridge: Cambridge University Press, 1984.

Antunes, Catia A. P. and Amelia Polonia. *Beyond Empires: Global, Self-Organizing, Cross-Imperial Networks, 1500–1800.* Leiden: Brill, 2016.

Appleby, John C. *Women and English Piracy, 1540–1720: Partners and Victims of Crime.* Woodbridge: Boydell Press, 2013.

Bahar, Matthew R. "People of the Dawn, People of the Door: Indian Pirates and the Violent Theft of an Atlantic World." *Journal of American History*, 101:2 (2014): 401–26.

Bahar, Matthew R. *Storm of the Sea: Indians and Empires in the Atlantic's Age of Sail.* New York: Oxford, 2019.

Benton, Lauren. "Legal Spaces of Empire: Piracy and the Origins of Ocean Regionalism." *Comparative Studies in Society and History* 47, no. 4 (2005): 700–724.

Benton, Lauren. *Search for Sovereignty: Law and Geography in European Empires, 1400–1900.* Cambridge: Cambridge University Press, 2010.

Benton, Lauren. "Toward a New Legal History of Piracy: Maritime Legalities and the Myth of Universal Jurisdiction." *International Journal of Maritime History* 12, no. 1 (2011).

Bialuschewski, Arne. "Pirates, Markets and Imperial Authority: Economic Aspects of Maritime Depredations in the Atlantic World, 1716–1726." *Global Crime* 9, nos. 1–2 (2008): 53–65.

Bialuschewski, Arne. *Raiders and Natives: Cross-Cultural Relations in the Age of Buccaneers.* Athens: University of Georgia Press, 2022.

Burwick, Frederick and Manushag N. Powell. *British Pirates in Print and Performance.* New York: Palgrave Macmillan, 2015.

Cathcart, Alison. "The Maritime Dimension to Plantation in Ulster, ca. 1550–ca. 1600." *Journal of the North Atlantic* 12, SP1 (2019): 95–111.

Chadwick, Mark. *Piracy and the Origins of Universal Jurisdiction: On Stranger Tides?* Leiden: Brill, 2019.

Coakley, John. "'The Piracies of Some Little Privateers': Language, Law and Maritime Violence in the Seventeenth-Century Caribbean." *Britain and the World* 13, no. 1 (2020): 6–26.

Coakley, John, C. Nathan Kwan, and David Wilson. "Introduction: Piracy and Occasional State Power." *The International Journal of Maritime History* 32, no. 3 (2020): 656–65.

Davies, J. D., Alan James, and Gijs Rommelse, eds. *Ideologies of Western Naval Powers, c. 1500–1815.* New York: Routledge, 2019.

Elliot, Derek L. "Pirates, Polities and Companies: Global Politics on the Konkan Littoral, c. 1690–1756." *Working Papers* 136, no. 10 (2010): 1–43.

Elliot, Derek L. "The Politics of Capture in the Eastern Arabian Sea, c. 1700–1750." *International Journal of Maritime History* 25, no. 2 (2013): 187–98.

Galvin, Peter. *Patterns of Pillage: A Geography of Caribbean-Based Piracy in Spanish America, 1536–1718.* New York: Peter Lang, 1999.

Gibbs, Joseph. "'A Certain False, Malicious, Scandalous and Famous Libel': Sir Henry Morgan's Legal Action against a London Publisher of Alexandre Exquemelin, 1685." *International Journal of Maritime History* 30, no. 1 (2018): 3–29.

Hahn, Steven C. "The Atlantic Odyssey of Richard Tookerman: Gentleman of South Carolina, Pirate of Jamaica, and Litigant before the King's Bench." *Early American Studies: An Interdisciplinary Journal* 15, no. 3 (2017): 539–90.

Hanna, Mark G. *Pirate Nests and the Rise of the British Empire, 1570–1740.* Chapel Hill: University of North Carolina Press, 2015.

Head, David, ed. *The Golden Age of Piracy: The Rise, Fall, and Enduring Popularity of Pirates.* Athens: University of Georgia Press, 2018.

Heller-Roazen, Daniel. *The Enemy of All: Piracy and the Law of Nations.* New York: Zone Books, 2009.

Jowitt, Claire. *The Culture of Piracy, 1580–1630: English Literature and Seaborne Crime.* Farnham: Ashgate, 2010.

Kempe, Michael. "'Even in the Remotest Corners of the World': Globalized Piracy and International Law, 1500–1900." *Journal of Global History* 5, no. 3 (2010): 353–72.

Kwan, C. Nathan. "'Barbarian Ships Sail Freely about the Seas': Qing Reactions to the British Suppression of Piracy in South China, 1841–1856." *Asian Review of World Histories* 8, no. 1 (2020): 83–102.

Kwan, C. Nathan. "'Putting Down a Common Enemy': Piracy and Occasional Inter-state Power in South China during the Mid-Nineteenth Century," *International Journal of Maritime History* 32, no. 3 (2020): 697–712.

Lane, Kris E. *Pillaging the Empire: Global Piracy on the High Seas, 1500–1750.* New York: Routledge, 2015.

Layton, Simon. "Hydras and Leviathans in the Indian Ocean World." *International Journal of Maritime History* 25, no. 2 (2013): 213–25.

Layton, Simon. "The 'Moghul's Admiral': Angrian 'Piracy' and the Rise of Bombay." *Journal of Early Modern History* 17 (2013): 75–93.

Lincoln, Margarette. *British Pirates and Society, 1680–1730.* Farnham: Ashgate, 2014.

Linebaugh, Peter and Marcus Rediker. *The Many-Headed Hydra: Sailors, Slaves, Commoners, and the Hidden History of the Revolutionary Atlantic.* Boston: Beacon Press, 2000.

Lipman, Andrew. *The Saltwater Frontier: Indians and the Contest for the American Coast.* New Haven: Yale University Press, 2015.

Macfarlane, C. Alasdair. "Pirates and Publicity: The Making and Unmaking of Early Modern Pirates in English and Scottish Popular Print." *Humanities* 9, no. 1:14 (2020).

Mancall, P. C. and C. Shamma. *Governing the Sea in the Early Modern Era: Essays in Honor of Robert C. Ritchie.* California: Huntington Library, 2015.

McDonald, Kevin P. *Pirates, Merchants, Settlers, and Slaves: Colonial America and the Indo-Atlantic World.* Oakland: University of California Press, 2015.

Pennell, C. R., ed. *Bandits at Sea: A Pirates Reader.* New York: New York University Press, 2001.

Rediker, Marcus. "'Under the Banner of King Death': The Social World of Anglo-American Pirates, 1716 to 1726." *William and Mary Quarterly* 38, no. 2 (1981): 203–27.

Rediker, Marcus. *Between the Devil and the Deep Blue Sea: Merchant Seamen, Pirates, and the Anglo-American Maritime World, 1700–1750.* Cambridge: Cambridge University Press, 1993.

Rediker, Marcus. *Villains of All Nations: Atlantic Pirates in the Golden Age.* Boston: Beacon Press, 2004.

Rennie, Neil. *Treasure Neverland: Real and Imaginary Pirates.* New York: Oxford, 2013.

Risso, Patricia. "Cross-Cultural Perceptions of Piracy: Maritime Violence in the Western Indian Ocean and Persian Gulf Region during a Long Eighteenth Century." *Journal of World History* 12, no. 2 (2001): 293–319.

Ritchie, Robert. *Captain Kidd and the War against the Pirates.* Cambridge, MA: Harvard, 1986.

Rubin, Alfred P. *The Law of Piracy.* Newport, RI: Naval War College Press, 1988.

Sicking, Louis. "The Pirate and the Admiral: Europeanisation and Globalisation of Maritime Conflict Managements." *Journal of the History of International Law,* 20 (2018): 429–47.

Starkey, David J., E. S. Van Eyck Van Heslinga, and J. A. De Moor, eds. *Pirates and Privateers: New Perspectives on the War on Trade in the Eighteenth and Nineteenth Centuries.* Exeter: University of Exeter Press, 1997.

Stein, Stanley J. and Barbara H. Stein. *Silver, Trade, and War: Spain and America in the Making of Early Modern Europe.* Baltimore: Johns Hopkins University Press, 2000.

Steinberg, Philip E. *The Social Construction of the Ocean.* Cambridge: Cambridge University Press, 2001.

Thomson, Janice E. *Mercenaries, Pirates, and Sovereigns: State-Building and Extraterritorial Violence in Early Modern Europe.* Princeton: Princeton University Press, 1994.
United Nations Convention on the Law of the Sea. December 10, 1982. Part VII, Article 101.
Wilson, David. "From the Caribbean to Craignish: Imperial Authority and Piratical Voyages in the Early Eighteenth-Century Atlantic Commons." *Itinerario* 42, no. 3 (2018): 430–60.
Wilson, David. "European Colonisation, Law, and Indigenous Marine Dispossession: Historical Perspectives on the Construction and Entrenchment of Unequal Marine Governance." *Maritime Studies*, 20 (2021): 387–407.
Wilson, David. *Suppressing Piracy in the Early Eighteenth Century: Pirates, Merchants and British Imperial Authority in the Atlantic and Indian Oceans.* Woodbridge: Boydell Press, 2021.
Zahedieh, Nuala. "The Merchants of Port Royal, Jamaica, and the Spanish Contraband Trade, 1655–1692." *The William and Mary Quarterly* 43, no. 4 (1986): 570–93.
Zahedieh, Nuala. "Trade, Plunder, and Economic Development in Early English Jamaica, 1655–89." *Economic History Review* 39, no. 2 (1986): 205–22.

## About the Authors

**John Coakley** is an historian of early America and the Atlantic world, focusing on maritime predation in the Caribbean. He received his Ph.D. from the University of Wisconsin–Madison. He is the author of '"The Piracies of some Little Privateers': Language, Law and Maritime Violence in the Seventeenth-Century Caribbean," *Britain and the World*, 13:1 (2020), 6–26.

**C. Nathan Kwan** teaches at the Education University of Hong Kong. His research focuses on Qing China's maritime relations with the West. He is the author of "'Barbarian Ships Sail Freely about the Seas': Qing Reactions to the British Suppression of Piracy in South China, 1841–1856," *Asian Review of World Histories*, 8 (2020): 83–102.

**David Wilson** is lecturer in maritime history at the University of Strathclyde. His research interests include early modern piracy, maritime law, and coastal communities. He is the author of *Suppressing Piracy in the Early Eighteenth Century: Pirates, Merchants, and British Imperial Authority in the Atlantic and Indian Oceans* (Woodbridge: Boydell & Brewer, 2021).

# Section I

## Jurisdiction

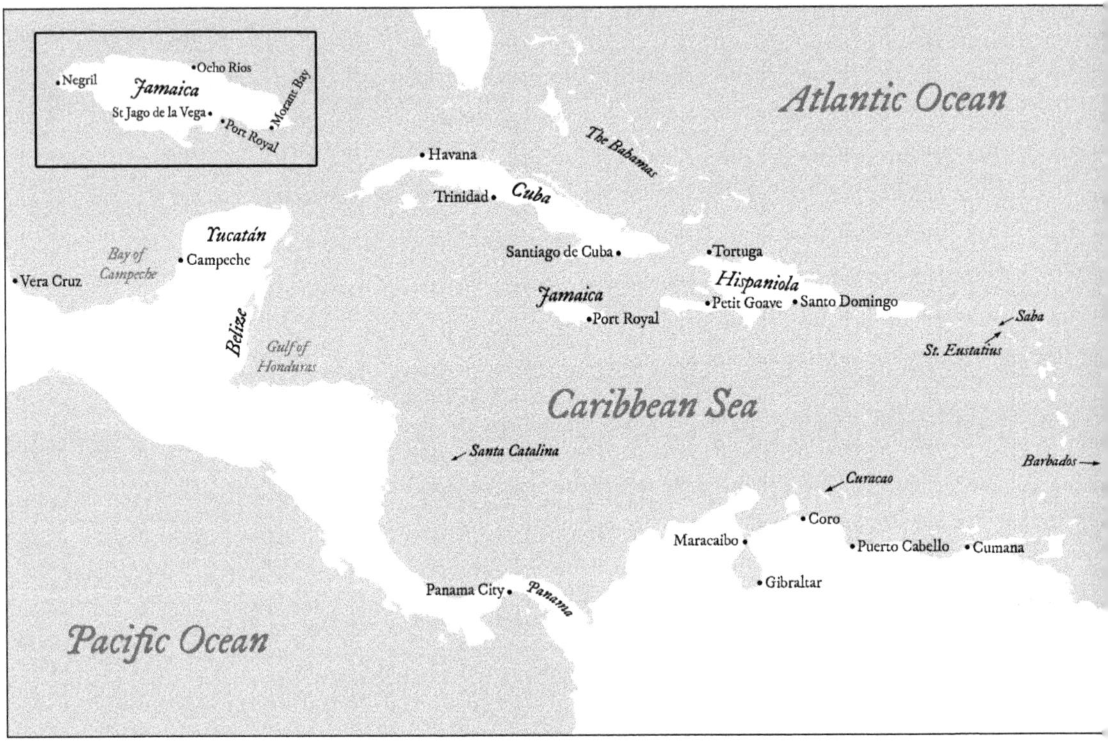
Negril
Ocho Rios
Jamaica
St Jago de la Vega
Port Royal
Morant Bay
Atlantic Ocean
The Bahamas
Havana
Trinidad
Cuba
Yucatán
Bay of Campeche
Campeche
Vera Cruz
Santiago de Cuba
Tortuga
Hispaniola
Jamaica
Petit Goave
Santo Domingo
Port Royal
Saba
St. Eustatius
Belize
Gulf of Honduras
Caribbean Sea
Santa Catalina
Barbados
Curacao
Coro
Maracaibo
Puerto Cabello
Cumana
Gibraltar
Panama City
Panama
Pacific Ocean

# 1. Local Maritime Jurisdiction in the Early English Caribbean

*John Coakley*

**Abstract**
In the seventeenth century, England claimed jurisdiction over sea corridors in the Caribbean, hoping to exercise those claims from Jamaica. The state was forced to rely on local elites who hired privateers for Jamaica's maritime security, thereby making England's maritime jurisdiction a local phenomenon. During the first two decades of the colony, the English state attempted but failed to exert its own control over the Caribbean and get around local actors. The boldest state attempt to subvert colonial jurisdiction was the 1670 Treaty of Madrid, but local conditions won out, and Jamaican governors continued to control maritime affairs. A gradual transition to state control began in 1675 and lasted another twenty years, ending the period of local jurisdiction.

**Keywords:** Jamaica; Privateers; Law; Treaty; Regionalism; Piracy

## Introduction

Early modern European states overflowed with ambition to control maritime space in the Americas and elsewhere, but they mostly lacked the resources to fulfil those dreams. By the seventeenth century, most states with imperial dreams had established colonies in the highly contested close quarters of the Caribbean region, all of which relied on maritime trade and defence for their survival. Owing to treaties and papal bulls dating back to the fifteenth century, however, Spain claimed jurisdiction over nearly all land and sea spaces in the Americas, making subjects of any other states mere interlopers

Coakley, J., C. Nathan Kwan and D. Wilson, *The Problem of Piracy in the Early Modern World: Maritime Predation, Empire, and the Construction of Authority at Sea*.
Taylor & Francis Group, 2024
DOI 10.5117/9789463720960_CH01

in a purported Spanish monopoly zone.[1] England, emerging as a main rival to Spain, aggressively pursued landed colonies and sea-based claims in the 1600s, seeking to puncture that Iberian bubble. As Lauren Benton demonstrates, maritime jurisdictional claims were important for empire building; although some believed in a concept of "free seas," international jurists agreed that a state could exert jurisdiction over sea corridors but could not own the seas.[2] The English state was keen to establish its own sea corridors in the Caribbean despite Spanish dominance, thereby cementing its presence. Only sizable navy fleets policing these corridors and protecting English trading interests could hope to turn the state's maritime ambitions into reality, but it was impossible to commit such a force that far from home during the early-modern period.

In the latter half of the seventeenth century, England pinned its hopes for some form of Caribbean marine control on its newly conquered and centrally located island colony of Jamaica. Begun as a military outpost, the colony was at first temporarily home to a dozen or so navy ships, centred around some mid-size frigates. Later, a handful of smaller state ships remained, but, by the 1660s, always with the assumption that their efforts would be buttressed by more numerous private ships operated by private crews.[3] The growing English empire therefore relied on seafaring private colonists sailing private vessels to control maritime affairs and prop up its ambitious jurisdictional claims. In the case of Jamaica, colonial leaders hired local privateers to ensure the island's maritime security.[4] However, both elite leaders and poor seafarers alike sometimes did not recognise state claims to maritime control or did not always act as the state agents they were assumed

1 By the mid-seventeenth century, the most recent example was the Anglo-Spanish Treaty of Madrid (1630), which allowed for trade between the empires, but only in locations that had enjoyed such commerce before Elizabeth I's war with Spain fifty years earlier. The treaty was written so as to disallow English entry into America, since England had had no legitimate business there prior to Elizabeth's war. Davenport, *European Treaties*, pp. 305–13. European claims to land colonies and sea space permeated popular cartography, shaping Europeans' visual understanding of the sea. See Benton, *Search for Sovereignty*, pp. 105–12. Such cartography was important to the way "the colonial state imagined its dominion." Anderson, *Imagined Communities*, p. 164.

2 Benton shows that, in developing international law, both Alberico Gentili and Hugo Grotius made several provisions for states to claim jurisdiction, but not actual ownership, over sea space. Benton, *Search for Sovereignty*, pp. 125–36; Benton, "Legal History of Piracy."

3 Pestana canvassed all relevant sources for a census of state ships on the station for the decade 1655–1665. She notes that the wind and waters around Jamaica were too dangerous for large ships, so mid-rated frigates and smaller ships prevailed and had to rotate regularly because of damaging conditions. She dates the transition to reliance on private ships to about 1662. Pestana, "Early English Jamaica."

4 Coakley, "Private Seafarers."

to be, and instead exerted a kind of local jurisdiction. Since English claims over Caribbean Sea space exceeded England's ability, the state could not enforce them without private assistance, a situation the local elites exploited. Rather than England's imagined control of Caribbean Sea corridors, in reality Jamaica controlled its own maritime affairs, revealing that maritime jurisdiction was therefore largely a local phenomenon in the region.

Some historians of Caribbean piracy and privateering assert otherwise, essentially reading England's fictional claims as fact. According to this argument, the English state always saw the sea-raiding activities of Jamaican privateers and the unrulier buccaneers as illicit, if not outright illegal. State administrators covertly allowed this predatory behaviour in the Caribbean during the seventeenth century only because it aligned with English state interests to undercut Spain. Imperial administrators therefore declared the region a sort-of free-for-all with the policy of "no peace beyond the line," hoping that the plunder brought in would benefit English colonial projects and hurt the Spanish empire and its purported monopoly. In this line of thinking, it is therefore natural that the English empire began suppressing Caribbean piracy in the early eighteenth century. To do so was simply to enforce maritime jurisdiction England had always held over certain Caribbean Sea corridors but had strategically declined to use. By the 1700s, according to this argument, piracy began to threaten England's cash-crop trading interests, so the royal navy drove out maritime predators and the empire prosecuted them as "enemies of all."[5] Only recently have some historians pushed back against this thesis in the case of Jamaica. Carla Pestana shows that the state navy actively encouraged and worked with private raiders while the colony was new during the 1650s and 1660s.[6] For the next decade or so, as Mark Hanna proves, privateers were integrated into Jamaican society and always found eager buyers and sellers of their goods.[7] In short, rather than choosing to allow local predation, it is clear the English state was unable to suppress it until at least 1700. Even then, as David Wilson

5 This theory has circulated in academic histories since the nineteenth century. Versions of it appear in Dunn, *Sugar and Slaves*; Rediker, *Villains of All Nations*; Ritchie, *Captain Kidd*; Thomson, *Mercenaries*; and more recently in Latimer, *Buccaneers of the Caribbean*. For a fuller discussion of this prevailing view, see Pestana, "Jamaica without Pirates," pp. 321–25. The idea of "no peace beyond the line" traditionally appeared in European treaties allowing peace in Europe but intermittent warfare in colonial spaces. See Gould, "Lines of Plunder." Although Gould sees a continuation of the concept in some form into the nineteenth century, Francis Davenport notes that the 1630 treaty with Spain effectively did away with the idea by making clear that the treaty would be in force beyond the line. Davenport, *European Treaties*, p. 313.

6 Pestana, "Early English Jamaica"; Pestana, *English Conquest*, pp. 184–88.

7 Hanna, *Pirate Nests*.

shows, England could only muster a scattershot approach at suppression in the Caribbean that mostly consisted of offering boilerplate pardons to raiders and deputising colonial actors to hunt down the unredeemed.[8] Depicting most maritime raiders as *hostes humani generis* aided the state in its war against the pirates in the early eighteenth century, but, as Benton points out, it was not consistently applied as a legal tool to prosecute pirates before that point.[9]

This chapter closely examines colonially controlled seafaring and jurisdiction in Jamaica. I argue that, during Jamaica's first twenty years as an English colony (1655–1675), its elite political leaders claimed control over the colony's own maritime affairs, and authorised poor private seafarers to manifest those claims and provide maritime security that the English state could not. Politicians and privateers made and enforced these claims in the name of the colony and not simply as agents of the empire, hence setting themselves in opposition to the state. A tense tripartite legal pluralism emerged, where the established Spanish empire claimed blanket jurisdiction, the growing English empire made strategic inroads into certain waters, but ultimately it was local English colonists that manifested those claims in ways that benefitted themselves. To exert its claimed power and cut out the colonists, England made greater and greater state claims to maritime jurisdiction in the Caribbean through orders to governors and the sweeping 1670 Treaty of Madrid, but still Jamaicans were the only ones who could maintain control over the island's maritime activities. Imperial leaders implicitly recognised Jamaica's local jurisdiction until the late 1680s, when European politics, warfare, and anti-piracy laws helped the state begin to achieve the maritime sovereignty it had long sought.

## Cromwellian Jamaica

England claimed the large, central Caribbean island of Jamaica starting in 1655, won as part of Lord Protector Oliver Cromwell's "Western Design" campaign to wrest the entire West Indies from Spanish hands. The island had been in Spanish hands since the late fifteenth century, though for a time it was owned directly by the Columbus family. Since the early 1500s, Spanish conquerors had destroyed the island's Indigenous Taíno communities, cleared land, set up plantations, built cities (including the recent

8 Wilson, *Suppressing Piracy.*

9 Benton, "Legal History of Piracy."

inland capital of St. Jago de la Vega), and imported livestock and enslaved labourers all in the name of economic and imperial expansion. But since the arrival of English settlers on the periphery of the Caribbean in the 1620s, the Spanish hold on Jamaica weakened. Raider William Jackson—operating with an outdated commission from the Puritan sea-raiding base Providence Island—briefly conquered Jamaica's southern harbour in 1643, signalling to England the tantalising potential of expansion into the central zone of this important maritime space. When Penn and Venables arrived in the region in 1655, they turned their attention to Jamaica after botching their planned invasion of Santo Domingo in nearby Hispaniola. Like Jackson before them, the Western Design leaders landed first at Jamaica's southern harbour (which would later become Port Royal) in May 1655. The soldiers and sailors of the Design, though dying from disease and malnutrition, spread out slowly from there, fighting the retreating Spanish inhabitants and their slaves.[10]

The English state nominally administered the colony as a military outpost, but real authority quickly devolved to a series of local leaders who, at first, embraced the state project of the Western Design: expand English power in the region at the expense of the Spanish. With a cadre of sailors at their disposal, Jamaica's leaders used maritime predation to grow and protect the island and to pursue the Design. The Design's admiral, William Penn, and general, Robert Venables, both left quickly after initial conquest, so their successors bore the burden of settling the struggling colony and interpreting distant orders from the English state. These were not just terrestrial commanders, since Cromwell always made provisions for control of English sea power.[11] With this backing, Jamaica's local leaders authorised, regulated, and participated in, sea raids against Spanish targets. Attempts to take the heavily armed treasure ships of the Spanish plate fleet predictably failed, despite an early promotional pamphlet boasting that "the Privateers from Jamaica are often fingering the Plate."[12] If not directly from passing ships, however, Jamaicans did get their hands on Spanish currency, goods, land, and slaves via several successful sea raids in the first five years of the colony. Under Design leaders, Jamaicans re-took the small nearby island of Tortuga, a former English possession that had been lost to the French in

10 Coakley, "Private Seafarers"; Pestana, *English Conquest*.

11 Colonel William Brayne in 1657 was the first Jamaican commander to assume control of both land and sea forces when his predecessors died or left. When Brayne himself died later the same year, he passed the same dual authority to his successor, Colonel Eward Doyley. See Barber, "Brayne, William"; Capp, "Goodsonn, William."

12 Hickeringill, *Jamaica Viewed*, p. 47.

1640.[13] Thousands of Jamaicans signed onto a series of raids under visiting Commodore Christopher Myngs, who led them to sack Coro, Cumaná, and Puerto Cabello on the Spanish Main and several smaller Cuban targets through 1659.[14]

The man who shouldered most of the responsibility for creating the colony and for its maritime raids was Colonel Edward Doyley, who quickly rose through the ranks simply because he did not succumb to the tropical diseases that killed most of his superiors and subordinates alike.[15] Doyley then survived the 1660 Restoration of the English monarchy, becoming the first royal governor of the island until his replacement arrived in 1662. During his tenure, he shied away from offering raiding commissions to anyone who asked, preferring to keep his soldiers working the land around the harbour, but he did allow some petty subsistence raids, led one of the plate fleet expeditions, and authorised the conquest of Tortuga. Despite authorising maritime predation, Doyley questioned the enterprise. He deeply disliked brash Commodore Myngs—even getting him recalled for "robbing the Merchant and the State" through embezzlement of prize goods in 1659—and grumbled about the soldiers who followed him.[16] But he realised Jamaican colonists were largely on their own in foreign maritime space with only sporadic correspondence from England. Exercising jurisdiction over the colony's maritime security, and therefore maritime predation, was a necessity during the colony's early years. It fell to civilian and military leaders on the island because the state was unequipped to run a colony.

During the Protectorate, a special Committee of the Council of State ran the fledgling Caribbean outpost from the metropole, but the inefficient body was unable to fulfil the colonists' basic needs or control their activities. For example, on their way to Hispaniola and Jamaica, the Western Design had stopped at England's established Barbados colony, hiring men and buying supplies, expecting Cromwell to foot the bill. By late 1655 these bills, and dozens more, faced an overwhelmed committee. By then, Doyley was beseeching

13 Lying just north of Hispaniola, Tortuga had been administered by the English Providence Island Company in the 1630s. Jamaica recaptured it late in 1655 and administered it through Colonel Elias Watts until it again fell under French control. The National Archives, UK (TNA), Colonial Office Papers (CO) 324/1/259–262. "The Condition of the Island of Jamaica at the Lord Windsor's Departure," October 20, 1662; Kupperman, *Providence Island*, ch. 9.

14 Dyer, "Christopher Myngs"; Firth, "Capture of Santiago," pp. 536–40. Pestana notes that these were almost all navy raids, part of Cromwell's project of naval warfare against Spain. Pestana, "Early English Jamaica."

15 See the discussion of Doyley and the death rates in Pestana, *English Conquest*, pp. 157–60.

16 National Library of Jamaica, MS 383. Edward D'Oyley to Commissioner of Admiralty, January 1659.

England to send more food and supplies from throughout the empire (the soldiers particularly asked for peas from New England) to keep his men alive, but the death rate mounted alarmingly.[17] That cost the state money, too; by 1656 the widows of the hundreds of perished soldiers started petitioning the committee for relief funds.[18] Though committee members declared Jamaica the "centre of the interests of his Highness and this Commonwealth in America," in 1658 they knew that managing the colony from across the Atlantic involved a great deal of time and money that the state did not have.[19] Despite the presence of the committee—evidence of an attempt at centralised authority—the state still lacked control over what it deemed its most important colony in the 1650s. This set a precedent for fierce local control of colonial, and especially maritime, affairs that Jamaica maintained for the next decade and a half.

## Restoration Jamaica

Upon officially taking the throne in 1660, King Charles II chose to retain Jamaica and assert maritime access to it, but to also pursue greater diplomacy with Spain, leaving the naval war in the Caribbean in a limbo state. Moreover, since Spain did not recognise any English jurisdictional claims in the Americas, territorial or maritime, all English colonies there were suspended in a murky legality. Pestana notes, "the positions of the two governments remained irreconcilable," and a conflict over sovereignty between the two empires loomed.[20] Centrally located and only recently conquered, Jamaica was the most problematic of England's possessions, especially since its inhabitants comprised the rump of Cromwell's military forces abroad. To deal with this problem, the king de-commissioned the Western Design, but left the soldiers on the island, ordering Doyley to pay them out with a paltry lump sum before recalling the governor, too, in favour of monarchists.[21] In an effort to unify state oversight of all colonies

17 British Library (BL), Egerton MS 2395, 146r. Edward D'Oyley, September 1657; BL, Additional MS 12423. "Coll. Edward D'Oyley's Journal of his proceedings during the time he held chief command in the island of Jamaica."

18 *Calendar of State Papers, Colonial: America and the West Indies* (CSPC), 1574–1660, October 13, October 20, December 18, 1657.

19 BL Egerton MS 2395, fol. 157. Povey et al. to Committee on Jamaica, January 11, 1658.

20 Pestana, "Early English Jamaica," pp. 334–35.

21 The sum was mostly in goods, not hard currency, but one official estimated the value of the goods would have covered only twenty-five weeks of service for the remaining soldiers. Huntington Library, MS BL 319. "Memorandum of the devident," c. 1660.

from the metropole, in the next few years Charles also erected the Lords of Trade and Plantations, a select group of the Privy Council that worked closely with the Secretaries of State on colonial issues, and which gradually morphed into the Board of Trade by the 1700s. At this early stage, however, the Lords mostly made recommendations, gathered data from colonial leaders, and wrote governor's commissions. As such, they prepared boilerplate commissions for the king's choice of leadership for the troublesome island, Thomas Windsor as governor and Edward Lyttleton as lieutenant governor. Both men arrived in the hastily renamed Port Royal harbour in 1662, but soon clamoured to return home lest they succumb to the tropical diseases that still killed much of the settler population; by 1664, both were back in England and the fledgling colony was again rudderless.

During that brief two-year period, the new island leaders did not strictly enforce state policy, instead they leveraged the vague language in their commissions to authorise (and profit from) maritime predation against Spanish targets. The struggling former soldiers and sailors that made up the bulk of the colonial population eagerly signed up for these expeditions, just as they had before. Christopher Myngs even returned to the island, now as an admiral, in 1662 and easily hired 1,300 of the men to loot St. Jago de Cuba, and even more to plunder the Yucatan city of Campeche the next year, a clear sign that England's naval war persisted despite the geo-political changes.[22] A few things looked different now. State interests and local leaders had both shifted, and the two were no longer in sync; the state now sought to maintain but not ignite the simmering tensions with Spain, and local leaders wanted to continue to secure the colony (and line their pockets) through maritime predation. The men who made the raids were the same as before, but because the Design was decommissioned, now they were private actors and only theoretically represented the English state as subjects. A new word—"privateer"—rose to prominence in connection with these Jamaican men and their ships during the early 1660s.[23] Jamaica's war with its Spanish neighbours was not over, and the privateers were the ones fighting it. But they did not fight just for Windsor nor for an imagined glory of England, they plundered for their own security and to ensure the future security of their island home. The currency, goods, and slaves brought in from these expeditions propelled the colony to begin sugar planting, which though a

22 Firth, "Capture of Santiago," p. 9. Beeston, "A Journal Kept by Col. William Beeston," pp. 276–77. Pestana estimates a couple state ships and fourteen private vessels taking part, Pestana, "Early English Jamaica," p. 332.

23 Coakley, "Little Privateers," pp. 6–26.

lucrative business, required a high initial investment; money that no one on the island would have had were it not for the privateers.[24] In Jamaica in the 1660s, maritime predation brought security, so privateers carried on this local enterprise, regardless of the king's reaction.

The king's reaction was mixed. Although privateering served Charles II's interests to some extent, too much of it imperilled ongoing treaty negotiations with Spain. So, in 1663 the king ordered a cessation of "such undertakings as have lately been set on foot."[25] The wording was intentionally vague and only sowed confusion on the island, one observer noting that it was not "well understood" what the king meant. Some surmised that the king simply wanted privateers to update their commissions from Doyley to Windsor and then to Lyttleton, since new commissions would also use the king's name and ensure that 10 per cent of the plundered loot would be owed to royal coffers.[26] There are two reasons for Charles's intentionally vague wording. First, imperial administrators continued to benefit from maritime predation because it brought security and economic boon to a colony that they knew the state could not fully support outright. Second, the state did not have the resources to wrest maritime jurisdiction out of local hands and enforce state policy directly using military or other means. A royal order was a good reminder of the state's jurisdictional claims over maritime activity, but the state's claim did not represent real control. Colonial leaders maintained actual control over colonial maritime predation.

## Modyford's Jamaica

The arrival of a new royal governor in 1664 brought the potential for a greater alignment of interests between the state, the island politicians, and the poor private sailors. Though such alignment did occur at first, those interests diverged further over the next seven years. Thomas Modyford, the king's choice for governor from 1664 to 1671, controlled Jamaica's maritime operations for his own (and the colony's) benefit more than any other colonial or imperial leader in the seventeenth century.

An ambitious man with a large ego, Modyford had already served as governor of Barbados before he came to Jamaica. He made a great deal of money in the sugar boom on the smaller island, but his rivals took the

24 Zahedieh, "Trade, Plunder, and Economic Development."

25 TNA, CO 1/17/23. King to Lyttleton, April 1663.

26 TNA, CO 1/18/110. "Extract of a Letter," 1664?.

governorship from him. Though once a Cromwellian favourite, Modyford had shrewdly maintained royalist connections, including with his relative George Monck, the Duke of Albemarle, who now convinced the king to give Modyford the top job in Jamaica.[27] He arrived in Port Royal in 1664 with Barbadian migrants, royal funds to jump-start the economy, and broad political powers in his commission from the Lords of Trade. He used the latter to oust potential rivals from positions of power, like a young Thomas Lynch, who sneered that Modyford "would have none to shine in this Hemisphere but himself and sons."[28] Modyford promised to bring Jamaica into line with the other trade-based and cash-crop territories, converting the Jamaican economy to a sugar plantation culture. Since he controlled land distribution on sparsely inhabited Jamaica, Modyford shaped the map of the island, patenting over 180,000 acres and allowing rich landowners to consolidate holdings to create vast estates. He and his sons kept over 9,000 acres of plantation land for their own, with more to other members of the Modyford clan.[29] He also oversaw some of the first official imports of enslaved Africans to Jamaica through the Company of Royal Adventurers (the precursor to the Royal African Company), of which he was an agent. Records reveal the arrival of at least 5,000 enslaved Africans to the island during his term as governor, but at least 1,500 others likely arrived illegally because of privateering raids.[30] Modyford personally owned 400 human beings in 1670.

Modyford pursued a bellicose foreign policy in the Caribbean Sea, using the privateers to fight local wars that dwarfed Windsor's, predominantly against Spanish colonies. As the governor saw it, Jamaica was "circled in with enemies countries" and he wanted to force those enemies to avoid making the island a target.[31] Though at first he pursued peaceful diplomacy—hoping to gain Spanish goods and currency through trading rather than raiding—Modyford soon realised that hiring the poor local sailors to prey on those enemies might be a quick way to realise his ambitions. Aside

27 Zahedieh, "Modyford, Sir Thomas."

28 TNA, CO 1/19/23. Lynch to Bennet, February 12, 1665.

29 TNA, CO 138/1 fols. 61–79. "Survey of Jamaica," September 1670; Dunn, *Sugar and Slaves*, 68, 81, 154.

30 This information taken from *Voyages: The Trans-Atlantic Slave Trade Database*, http://www.slavevoyages.org (accessed August 31, 2012). The 1,500 estimate is based on slaves known to have been captured during Edward and Henry Morgan's raids for Modyford. For the practice of privateer slaving, see Heywood and Thornton, *Central Africans*, pp. 20–38.

31 "The governor of Jamaica's answers to the inquiries of his Majesty's commissioners" in *Journals of the Assembly of Jamaica* (Jamaica, 1811), appendix, 23.

from petty raids, which he commissioned liberally, Modyford's first major campaign at sea was another Western Design, this time meant to "make a clear board" in the Caribbean by crushing the French and the Dutch, both of which had recently made war on England.[32] He appointed deputy-governor Edward Morgan to lead the expedition, and they raised over 500 local private men to be the fighting force.[33] The results disappointed Modyford—no land permanently changed hands, plunder was minimal, and Morgan himself died in battle—and mostly they exposed his early inability to manifest his ambitions. This second Design was neither a state-sponsored nor a state-run expedition; it was sponsored by Modyford himself. As governor, he took it upon himself to exert the state's tenuous claims to land and sea in the region, but he did so not simply for state purposes, rather to swell his own and the colony's power in the region. He also had only a tenuous hold over the private sailors acting as his agents in the field, who, though they extended Jamaica's sovereignty into maritime space, also primarily acted for their own benefit.

Maritime predation from Jamaica increased through 1668 as Modyford allowed the privateers to target Spain once again, hoping to make good on his ambitions that way. An expedition under Edward Mansfield in 1666 re-conquered Providence Island, which Spain had taken from England more than two decades earlier. Although Mansfield made his conquest under Modyford's local jurisdiction, because it resulted in the successful capture of new territory, it aligned with imperial interests; King Charles appointed Modyford's younger brother Sir James Modyford to act as governor of Providence Island.[34] This success did not last long—before Sir James could reach Providence, it was already back in Spanish hands—but he settled in Jamaica where his elder brother took care of him, appointing him Lieutenant-Governor of the island, Governor of Port Royal, and sole Admiralty judge. Thomas Modyford neither sought nor received royal approval, he simply informed the king of his brother's new posts.[35] In the defeat at Providence, the loss of the territory was probably not as significant to Modyford and the king as the loss of the key sea route between Jamaica and Providence. The ability to claim jurisdiction over a maritime corridor leading toward the natural resources of the Bay of Honduras would have been a win for the state, the local leaders, and the privateers. All three gained something

32 TNA, CO 1/19/29. Modyford to Arlington, April 1665.

33 TNA, CO 1/19/27. Modyford to Arlington, February 20, 1665.

34 TNA, CO 1/20/166. "Commission appointing Sir James Modyford," November 1666.

35 TNA, CO 140/1/168. Jamaica Council Minutes, February 22, 1666.

nonetheless from Mansfield's outing. The state benefited from actions that destabilised Spain without starting an all-out war; Modyford got to exert his own and the island's power; and the sailors got plunder and the governor's good graces. In that moment, having private sailors attack Spain in the Caribbean served all three interests.

Ever ambitious, Modyford then sought to cement his hold over the sailors and to use them to increase Jamaica's maritime control. Henry Morgan, perhaps the most famous of the privateers, emerged in this time as both a leader of the men and something of a politician himself, acting as an intermediary between the Modyfords and the sailors. Sensing Morgan's ambition and having noticed his skills on Mansfield's outing, Thomas Modyford eventually promoted the young privateer into a new position: Admiral of Jamaica. The creation of this role, for which Modyford again neither sought nor received royal approval, clearly indicates that Jamaica's jurisdiction over its own maritime affairs in this time was complete in the minds of its leaders, and at least tacitly accepted by imperial administrators. Admiral Morgan did not work for the Royal Navy, but he did work for the governor of Jamaica. In this unique position, Morgan made three significant expeditions—all using privately-raised sailors and ships from Jamaica and Tortuga—in 1668 and 1669 that brought in shiploads of new currency, goods, and slaves to the island in a brief economic boom. All had been stolen quite violently from nearby Spanish colonies, some of which lay in ruins, their populations brutalised, from the privateers' ruthless tactics. Retribution from Spanish colonial leaders came swiftly; more English sailors sat in Spanish prisons and rumours abounded of Spanish troops massing nearby.[36]

After Morgan's 1669 attack on Maracaibo and Gibraltar, Modyford declared Jamaica at peace with Spain, believing that Spanish colonies had been appropriately subdued.[37] This also relieved Modyford of pressure from the king, who now had a bigger problem with privateering than he had had in the past. English negotiators had been working on a new treaty with Spain, that by now was close at hand, so tensions in the Caribbean no longer served English state interests. Based on the success of a 1667 commercial treaty that allowed England to trade with Spain in Europe, the new treaty, which Charles II soon hoped to sign, would be a first-of-its-kind Anglo-Spanish peace pact specifically covering actions in the Americas. He and many members of his court even hoped for a "general trade with the Indies"—an

36 See the discussion of Morgan's raids in Coakley, "Private Seafarers," pp. 36–38.

37 "James Modyford to Andrew King, June 29, 1669, Westminster Abbey Muniments 11929" in Thornton, "Modyfords," p. 56.

extension to the Americas of the trade benefits of the earlier treaty—that would allow legal English trade into the closed-off Spanish colonial world.[38] Modyford's ceasefire in the Caribbean provided a welcome impetus to finalise the document, but the tempting détente did not last long.

A year later, Modyford cancelled the peace and got the governor's council to declare a new war on Spain, based on evidence that Spain was now allowing maritime retaliation against English hostilities and direct reports of Spanish attacks on Jamaica's poorly defended north side. The governor ordered Morgan ready for war again; with the promise of more plunder than previous missions, the admiral raised nearly 2,000 private sailors from Port Royal and surrounding English and French colonies. Each man dreamed of amassing his own personal fortune of Spanish silver from the rich cities they would loot on Morgan's big raid. With no target set initially, the admiral's council of war debated such large prizes as Havana before finally settling on the city of Panama itself. The impressive fleet weighed anchor and set off in August of 1670. After first gathering supplies, then fighting their way onto and across the Darien isthmus, by early 1671, Morgan and his now-depleted men reached the Pacific Ocean and Panama, two places they likely never dreamed of seeing. The privateers hunted fruitlessly for the promised mountains of treasure, torturing inhabitants, and even letting the city burn after a blaze started downtown. After a month as de-facto conquerors of this richest of Spanish American cities, the men trekked and floated back to Port Royal, disappointed in the meagre pay-out that was coming, and convinced the admiral was cheating them out of more loot.[39]

The Panama expedition exemplified the near-total control island leaders now had over Jamaica's maritime activities and therefore the maritime zones in the area over which England claimed jurisdiction. Morgan was the lynchpin; through him, Modyford had more control over the privateers than before. But his authority as admiral only existed because there was no royal navy presence, laying bare the fact that the state lacked the ability to exercise jurisdiction over maritime space or even over many Jamaican colonists.[40] King Charles and the secretary of state ordered Modyford to cancel the raid and the governor even received a copy of the newly-signed Treaty of Madrid in December 1670, but the governor claimed the orders arrived too late to

38 "The Lord Sandwich's Discourse of what Advantages his Majesty may father have from Spain, by a nearer League" in *Hispania Illustrata*, pp. 93–105.

39 Coakley, "Private Seafarers," p. 38. See also Earle, *Sack of Panama*.

40 One navy ship, HMS *Oxford*, was on the way to Jamaica in 1668, but never arrived. The ship blew up at anchor off Hispaniola while Morgan held a council of war on board, planning a raid. CSPC, 1675–1676, addendum 1669, no. 1207. Richard Browne to Williamson, January 1669.

reach the admiral at sea and the raid continued as planned.[41] The king was embarrassingly on the back foot; there was little he could do in that moment to control the actions of his own subjects in the Americas. This was because, for the decade of his administration, colonial maritime activity and England's claimed maritime corridors through Spain's purported monopoly in the Caribbean were not under his control. England claimed sea space it could not manage, so actual authority fell to local colonies. But lack of imperial control did not indicate that the Caribbean was a lawless world of sea-robbers. As the case of Jamaica, Morgan, and the Modyfords makes clear, elite colonial leaders acting for themselves enforced jurisdiction over their own maritime affairs, using bands of armed private sailors seeking money and security.

## The Treaty's Jamaica

Despite these realities, the empire continued to assert jurisdiction over its Caribbean colonies' maritime affairs, and those assertions reached new heights with the 1670 Treaty of Madrid, the document that had so occupied the king while Morgan attacked Panama. The first European treaty to cover exclusively American land and sea space, the treaty was the result of years of compromise between the two major powers of Spain and England, but only brought tenuous peace at best to the region. For English negotiators, the treaty was successful in one major area: it recognised the English king's claims to all American lands that "His Subjects do at present hold and possess." But in a huge blow to the trade-oriented politicians dominating Restoration English politics, the treaty forbade trade between English and Spanish colonies in the Americas, making only a few concessions for small future transactions.[42]

In the way it treated regional maritime predation, the treaty insisted that maritime jurisdiction in the Caribbean rested solely with imperial leaders in Europe. It declared that all prior attacks between English and Spanish subjects in the Americas were to be "expunged out of remembrance, and buried in Oblivion, as if no such thing had ever past."[43] The treaty also forbade all future acts of maritime predation between the two, legally converting any such act into "piracy." Though the restriction seems obvious in a treaty of peace, other treaties of the time still allowed for individual small-scale

41 TNA, CO 1/25/103. Modyford to Arlington, December 18, 1670.

42 *A Treaty for the Composing of Differences*, secs. VII–IX.

43 *A Treaty for the Composing of Differences*, sec. VII.

raids under the medieval rubric of reprisal. Once common throughout European waters, reprisal allowed a local leader to authorise a single captain to plunder a foreign ship as a way to recoup economic losses inflicted by those same foreigners during a previous trading mission.[44] Negotiators for the Treaty of Madrid openly recognised that such "private injury" may still occur, but the treaty forbade reprisal as an option, instead essentially forcing injured parties to apply to their colonial governors who could apply to their king who could try to get reparation.[45] In limiting colonists' behaviour and curtailing their traditional use of local maritime enterprise to solve local problems, the Treaty of Madrid is the strongest seventeenth-century assertion of imperial (rather than colonial) jurisdiction over American maritime affairs. But these assertions were simply that, and they did not recognise regional realities. The treaty failed in its too-ambitious efforts to reshape Caribbean maritime activities.

The treaty also upended Jamaican politics. The king sent Thomas Lynch to be Jamaica's new leader, ordering him to arrest Modyford for violating the treaty. Anticipating an island-wide rebellion at the shocking orders, the state sent Lynch with two navy ships that could be used to quell protests and force Modyford into custody. He did not need the firepower, instead luring the governor onto the *Assistance* frigate, and imprisoning him on the spot.[46] Lynch delayed a year fulfilling another order to arrest Morgan, preferring to keep the navy ship on the station rather than waste it on transporting the imprisoned admiral.

Like Modyford before him, Thomas Lynch began his tenure by outlawing maritime raiding against Spanish targets, in line with the treaty's provisions. Lynch tried both to cajole and to coerce the privateers (a word he still used to refer to Jamaica's sizable cadre of private sailors despite being ready to prosecute them as "pirates"), but neither method proved effective. He allowed a grace period and a generous pardon for the private sailors who surrendered their commissions and ships willingly, but though some came in, few trusted the new lieutenant-governor enough to comply.[47] To quell the privateer population, Lynch recognised that he must buck the treaty and utilise the island's de-facto jurisdiction over its own maritime affairs. His orders were more circumscribed than Modyford's, meaning he was forbidden

44 Clark, "The English Practice"; Starkey and McCarthy, "Persistent Phenomenon."

45 See TNA, State Papers (SP) 103/66/1. See also *A Treaty for the Composing of Differences*, sec. XIV.

46 TNA, CO 1/27/22. Lynch to Arlington, August 20, 1671. The discussion of Lynch below follows a similar interpretation to that found in Coakley, "Private Seafarers," 39–40.

47 The pardon is in TNA, CO 1/25/107. Instructions for Thomas Lynch, December 31, 1670. That few took advantage of it is mentioned in TNA, CO 140/1/223. Council Minutes, August 12, 1671.

from giving orders to royal navy fleet admirals in the Caribbean.[48] Despite these limitations, Lynch replaced the captain of the *Assistance* and then made a lieutenant out of a man named Prynce, "one of the most famous of the privateers."[49] He sent it out pirate-hunting in consort with the *Lilly* under captain Morris, a convicted pirate whom Lynch had just released from custody for the mission.[50]

Lynch was keenly interested in the channel that ran between Jamaica's north side and the south side of neighbouring Cuba; he made that the focus of his policing efforts, even sending four private ships and 500 men to the area.[51] This policing cleared the way for establishing an illicit trade between the two islands. The many small cays that dotted the channel were ideal for clandestine smuggling transactions, which Lynch now openly encouraged to "steal into a trade with the Spaniards," largely in "hides and cattle."[52] Hiring private sailors to hunt down their former compatriots and using the cleared seaways to promote smuggling went beyond Lynch's orders and the treaty. They did, however, both feed English desires to tap into the Spanish-American market, so his superiors tacitly allowed him to continue local jurisdiction of maritime affairs even though the treaty seemingly banned it.

Lynch then pushed his local jurisdiction further, using the privateers to ignite an undeclared war with Spain over the lucrative logwood trade, and his ambitions got him recalled. Logwood, a tree that grew in brackish lowland Central America, had for decades been a source of rich dyes for the European textile industry. It commanded a good price at the time, so the stakes were high: whomever controlled the wood controlled the Caribbean economy.[53] The English state wanted more of the trade and tried to use the treaty to take it. Although Spain claimed jurisdiction over the territory where the trees grew, there had been English woodcutting camps in the region for decades. State authorities pressed legal arguments that the presence of the camps gave England jurisdiction instead.[54] Seizing the chance to gain profit and power, Thomas Lynch encouraged Jamaican ships to patronise the logging camps,

48 CSPC, 1669-74, no. 386.

49 TNA, CO 140/1/223. Lynch to Arlington, December 17, 1671.

50 TNA, CO 1/28/8. Lynch to Williamson, January 27, 1672.

51 TNA, CO 1/28/35. Lynch to Don Francisco Rodrigues, April 5, 1672.

52 TNA, CO 1/29/6. Lynch to Williamson, July 5, 1672; TNA, CO 1/30/19. Lynch to Council for Trade, April 4, 1673.

53 On logwood generally, see Botella-Ordinas, "Debating Empires"; Campbell, *Becoming Belize*; Offen, "British Logwood Extraction." For another side of the logwood trade at the time, see Steven J. Pitt's contribution to this volume.

54 Campbell, *Belize*; Botella-Ordinas, "Debating Empires."

increasing the trade. Official reports note that Jamaicans exported 920 tons of the wood during Lynch's tenure, but he estimated that English cutters took at least 2,000 tons all told, and most went through the island.[55] It was enough to focus the ire of Spanish governors, who commissioned private seafarers to attack and imprison English traders. One, Philip Fitz-Gerald, an Irishman working out of Havana, gained fame for his brutal attacks, pledging to drink English blood "as freely as water" in 1673.[56] Jamaican sailors, like Captain Yallahs, switched allegiances too; he captured more than a dozen English ships from the Spanish port of Campeche. Lynch fought back, but knowing he could not issue reprisals, he treated Yallahs as a rogue captain and sent the *Lilly* and the *Assistance* to hunt down the pirate.[57]

This warfare over logwood reached greater proportions as Lynch supported the pillaged traders. He insistently pressed the Lords and others to work on getting compensation from the Spanish government, and most importantly to give him the power to issue reprisal commissions. Reprisals and restitution never came, and Lynch fumed that his logwood traders would see justice only "in the world to come."[58] Once again, he took matters into his own hands, letting the privateers seek justice for Englishmen in their typical blunt manner. By the end of 1674, Lynch already knew he was to be replaced in the coming year, so he reported with some smugness that up to 600 private seamen were gathering at Tortuga, plotting a deadly campaign against the Spanish, and that "the better half" of their number were angry Englishmen.[59] He carefully avoided issuing reprisal commissions himself, but they could be easily acquired from French governors. Lynch was disappointed if he hoped the privateers would make an expedition on the scale of the Western Design or the Panama raid, because nothing like that happened this time; he would have to wait until the 1683 looting of Vera Cruz for that. But the fact of their gathering said enough. In just four years, the treaty was proved to be toothless and Jamaica reverted to the status quo. Despite the treaty's assertion that English maritime jurisdiction in the region belonged to the state, like his predecessors before him, Lynch proved that in fact for the moment, Jamaicans ruled those waves.

55 TNA, CO 1/43/59 in Zahedieh, "Trade, Plunder, and Economic Development," 207; TNA, CO 1/29/43. Lynch to H. Slingesby, November 5, 1672.

56 TNA, CO 1/31/12. Petition of Thomas Jarvis et al. to the King, February 27, 1674.

57 TNA, CO 1/29/8. Hender Molesworth to Thomas Duck, July 8, 1672; TNA, CO 1/28/7. Lynch to Williamson, January 27, 1672. Morris may have used the trip as an excuse to get logwood himself. TNA, CO 1/29/33. R Browne to Williamson, September 28, 1672.

58 TNA, CO 1/30/58. Lynch to Worsley, August 12, 1673.

59 Lynch stated that he prevented Jamaicans from going to Tortuga for the mission. TNA, CO 1/31/77. Lynch to Williamson, November 20, 1674.

## Conclusions

During the island colony's first twenty years, Jamaicans proved remarkably tenacious—and successful—at maintaining local jurisdiction over maritime affairs, regardless of imperial assertions to the contrary. Maritime jurisdiction began to shift in favour of the English state in 1675, a lengthy process lasting at least the next two decades. When Lynch was recalled, John, Lord Vaughan, arrived as the new governor, bringing with him the freshly knighted Sir Henry Morgan as his lieutenant. The inexperienced Vaughan proved unpopular. By attempting to work with both rival political factions in the colony's political system, he infuriated both and drew the ire and scrutiny of the king and the lords. He prosecuted the popular Morgan and his brother-in-law for encouraging Jamaican privateers to work for French governors, and then went against the lieutenant-governor again by sentencing a pirate to death in an illegal trial. Finally, he allowed his council to re-draw Admiralty Court jurisdictions for individual parishes to include immediate coastal waters, in order to prosecute a slave trade interloper captured in port. Meanwhile, the colony's assembly asserted its own political authority independent of the king's.[60]

Imperial officials took alarm at these actions, especially as they all seemed to come at once and amid other troubles in the empire in the mid-1670s: King Philip's War in New England, Bacon's Rebellion in Virginia, and a slave revolt in Barbados. Attempting to solve this Jamaican constitutional crisis in 1679, the king and lords opted not simply to assert control over the island, but to enforce it as well, by authoring a special version of Poyning's Law demanding adherence to royal rules for the colony. Although the law failed and further compromises were reached with the Jamaican assembly, the island's politicians and its private seafarers never again enjoyed the kind of de-facto local control that they had in previous decades. Morgan and the Modyford family remained significant figures throughout the 1680s, and Lynch even returned as governor one more time, but they were forced to adapt to a new imperial reality that hindered local control of maritime affairs. During the late 1680s, King James II asserted personal control over pirate-hunting operations in the area.[61] The Glorious Revolution and the Port Royal earthquake of 1692 then aided the transition further. In the wake

60 Hanna, *Pirate Nests*, pp. 126–32; Whitson, *Constitutional Development*.

61 James II, *A Proclamation for the more Effectual Reducing and Suppressing of Pirates and Privateers in America* (London, 1687).

of the quake's devastation, governor William Beeston—who had held many local offices since his arrival in Jamaica in 1660—found his effectiveness curtailed by the new monarchs' interpretation of Admiralty Law. When the French invaded in 1694, he mounted remarkable and successful ground defences but had to wait months after the invaders left for a Royal Navy fleet to arrive; when it did, he fumed that his orders prevented him from working effectively with the commanders.[62]

Eventually the English state achieved with legal precision by the end of the century what it had not been able to with the blunt instrument of the Treaty of Madrid in 1670: wresting maritime jurisdiction from colonial elites and installing it as a cornerstone of state authority. Metropolitan jurisdiction over Caribbean colonial waters was a long time coming. It had been asserted since at least the Civil Wars, when attempted enforcement arrived in the form of fleets from each side. The Western Design had seemed to usher in a new era of Cromwellian state authority in the region, but it proved a fantasy. The treaty was the boldest written assertion, but as Lynch and the logwood traders discovered, words did not translate into action. Action, and therefore effective control for the state, only came in the 1690s, ironically perhaps when Jamaica could have benefited the most from local control over its maritime affairs. But the state was not yet able to extend its authority completely to the private seafarers. Those who escaped prosecution and natural disaster continued maritime predation, mostly for the time being in the Indian and Pacific Oceans, reminding European empire builders that despite their best dreams and documents, no one owned the seas.

## Bibliography

### Manuscript Sources

Colonial Office. *America and West Indies, colonial papers*, 1663–1674. The National Archives, UK (TNA). Colonial Office (CO) Series 1/17-1/31.

Colonial Office. *General Entry Books, Series I: Commissions, Instructions, Petitions, Grants, Orders in Council, Warrants, Letters, etc*, 1662–1674. TNA, CO Series 324/1.

Colonial Office. *Jamaica, Entry Books: Commissions, Instructions, Board of Trade Correspondence*, 1661–1674. TNA, CO Series 138/1.

62 TNA, CO 137/1/193–5. William Beeston, "Brief Account of what happened," August 1694; Buisseret, "French Invasion," pp. 31–33.

Colonial Office. *Jamaica, Original Correspondence: Board of Trade*, 1694–1695. TNA, CO Series 137/1.
Colonial Office. *Jamaica, Sessional Papers: Council Minutes*, 1661–1672. TNA, CO Series 140/1.
"Memorandum of the devident." c. 1660. Huntington Library, Manuscript Collection. No. BL 319.
National Library of Jamaica, Manuscript Collection.
*Official papers relating to the English settlements in America and the West Indies*, British Library (BL), Egerton Manuscripts 2395.
*Original Journal of the proceedings of Colonel Edward D'Oyley, during the time he held the chief command in the island of Jamaica*, BL, Additional Manuscripts 12423.
Secretaries of State. *State Papers Foreign, Treaty Papers: Spain*, 1667–1719. TNA, State Paper Foreign (SP) Series 103/66.

## Published Primary Sources

Beeston, William. "A Journal Kept by Col. William Beeston." In *Interesting Tracts, Relating to the Island of Jamaica*. St. Jago de la Vega, Jamaica, 1800.
Davenport, Francis Gardiner, ed. *European Treaties Bearing on the History of the United States and its Dependencies to 1648*. Clark, NJ: Lawbook Exchange, 2004.
*Hispania Illustrata Or, the Maxims of the Spanish Court, and Most Memorable Affairs*. London, 1703.
Hickeringill, Edmund. *Jamaica Viewed with All the Ports, Harbours, and Their Several Soundings, Towns, and Settlements Thereunto Belonging together* [...], 2nd ed. London: John Williams, 1661.
James II. *A Proclamation for the more Effectual Reducing and Suppressing of Pirates and Privateers in America*. London, 1687.
*Journals of the Assembly of Jamaica*, Vol. I: 1664–1709. Jamaica: Alexander Aikman, 1811.
Sainsbury, W. Noel, ed. *Calendar of State Papers, Colonial Series, America and West Indies* (CSPC), 1574–1660. London: H. M. Stationery Office, 1860.
Sainsbury, W. Noel, ed. *Calendar of State Papers, Colonial Series, America and West Indies* (CSPC), 1669–1674. London: H. M. Stationery Office, 1884.
Sainsbury, W. Noel, ed. *Calendar of State Papers, Colonial Series, America and West Indies* (CSPC), 1675–1676. London: H. M. Stationery Office, 1893.
*Treaty for the Composing of Differences, Restraining of Depredations and Establishing of Peace in America: Between the Crowns of Great Britain and Spain*. London, 1670.
Westminster Abbey Muniments. In A. P. Thornton. "The Modyfords and Morgan." *Jamaican Historical Review* 2 (1952): 36–60.

## Secondary Sources


Anderson, Benedict. *Imagined Communities: Reflections on the Origin and Spread of Nationalism*, Revised Edition. London: Verso, 1991.

Barber, Sarah. "Brayne, William." *Oxford Dictionary of National Biography*, 2004. Accessed July 16, 2022.

Benton, Lauren. *A Search for Sovereignty: Law & Geography in European Empires, 1400–1900*. New York: Cambridge University Press, 2010.

Benton, Lauren. "Toward a New Legal History of Piracy: Maritime Legalities and the Myth of Universal Jurisdiction." *International Journal of Maritime History* 23, no. 1 (2011): 225–40.

Botella-Ordinas, Eva. "Debating Empires, Inventing Empires: British Territorial Claims against the Spaniards in America, 1670–1714." *Journal for Early Modern Cultural Studies* 10, no. 1 (2010): 142–68.

Buisseret, David. "The French Invasion of Jamaica—1694." *Jamaica Journal* 16, no. 3 (1983): 31–33.

Campbell, Mavis Christine. *Becoming Belize: A History of an Outpost of Empire Searching for Identity, 1528–1823*. Kingston, Jamaica: University of the West Indies Press, 2011.

Capp, Bernard. "Goodsonn, William." *Oxford Dictionary of National Biography*, 2004. Accessed July 16, 2022.

Clark, Grover. "The English Practice with Regard to Reprisals by Private Persons." *American Journal of International Law*, 27 (1933): 694–723.

Coakley, John. "Jamaica's Private Seafarers: Politics and Violence in a Seventeenth-Century English Colony." In *The Golden Age of Piracy: The Rise, Fall, and Enduring Popularity of Pirates*, edited by David Head. Athens, GA: The University of Georgia Press, 2018.

Coakley, John. "'The Piracies of Some Little Privateers': Language, Law and Maritime Violence in the Seventeenth-Century Caribbean." *Britain and the World* 13, no. 1 (2020): 6–26.

Dunn, Richard S. *Sugar and Slaves: The Rise of the Planter Class in the English West Indies, 1624–1713*. New York: Norton, 1972.

Dyer, Florence E. "Captain Christopher Myngs in the West Indies." *Mariner's Mirror* xviii (1932): 168–87.

Earle, Peter. *The Sack of Panama: Captain Morgan and the Battle for the Caribbean*. New York: St. Martin's, 1981.

Firth, Charles H. "The Capture of Santiago, in Cuba, by Captain Myngs, 1662." *The English Historical Review* 14, no. 55 (1899): 536–40.

Gould, Eliga. "Lines of Plunder of Crucible of Modernity? The Legal Geography of the English-Speaking Atlantic, 1660–1825." In *Seascapes: Maritime Histories,*

*Littoral Cultures, and Transoceanic Exchanges,* edited by Jerry H. Bentley, Renate Bridenthal, and Kären Wigen, 105–20. Honolulu: University of Hawai'i Press, 2007.

Hanna, Mark G. *Pirate Nests and the Rise of the British Empire, 1570–1740.* Chapel Hill: University of North Carolina Press, 2015.

Heywood, Linda M. and John K. Thornton. *Central Africans, Atlantic Creoles, and the Making of the Foundation of the Americas, 1585–1660.* New York: Cambridge University Press, 2007.

Kupperman, Karen Ordahl. *Providence Island, 1630–1641: The Other Puritan Colony.* Cambridge: Cambridge University Press, 1993.

Latimer, Jon. *Buccaneers of the Caribbean: How Piracy Forged an Empire.* Cambridge, MA: Harvard University Press, 2009.

Offen, Karl H. "British Logwood Extraction from the Mosquitia: The Origin of a Myth." *Hispanic American Historical Review* 80, no. 1 (February 2000): 113–35.

Pestana, Carla Gardina. "Early English Jamaica without Pirates." *The William and Mary Quarterly* 71, no. 3 (2014): 321–60.

Pestana, Carla Gardina. *The English Conquest of Jamaica: Oliver Cromwell's Bid for Empire.* Cambridge, MA: Belknap, 2017.

Rediker, Marcus. *Villains of All Nations: Atlantic Pirates in the Golden Age.* Boston: Beacon Press, 2004.

Ritchie, Robert C. *Captain Kidd and the War against the Pirates.* Cambridge, MA: Harvard University Press, 1986.

Starkey, David J. and Matthew McCarthy. "A Persistent Phenomenon: Private Prize-Taking in the British Atlantic World, c.1540–1856." In *Persistent Piracy: Maritime Violence and State Formation in Global Historical Perspective,* edited by Stefan Amirell and Leos Mu̍ller, 131–51. London: Palgrave Macmillan, 2014.

Thomson, Janice. *Mercenaries, Pirates, and Sovereigns: State-Building and Extraterritorial Violence in Early Modern Europe.* Princeton: Princeton University Press, 1994.

Whitson, Agnes. *The Constitutional Development of Jamaica, 1660–1729.* Manchester, 1929.

Wilson, David. *Suppressing Piracy in the Early Eighteenth Century: Pirates, Merchants and British Imperial Authority in the Atlantic and Indian Oceans.* Woodbridge: Boydell Press, 2021.

Zahedieh, Nuala. "Trade, Plunder, and Economic Development in Early English Jamaica, 1655–89." *Economic History Review* 39, no. 2 (1986): 205–22.

Zahedieh, Nuala. "Modyford, Sir Thomas, First Baronet (c. 1620–1679), Planter and Colonial Governor." *Oxford Dictionary of National Biography*, 2004. Accessed February 4, 2022.

## About the Authors

**John Coakley** is an historian of early America and the Atlantic world, focusing on maritime predation in the Caribbean. He received his Ph.D. from the University of Wisconsin–Madison. He is the author of '"The Piracies of some Little Privateers': Language, Law and Maritime Violence in the Seventeenth-Century Caribbean," *Britain and the World*, 13:1 (2020), 6–26.

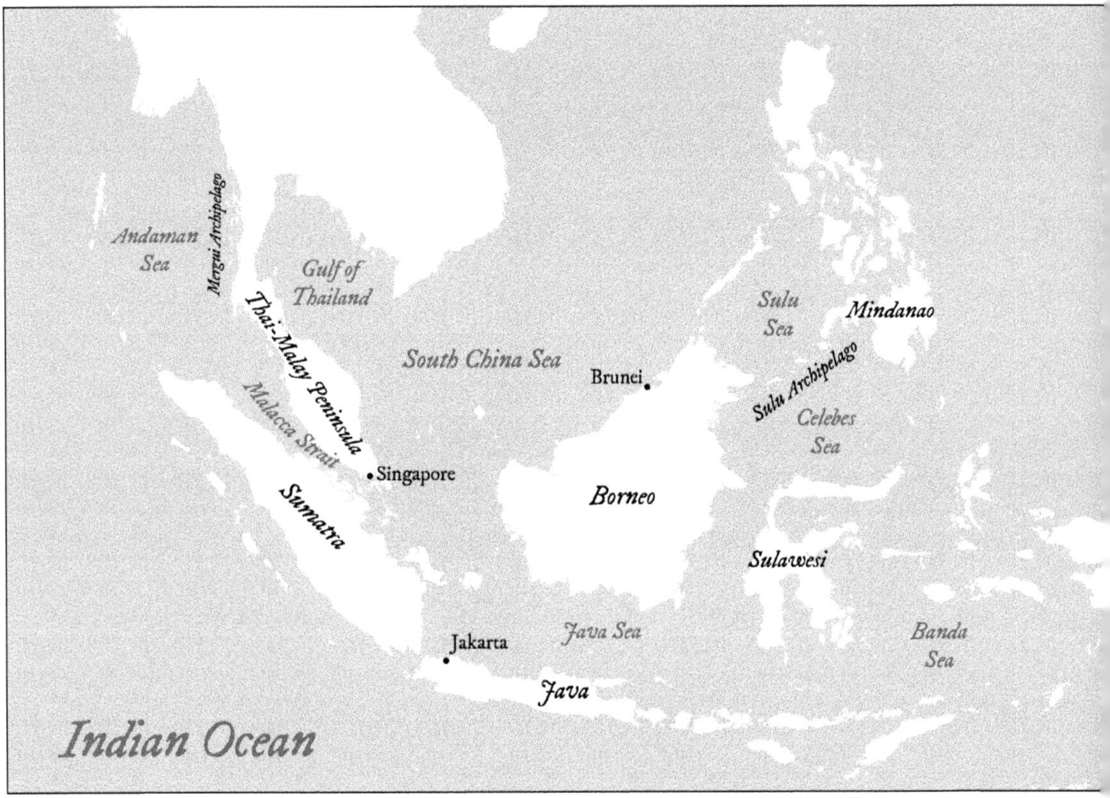
Andaman
Sea
Mergui Archipelago
Gulf of
Thailand
Thai-Malay Peninsula
Malacca Strait
South China Sea
Sulu
Sea
Mindanao
Brunei
Sulu Archipelago
Celebes
Sea
Singapore
Sumatra
Borneo
Sulawesi
Jakarta
Java Sea
Banda
Sea
Java
Indian Ocean

# 2. Primitive, Peregrinate, Piratical: Framing Southeast Asian Sea-Nomads in Nineteenth-Century Colonial Discourse and Imperial Practice

*Martin Müller*

**Abstract**

This chapter examines the epistemic and violent processes through which Southeast Asian non-sedentary communities on sea, land, and along the littoral were assessed, criminalised, and rendered governable by colonial regimes during the nineteenth century. It hypothesises that the concept of territoriality proved pivotal in defining the subjects of colonial rule. Territoriality determined the status of members of these itinerant cultures, notwithstanding whether they were potentially "domesticable" within a capitalist world system, or decried as pirates, dacoits, or bandits. The chapter juxtaposes attitudes towards maritime and terrestrial nomads, exposing how physical spatiality influenced how territoriality was conceptualised, laws and legal procedures implemented, and how this affected the governmental strategies of othering the inhabitants inhabiting and of bringing order to them.

**Keywords:** Malay; Non-Sedentary; British Empire; Territoriality; Frontier; Piracy; Law

The entry on "Piracy and Pirate" in John Crawfurd's (1783–1868) "A Descriptive Dictionary of the Indian Islands and Adjacent Countries" (1856) opened with these words:

> There is no name in Malay and Javanese, or indeed in any other native language for piracy or robbery on the high seas. There is, in fact, no word to

Coakley, J., C. Nathan Kwan and D. Wilson, *The Problem of Piracy in the Early Modern World: Maritime Predation, Empire, and the Construction of Authority at Sea.*
Taylor & Francis Group, 2024
DOI 10.5117/9789463720960_CH02

distinguish the element on which the act of plunder is committed, a thing natural enough with a people who live as much on the sea as on the land.[1]

According to Crawfurd—who by then was broadly recognised as the authority on Southeast Asia in a British context[2]—the notion, if not the actual practices, of piracy was an unknown phenomenon in Southeast Asia before Europeans entered its waters and introduced this category as a potent politico-legal signifier. During the nineteenth century, piracy came to constitute a key category in the ideological idiom[3] that underwrote the imperial projects carried out by European colonial regimes in Southeast Asia and the Indian Ocean region with the ostensible aim at bringing (their) trade, order, and, in their own optics, progress in the broadest sense of the word to these seas and their adjacent terrestrial regions.

Despite this an alleged semantic absence of what Europeans called piracy, acts normally associated with this category were, nonetheless, deemed to be rampant in the region. To borrow J. L. Anderson's typological classification, piracy was not seen as merely an *episodic* phenomenon in insular and maritime Southeast Asia or, indeed, a *parasitic* function of trade; rather it formed an *intrinsic* part of the economic structure and political organisation in many smaller as well as larger coastal polities.[4] At least it had been so ever since Europeans first gained a foothold in the region and then slowly began extending and intensifying their influence, a process culminating when the last indigenous state formations succumbed to European rule during the latter half of the nineteenth century. This meant that internecine depredation and slave raiding were everyday reality on the high seas and along the littoral, and they formed the primary means through which warfare or other forms of violent competition between coastal polities were carried out.[5] Hence, these so-called "pirates" were, as pointed out recently by Sebastian Prange, "not outlaws but in-laws, respected members of society and a political force that rulers needed to reckon with."[6] This did not imply that these seas represented a lawless space; rather, in the words of Renisa Mawani and Sebestian Prange, it made up an unruly oceanic space "in which the state appears not as the sole purveyor of legitimacy but as one actor among many, in a multi-vocal and plural legal environment where

1 Crawfurd, *Descriptive Dictionary*, p. 353.
2 See Knapman, *Race and British Colonialism*; Müller, "Civilization, Culture and Race."
3 Antony, "Turbulent Waters," p. 24.
4 Anderson, "Piracy in the Eastern Seas," pp. 87–89.
5 See, for instance, Barnard, "Malay Warfare."
6 Prange, "Asian Piracy."

a diversity of religious laws and legal cultures intersected, coexisted and competed."[7]

With these larger issues in mind, this chapter examines the epistemic and violent processes through which especially Southeast Asian non-sedentary communities on sea and along the littoral were assessed, criminalised, and rendered governable by British and other colonial regimes. Focusing on an pivotal epoch in imperial territorialisation and intensified colonisation in Southeast Asia (c.1770–1850), it traces the range of especially British attitudes towards these communities through the framework of "piracy" with the aim at exposing how different types of geographical spatiality influenced the manner in which territoriality and a corresponding "maritory"[8] were conceptualised, laws and legal procedures framed and implemented, and how this affected governmental strategies of defining and ordering non-sedentary inhabitants in coastal and maritime contexts.

As a way to illuminate these foundational issues which only seldom explicitly appear in the historical source material, this chapter provides a rudimentary sketch of the epistemic and governmental processes through which Southeast Asian maritime based, non-sedentary communities, or so-called "sea nomads," were approached, analysed, assessed and in the end rendered governable by European colonial regimes during the nineteenth century, and in which the piracy category played a crucial role. In present day scholarly literature,[9] these peoples are usually classified into three geographically divided and culturally distinct ethnic groups who shared—and to a certain, albeit lesser, degree still share—a non-sedentary and maritime based mode of life within essentially identical physical environments but placed in different political and historical contexts. These are composed of the *Moken* who inhabit the coastal fringes of the Mergui Archipelago on the eastern coast of the Andaman Sea in what is nowadays Myanmar and Thailand, the *Orang Laut* (literally "People of the Sea") who are native to the Malacca Strait with its many islands, and lastly the *Sama-Bajau* scattered around eastern parts of the Indonesian

7 Mawani and Prange, "Unruly Oceans." Grounded in a comparative framework of historical sociology, Ota Atsushi, on the other hand, cautions us against uncritically equating state formation and the extension of territorial (and maritime) sovereignty with an inherently violent European colonial expansion, as it is often assumed in approaches steeped in postcolonial theory. Ota, "Introduction," pp. 5–10 and pp. 14–15.

8 "A specific [aqueous] territory of maritime exploitation." Hoogervorst, "Aquatic Lifestyles," p. 255.

9 For recent studies see the contributions in Bellina, Blench and Galipaud, *Sea Nomads of Southeast Asia*. Problems and perspectives regarding how to define "sea nomads" are discussed in their introductory chapter, pp. 3–9. See also Chou, "Research Trends."

Archipelago. This division has a long history, and there appears, as stressed by Barbara Watson Andaya, "to be a scholarly consensus that it was (and is) difficult to create a general history of sea peoples as a distinctive group that shares features in common."[10] However, during the historical period of interest here, especially the two latter terms were used rather interchangeably and frequently overlapped each other; often they seemed not to possess a well defined and meticulously demarcated ethnological meaning.[11]

The central hypothesis in the chapter is that the concept of territoriality proved to be pivotal in defining the subjects of colonial rule[12] and foundational in conceiving and moulding the colonial spaces and accompanying governmentalities.[13] As a concept, territoriality did not merely manifest itself in terms of geographic demarcations, but it also served as an inherently politically charged as well as a continuously contested legal concept and, ultimately, as an epistemic approach that grounded the political, cultural, and lived places within fixed spaces. Imperial spaces tended to be bounded by zonal frontiers rather than by linear borders, something which rendered these liminal spaces bi-directionally porous, overlapping and utterly dynamic.[14] As a consequence the frontier becomes an unstable category of space, and hence it seems more productive, as stressed by Benjamin D. Hopkins, "to think of frontiers as conceptual rather than physical spaces"; so, "rather than identifying specific places as 'frontiers', it is more useful to identify particular practices constituting frontiers."[15] Practices that inscribed

10 Andaya, "Recording the Past," p. 10.

11 Many instances of this exist. See for example Forrest, *Voyage to New Guinea*, p. 372 ("Oran Badjoo" [Bajau] conflated with the Orang Laut in the Straits of Malacca); Forrest, *Voyage from Calcutta*, p. 71 ("Boadjoos" [Bajau] are the same as "Oranlout [Orang Laut]); Raffles, "Short Account," p. 43 ("Badju, or *Orang Laut* tribe"—referring to present- day Bajau); Hamilton, *East-India Gazetteer*, vol. II, p. 573 (equating "Bajoo" [Bajau] and Orang Laut); Keppel, *Expedition to Borneo*, vol. II, pp. 146–49 (conflating Orang Laut and "Badjow" [Bajau]), Keppel, *Visit to the Indian Archipelago*, vol. I, p. 45 (equating "Bajow" with Sulu pirates); Low, *Sarawak*, p. 344 (The "Badjus" [Bajau] along the eastern coast of Borneo are seen as Orang Laut who have immigrated to here from Johore in the Malacca Strait) and Earl, *Eastern Seas*, pp. 135–36, 334–36 (Earl distinguishes between the Orang Laut and the "Orang Badju" [Bajau] but he sees them as almost identical).

12 That is, *subjects* in a triple sense of the term: as subjects of colonial knowledge production, as subjected to colonial rule, and also as subjects endowed with agency. Pels and Salemink, *Colonial Subjects*, p. 3.

13 Scott, "Colonial Governmentality." For further thoughts on the importance of these underling aspects of territoriality in the context of nineteenth-century imperialism see also Maier, "Transformations of Territoriality," p. 35; Maier, *Once Within Borders*, pp. 19–20; Osterhammel, *Transformation of the World*, p. 107

14 Maier, *Once Within Borders*, pp. 19–20.

15 Hopkins, *Ruling the Savage Periphery*, p. 15.

and performed, and which, in this case, were articulated in discourses steeped in the category of the pirate and enacted through the anti-piratical measures against the seemingly itinerant, maritime communities whose culture and lifestyles could not be contained within the expanding and intensified frontier governmentality[16] exerted by the colonial powers in maritime Southeast Asia.

It has to be stressed that this chapter does not claim to provide a thorough analysis of the structural realities on the ground, as well as on the seas, in maritime Southeast Asia neither during nor before this period. Before the nineteenth century these non-sedentary maritime communities such as the Orang Laut often played a pivotal role, as shown by Leonard Y. Andaya, as guardians of the maritory claimed by the Malay ruler, as collectors of maritime produce and as skilled pilots. Their allegiance to the Malay ruler lent legitimacy to their patrolling of these seas and extracting "revenues," whereas their high mobility made it relatively easy for them shift their loyalty, if they felt mistreated or it was deemed in their best interest.[17] Jennifer Gaynor has staked a similar claim in the context of the Sama-Bajau along the coasts of Sulawesi.[18] Thus it does not make much sense to perceive the mobile dwellers on the seas and along the littoral as marginal and marginalised inhabitants of an unregulated "watery Zoomia," to borrow James C. Scott's concept,[19] before the nineteenth century with the establishment of colonial rule and the introduction of new economic structures.[20]

Summing up, this chapter will not so much address who were the pirates as how they came to be approached, analysed and assessed as such. This implies analytical scrutiny of the geographical reconfigurations that followed with colonial rule and global capitalism and which both connected and divided the region's oceanic and terrestrial spaces,[21] as well as of the spatial imaginaries that gave form and ascribed meaning to these, in all the senses of the word, powerful configurations.[22]

16 Ibid., pp. 16–18.

17 Andaya, *Leaves of the Same Tree*, pp. 173–201.

18 Gaynor, *Intertidal History*, especially pp. 107–65. See also Barnard, "Celates, Rayat-Laut, Pirates"; Ota, "Pirates or Entrepeneurs." Early assessments of the relationships between the Bajau and Brunei and Sulu can be found in Brown, "Brunei and the Bajau"; Sather, "Sulu's Political Jurisdiction."

19 Scott, *Not Being Governed*, pp. 1–39.

20 Andaya, "Recording the Past," p. 8.

21 Campling and Colas, *Capitalism and the Sea*, p. 69.

22 "Spatial imaginaries are socially held stories, ways of representing and talking about places and spaces" that "help shape material practices moulding geographies through their linguistic circulation and embodiment." Watkins, "Spatial Imaginaries," p. 509.

For our purposes, the quote from Crawfurd's "Descriptive Dictionary" which initiated this chapter raises two salient questions that will be discussed in the following; in conjunction they address the tacit assumptions inherent in the spatio-temporal configurations through which maritime and insular Southeast Asia was observed, organised and ordered in and through the European gaze. The first question revolves around the prevailing idea of the *notion* of piracy being nonexistent in this region before the advent of the Europeans and what this entailed when placed within a civilisational framework; it addresses how the historical and political space of maritime Southeast Asia was in colonial (and especially British) discourses conditioned by the alleged inability of its inhabitants to *come to terms* with the omnipresence of piratical activities in the region. Then the focus is shifted towards the spatial demarcations upon which the concept of piracy relies, and how these were seemingly transgressed time and again when actually combating piracy; what can this reveal about the instability of the spatial entities of "land" and "sea" and the futility of attempting to define these as binaries without recognising spaces in between? The last part of the chapter then deals with the register of "solutions" proposed by British, Spanish and Dutch proponents of colonial order to address and eradicate the problem posed by non-sedentary maritime peoples with their supposedly inherent piratical inclinations.

## Coming to Terms with Piracy: Temporalising Modes of Violence at Sea

First, the purported absence of any terminology referring specifically to illegal maritime predation in Southeast Asian languages[23] seems to suggest that a categorical, clear-cut land-sea divide did not exist as a foundational and ordering principle neither within nor in between insular Southeast Asian polities before being introduced from the outside.[24] This was, according to Crawfurd, hardly surprising for "a people who live as much on the sea as on the land." Crawfurd further explained this apparent anomaly by pointing to both the region's geography and to the level of civilisation among its dwellers: "In a region like the Malay and Philippine Archipelagos, abounding in narrow

23 For general assessments of the cross-cultural complications involved in labelling maritime violence and the political and moral connotations therein see Amirell, *Pirates of Empire*, pp. 8–13; Risso, "Cross-Cultural Perceptions," pp. 296–302.

24 For a critique of this essentially orientalist binary see Ota, "Introduction," pp. 1–15.

seas, rivers, creeks, coves, and mangrove swamps, and often inhabited by rude and lawless tribes of fishermen, piracy must have existed as early as there was anything to plunder."[25] By invoking an explanatory mode anchored in geographical determinism and then inscribing it into a framework of stadial progress of civilisation, Crawfurd simultaneously normalised[26] and circumscribed the prevalence of piracy in the region. Given the geographical features of maritime Southeast Asia and the supposedly less advanced stage of society of its inhabitants, things could hardly have been different, and endemic piracy was seen as a natural outcome of this confluence of space and state of society. Yet it was also ascribed the quality of being something pertaining to the past, and which would be overcome once modern order was established in the region through either indirect or direct imperial influence in the shape of marketisation and/or outright colonial rule.[27] Even though this explanation of the origin and "naturalisation" of piracy in maritime Southeast Asia had to compete with other contemporary explanations, such as the so-called "decay theory" championed by Sir Thomas Stamford Raffles whenever deemed convenient,[28] it resonated particularly well with the approaches to and conceptualisations of those groups of people in the region who seemed to spend their entire life roaming the seas without any fixed abode on land; that is, to the vexatious category of "maritime nomads." To the European eye, the utter otherness, perhaps more ascribed than real, of their alleged erratic way of life epitomised this extremely heterogeneous, maritime region that resisted all attempts at neat classification.[29]

By 1856 Crawfurd entertained no doubt about the origins of these peoples: "This singular people, who are in fact maritime nomads, are, wherever found, in race and language genuine Malays."[30] The "singular" aspect in particular referred to the conjunction of a nomadic lifestyle with the oceanic realm.

25 Crawfurd, *Descriptive Dictionary*, p. 353.

26 "[T]he utter extirpation of Malayan piracy is as hopeless as that of theft and burglary in the best ordered states of society." Crawfurd, *Descriptive Dictionary*, p. 355.

27 Crawfurd repeated this argument throughout his entire career, both in his scholarly productions, as this one, and in his more polemical tracts published in newspapers. On the latter, see Crawfurd, "Malay Pirates"; Crawfurd, "Piracy in the Eastern Archipelago."

28 Amirell, "Civilizing Pirates," pp. 28–29; Warren, *Sulu Zone*, pp. 147–48.

29 Farish A. Noor recently has asserted that the floating signifier of the pirate during the nineteenth century was primarily seen as "a sign of decay and collapse, symptom of native degeneration." Noor, *Long Shadow*, p. 114. This essay, however, demonstrates that this was not the case; this interpretative framework partially relied and always had to compete with the idea that "piracy"—in practice though not in name—on the contrary represented the original state of affairs in maritime Southeast Asia.

30 Crawfurd, *Descriptive Dictionary*, p. 314.

This, indeed, seemed to constitute a deviation from the expected order; nomadism was usually associated with the plains or the desert, smooth and unbounded spaces too, but normally situated deep within the continental heartlands and far away from the wet element.[31] A nation composed of maritime nomads, on the other hand, constituted an ontological aberration; this triggered curiosity and begged further explanation.

These nineteenth-century ways of approaching and assessing Southeast Asian sea nomads drew upon a longer genealogy which from the very onset had endeavoured to render these people and their errant ways intelligible through comparisons to better known peoples or by inscribing into familiar categories such as that of piracy. Whereas William Dampier only made scant references to a brief encounter near "Pulo Nuttee" in the Strait of Singapore,[32] his contemporary Alexander Hamilton, a Scottish Country trader, emphasised the more bellicose aspects of sea nomadic activity in his *A New Account of the East-Indies* published in 1727.[33] This approach echoed earlier Portuguese and contemporary Dutch denunciations of the piratical ways of these so-called "Celates," a term referring to roughly the same people as those later labelled as sea-nomads.[34] Yet neither Alexander Dalrymple nor Thomas Forrest mentioned piracy in the often quoted decriptions of their first hand encounters with sea-nomads. Dalrymple and Fiorrest were probably the two most important figures in promoting the East India Company's interests in maritime Southeast Asia during the latter half of the eighteenth century. Writing on the Sama-Bajau of the eastern part of the archipelago Dalrymple as well as Forrest would instead emphasise their contributions as gatherers of maritime goods, especially of trepang (sea slugs), a coveted product on the Chinese market.[35] So, by the end of the eighteenth century three key features stood out in their descriptions of the non-sedentary, maritime based peoples of the region, viz. those of an itinerant way of life, of piracy and of a potential as gatherers of goods within a capitalist system. These would persist and see much elaboration throughout the next century.

31 On smooth and striated spaces in a maritime context, see Deleuze and Guattari, *Thousand Plateaus*, pp. 478–82.

32 Dampier, *Collection of Voyages*, vol. II, p. 4.

33 Hamilton referred to both Moken people and Orang Laut as "Salleiters" [Celates] who were notorious "freebooters." Hamilton, *New Account of the East-Indies*, vol. II, pp. 68–69, 159.

34 For an analysis of early European descriptions of sea nomads see Sopher, *Sea Nomads*, pp. 295–339. Hamilton is not mentioned by Sopher. See also Amirell, "Malay Pirate."

35 Dalrymple, *Extending the Commerce*, p. 68; Forrest, *Voyage to New Guinea*, pp. 373–74; Forrest, *Voyage from Calcutta*, p. 71.

More than forty years before Crawfurd characterised this "singular people" as maritime nomads in 1856, he had, in one of his first published texts on Southeast Asian matters, tried to encapsulate the essence of these "men of the sea," or Orang Laut, in the Malacca Strait by linking them to the well known ancient category of ichthyophagi, or fish eaters.[36] Quoting directly from Herodotus's description of the Padaoi, or "Indian ichthyophagi," John Anderson (1795–1845), another scholar-administrator and Malay translator would reiterate this comparison and, unlike Crawfurd, furthermore emphasise their "nomadic habits."[37] Both agreed on the Orang Laut resembling modern Malays in all essentials in terms of race and language, two of the most common indications of a shared lineage.[38] Perhaps more surprisingly, they both stressed a remarkable prevalence of cutaneous disfigurements in the shape of "scorbutic eruptions" among them, maladies primarily attributed to their diet as well as squalid and abject living conditions.[39] This latter aspect would in time evolve into one of the stable topoi used to simultaneously describe and condemn the mode of life associated with the sea nomads.[40] Anderson's description of the Orang Laut revolved around the notion of savagery most manifestly incarnated in an assumed former existence of cannibalism among them,[41] whereas Crawfurd by 1814 focused on their "simple and inoffensive" manners.[42] This tension between ascribed savage violence or primitive placidity would form a recurring theme in

36 Crawfurd, "History and Languages," pp. 161–62. Although not stated explicitly, it is indirectly suggested that Crawfurd may have had a first-hand knowledge of these people while stationed at Penang between 1808 and 1811. The article thus contains a word list containing "the language of the Orang Laut" (p. 162) which seems to be the very first printed in a European language. British Library (BL), India Office Records (IOR). Private Papers, Mack Private 85.

37 Anderson, *Considerations*, appendix, pp. xxxv– xxvii. See also Newbold, *Political and Statistical Account*, vol. II, pp. 414–16; Manickam, *Taming the Wild*.

38 See Crawfurd, "History and Languages," p. 162; Anderson, *Considerations*, appendix, p. xxxv.

39 Crawfurd, "History and Languages," p. 162; Anderson, *Considerations*, vol. I, pp. 34–35.

40 See Begbie, *Malayan Peninsula*, p. 9; Hamilton, *East-India Gazetteer*, vol. II, p. 21, 129; Horsfield, "Island of Bangka," p. 329; Newbold, *Political and Statistical Account*, vol. II, p. 413; Thomson, "Eastern Coast of Johore," p. 142.

41 Anderson's argument was hinged upon a connection suggested by John Leyden between the Battas [Batak], the notorious "lettered cannibals of Sumatra" and the ichthyophagi of Herodotus. Leyden, "Indo-Chinese Nations," p. 203. This was then extrapolated by Anderson to include the present-day ichthyophagi. Begbie debunked this linkage, and hence the anthropophagous past of the Orang Laut by pointing to Anderson's misreading of Leyden. Begbie, "Malayan Peninsula," p. 9–14.

42 Crawfurd, "History and Languages," p. 161. In the entry on the Orang Laut along the littoral of the island of Lontar, Hamilton would repeat these descriptions almost verbatim calling them "mild and inoffensive." Hamilton, *East-India Gazetteer*, vol. II, p. 129.

most discourses on these people throughout the nineteenth century. Most remarkable was, however, the glaring absence of any reference to piracy; this was not to last long.

The diplomatic mission to Siam (Thailand) and Cochin China (Southern part of Vietnam) in 1821–1822 yielded two descriptions of the Orang Laut living in the vicinity of the recently established colony of Singapore—one by the head of the mission, Crawfurd, and another by George Finlayson (1790–1823), the expedition's physician and naturalist. Finlayson focused almost entirely on their physiognomy, which he found superior to that of other Malays whose mode of living subjected them to less hardship.[43] Crawfurd's account was probably the first longer description with a distinct ethnological approach and based on first hand observations.[44] "I had no conception that any of the tribes bearing the Malay name were in so low a state of civilisation as these people are. By far the greater number of them are born, live, and die in their miserable canoes, and the few who live occasionally on land are scarcely more comfortable situated. These are ignorant of the culture of rice."[45] In other words, the sea nomads epitomised the "navigators and fishermen" who could be seen as representing the very first steps towards a more civilised state.[46] Once again the low level of civilisation of these erratic sea nomads would be framed in terms of being unspoilt: in character they were perhaps "indolent, improvident, and defective in personal cleanliness"; yet it was also a fact that they did not possess any "selfish, cunning, or mendacious traits."[47] The most important contributions of ethnological material in English on the sea nomads during the next decade were Peter James Begbie's "The Malayan Peninsula" (1834) and "Political and Statistical Account of the British Settlements in the Straits of Malacca" (1839) by Thomas John Newbold. Focusing on Orang Laut, these added new details to the understanding of their conditions, life and culture. However, in essence they followed the established pattern and described them as poor, shabby, primitive, geographically and politically marginalised but in most aspects peaceful and industrious too.[48] Later, Stamford Raffles' former scribe Abdullah bin Abdul al Kadir, also commonly known as Munshi

43 Finlayson, *Mission to Siam*, p. 75.

44 Crawfurd, *Journal of an Embassy*, p. 52–55.

45 Ibid., p. 53.

46 For more on these early nineteenth-century British debates regarding the necessary and sufficient conditions for the origins of a Malay civilisation, see Müller, "Manufacturing Malayness."

47 Crawfurd, *Journal of an Embassy*, p. 53.

48 Descriptions of groups can be found in Begbie, "Malayan Peninsula," pp. 9–12; Newbold, *Political and Statistical Account*, vol. II, pp. 411–16.

Abdullah (1796–1854), would express his disdain for these Orang Laut in quite undiluted terms in his autobiography, the "Hikayat Abdullah" published in 1849.[49] To what degree his denigratory assessments were influenced by nineteenth-century European ideas on race and civilisation begs further study, but Sandra K. Manickam suggests this possibility.[50] These descriptions represented some of the first and foundational instances of how sea nomads, and especially the Orang Laut in the Strait of Malacca, came in ethnographically oriented discourses to be seen as a sort of proto-Malays who embodied all the characteristics of the Malay civilisation in its infant stages.[51] They were living vestiges of a distant past, and as such the study of these could reveal otherwise hidden truths about the "original nature" of the Malays. Was this, then, inherently piratical?

By the time James Richardson Logan (1819–1869) began publishing *The Journal of the Indian Archipelago and Eastern Asia* in 1847, the link between piracy and a sea nomadic life seemed firmly established. Indeed, piracy appeared almost innate in these roaming people. John Turnbull Thomson had, among others, been instrumental in rendering this perception into an ethnological commonplace through an often quoted article in which he emphasised both the present-day "occasional quiet piracies" and the not that distant times when "these sea-gypsies were notorious for their piratical propensities."[52] Yet this association with piracy was already brought to the fore in Crawfurd's description of his experiences with these people in 1822, and Munshi Abdullah had also written about how he with his own eyes had seen that the "Sea Gypsies" of Singapore in 1819 collected "the skulls of men who were robbed at sea."[53] Crawfurd had sweepingly declared that "sometimes called the Orang-salat, or 'men of the straits', under which appellation they have been stigmatised for their piracy as long ago as the time of John [Jõao] De Barros, whose work was composed in the sixteenth century."[54] Also known as "Ryots" [Rayats], or subjects to

49 Abdullah, "Hikayat Abdullah," p. 129.

50 Manickam, *Taming the Wild*, p. 55. See pp. 43–71 for an analysis of how European anthropological ideas merged with Malay notions in Munshi Abdullah's texts on these issues.

51 Although not necessarily referring to the "Malay nation" as historically described in their own annals, or the "Malay race" in a more biological sense. For more on the entanglements of and intersections between these concepts in British nineteenth-century thought, see Müller, "Manufacturing Malayness."

52 Thomson, "Eastern Coast of Johore," p. 141.

53 Abdullah, "Hikayat Abdullah," p. 130.

54 Crawfurd, *Journal of an Embassy*, p. 53. For more on the exploits—ascribed and real—of the so-called "Celates" (Crawfurd's "Orang-salat") by De Barros and later Portuguese and Dutch writers, see Sopher, *Sea Nomads*, pp. 314–39.

the Sultan of Jehor [Johore], their piracies had made the use of "the term Jehor as synonymous with that of pirate or robber" among the Malays.[55] It is remarkable that Crawfurd here relied solely on historical sources and hearsay of Portuguese and Malay provenance; nowhere did he refer to his own observations or invoke contemporary knowledge as evidence. The inherent piratical inclination of all sea nomads—as an inevitable result of their ascribed primitiveness and errant mode of life—was, in other words, always more presumed than actually proved; yet this did not prevent it from becoming a ubiquitous epithet inextricably associated with Southeast Asian sea nomads in nineteenth-century discourses.[56]

Apart from being perceived as inveterate pirates in their own right and resulting from innate qualities ascribed to their primitive mode of living, the sea nomads would also be seen as subordinated abetters of piracy through their services to or alliances with other pirates. This represented a less developed, but actually much more frequently invoked, theory; this linked their piratical activities to collaboration with—or subordination to—other, more powerful and politically established bands of pirates such as the neighbouring Malay polities or the Lanuns or Illanuns [Iranun] from the Sulu region.[57] One of the first and most detailed descriptions hereof can be found in the report that American naturalist Thomas Horsfield wrote in 1814 on the island of Bangka to Sir Thomas Stamford Raffles, then lieutenant-governor of the conquered Dutch East Indian possessions.[58] Although not published until 1848,[59] it was certainly read by Raffles who referred to its content on several occasions.[60] Writing on the disruptive influence of pirates and mercenaries in recent history of this island, Horsfield distinguished between two kinds of pirates: Lanons [Iranun] and the Orang Laut, or "Rayads" [Rayats].[61]

55 Crawfurd, *Journal of an Embassy*, p. 53. For the historical background, see Barnard, "Celates, Rayat-Laut, Pirates."

56 See Skeat and Ridley, "Orang Laut," p. 249; Verschuer, "De Badjo's." For modern treatments see Osborn, *Quedah*, pp. 254–55; Sopher, *Sea Nomads*, pp. 253–60.

57 As mentioned in Horsfield, "Island of Bangka," p. 328. See also Amirell, *Pirates of Empire*, pp. 42–160; Tarling, *Piracy and Politics*; Warren, *Iranun and Balangini*; Teitler, van Dissel, and á Campo, *Zeeroof en Zeeroofbestrijding*; Warren, *Sulu Zone*; Angster, *Erdbeeren und Piraten*, pp. 247–64.

58 For more on Horsfield's sojourn to Bangka, his report and the contexts in which this was written and received, see Bastin, *Raffles and Some of His Friends*, pp. 90–157, 250–67, esp. pp. 140–42; Harrington, "Dr. Horsfield's Report"; Teitler, van Dissel and á Campo, *Zeeroof en Zeeroofbestrijding*, pp. 38–41.

59 Horsfield, "Island of Bangka," p. 299.

60 Raffles, "Discourse," pp. 9–11; Raffles, *Memoir of the Life*, pp. 149–51.

61 Horsfield, "Island of Bangka," pp. 314–15.

The Illanuns were described as the most active and intrepid, whereas the opportunistic Rayats were "of a disposition equally cruel and rapacious but less vigorous and enterprising."[62] Horsfield explained this behaviour by taking recourse to their general character and to the encouragement they received from powerful groups:

> Their character is by no means favourable delineated; the accounts uniformly agree in representing them as treacherous and very indifferent to as to the crime of murder. They possess a considerable degree of personal courage and enterprise: qualities which are improved by their mode of life exposing them to unceasing vicissitudes and hardships. But they employ these qualities often unlawfully to the detriment of others. If they can surprise a prow [prahu] of inferior force without the probability of being discovered, the contents are plundered and every person on board put to death. Accidents have several times led to the discovery of these atrocities many years after they were committed. But piracy were they at liberty to indulge it, is their more regular profession, and they have been too much supported by the princes and other inhabitants of Linga and Rhio, who have profited from their exploits. Advances have often been made them in money on condition of dividing with them the spoil, by which their piracies have been sanctioned and encouraged.[63]

What Horsfield offered here was a blend of these two prevalent modes of explanation that ascribed their piratical inclinations to an innate character derived from their mode of living as well as to the structural conditions prevalent in the region. These latter simultaneously excluded and included the sea nomads; on the one hand marginalised and derided as outcasts, yet their "piratical activities" formed an integrated part of Malay politics in the region. Horsfield never attempted to historicise the "custom of living on water" or to inscribe it into any deep history framework of primitiveness and civilisation.[64] Instead, he focused on a binary between those who lived solely on water "by necessity" and those who did it "by choice" despite having "possibilities which would afford a more comfortable mode of life."[65] This seems to be one of the earliest recognitions of how the geographical mobility

62 Ibid., p. 317.

63 Ibid., p. 330.

64 Horsfield's report primarily concerned itself with the more recent history of the island and its surroundings from mid-eigteenth century and onwards. It relied chiefly on oral testimomy obtained from "the most respectable persons" among the local populace. Ibid., p. 301.

65 Ibid., p. 328.

allowed by this mode of life could be converted into political independence, and this could be favoured by some even if it implied a materially poor and harsh life. Primitiveness seemed here more a deliberate choice than an inescapable condition.

The focus on own volition stood in contrast to what could be labelled the "poverty interpretation," and according to which the piratical activities of the sea nomads should first and foremost be attributed to their dismal material situation. Examples of the usage of this approach can be found in Dutch colonial administrator Cornets de Groot's "Paper Relating to the Piracies Committed in the Indian Archipelago, and to the Measures Adopted by the Netherland Government, in the Year 1815 to 1845, for Their Repression." An extract from a report written by the Commissioner of Borneo, M. J. H. Tobias, in 1822 thus stated in unequivocal terms that the "pirates (Rayats) of these districts are only so from poverty."[66] Another report written in Malay by Pangaran Said Hassan Alhabashy in 1830, and likewise quoted by de Groot, traced one of the ways in which sea nomads became pirates due to their unwillingness to subject themselves to the laws of Malay rulers in combination with non-sedentary lifestyle; this prompted them to elope whenever they were not satisfied with the treatment they received.

> When, contrary to usage, other penalties [than those complying with their own traditions, and which they are willing submit to] are applied to the Rayats, they take to flight; whole *soekoes*[67] abandon the territory of their Princes, and go to seek refuge at Borneo or Sumatra, or other islands, where they hope to obtain the means of living. In time all these fugitives become pirates. They seek in preference those who show a readiness to assist them; it is thus that they obtain from the petty Princes of Sumatra and Borneo rice, stores, and arms, on condition that the booty they make shall be divided into three parts, two for the pirates, and the third for those who have supplied them with necessaries.[68]

Just like Horsfield, Alhabashy stressed their active agency and capabilities to exploit the advantages of geographical mobility to navigate dexterously in the intricate world of maritime Southeast Asian politics.[69] In addition, he

66 Groot, "Piracies," p. 74.

67 Organisational subdivisions of the Rayats according to Alhabashy. These are led by a chief, called a "Ratin."

68 Groot, "Piracies," p. 85.

69 For more on the different, and often conflicting, views on these groups when seen from different perspectives and assessed in different frameworks of international relations, see Ota,

seemed to suggest a reversal of the power relation between sedentary and itinerant groups in the region; just as much as resulting from marginalisation, their freedom of movement accorded the sea nomads a propitious position of negotiation vis-á-vis sedentary rulers. This in the end facilitated their survival as discrete ethnic groups with the ability to act as autonomous political units.[70]

## Placing Piracy: Territorialising Violence on and beyond the "High Seas"

The second issue raised in the initial Crawfurd quote is of a fundamentally spatial nature; it results from the fact that piracy was defined as "robbery on the *high seas.*"[71] This reflected a well established European legal tradition,[72] notwithstanding discrepancies in the exact definition of what constituted the high seas and whether it should, for instance, be positively or negatively defined.[73] Especially in the context of piracy it proved difficult to find consensus on a viable definition of the "high seas." Writing in 1874, A. T. Whatley claimed that the "'high seas' must be interpreted, not in its general sense of all the sea beyond an imaginary line drawn from point to point on the coast, but so as to include bays, harbours and rivers up to the first bridge, which are the limits of the jurisdiction of the admiralty."[74] Although this spatial setting of the high seas may serve as a legal prerequisite in defining an act as piracy, the violent activities of pirates were only seldom confined to the watery realm, and some legal definitions of piracy explicitly included attacks "by descent from the sea on the coast"; yet others, like Whatley's, strongly contested the validity of such a claim.[75] In the context of alleged piracy in a Southeast Asian setting this was even more so the case, both with regard to depredations on coastal shipping and to the just

"Pirates or Entrepreneurs"; Ota, "Trade, Piracy, and Sovereignty."

70 Andaya, *Leaves of the Same Tree*, pp. 199–200.

71 Crawfurd, *Descriptive Dictionary*, p. 353.

72 See Kempe, "Remotest Corners"; Kempe, *Fluch der Weltmeere*; Heller-Roazen, *Enemy of All*; Policante, *Pirate Myth.*

73 Defining the sea negatively as something that could never be seized due to its inexhaustible size and unfenceable nature, Grotius came to the conclusion that it logically had to remain common property for all. Fitzmaurice, *Sovereignty*, p. 92.

74 Whatley, "Historical Sketch," p. 551.

75 Whatley, on the contrary, emphasised that piracy should first and foremost be defined in terms of the place where the deed happened rather than from whence it proceeded. Ibid.

as frequent raids on land.[76] This apparent discrepancy between definition and reality on the ground did not, however, imply that the former was totally devoid of practical influence, or that the relationship between the two spheres can be construed as a binary composed of fiction versus fact. On the contrary, seeing and perceiving with and through the pirate category exerted a profound impact upon how insular Southeast Asia and the vast heterogeneity of the polities and peoples in and along its maritime zones were conceptualised, compared and, ultimately, controlled. As such, looking for pirates everywhere both prefigured the geopolitical approaches to the region and prescribed the range of policies to be adopted in regulating and ruling it.[77] In terms of spatial scope too, the piracy category involved more than merely hunting pirates on the high seas; from the very onset it transgressed the imaginary boundary between the high seas and the littoral with its innumerable inlets, rivers, mangrove swamps, etc. British, Dutch and Spanish authorities all extended their anti-piratical policies to include punitive expeditions and amphibious campaigns against alleged pirate haunts on land. These were often composed of entire communities whose inhabitants were collectively branded as pirates through "ethnologising" discourses that ascribed piratical propensities to pertain to entire groups as a result of their race, stage of civilisation or as an intrinsic part of their specific culture.[78] The sanguinary campaigns against the so-called Sea Dyaks along the northern coast of Borneo in the years between 1843 and 1849 conducted by Royal Navy commanders at the behest of James Brooke, the self-proclaimed Rajah of Sarawak, are probably the best known examples of such policies due to the intense political controversies they stirred back in Britain.[79] But neither the Spanish nor the Dutch desisted from applying similar measures, as illustrated by, for instance, the Spanish expedition against Balanguingui in 1848 and the conquest of Jolo in 1876,[80] as well as by the numerous large and small punitive expeditions undertaken by the Dutch against denounced piratical dens.[81]

76 As also stressed in Gaynor, "Piracy in the Offing," p. 841; Rubin, *Law of Piracy*, p. xiii.

77 For more on these issues, see Amirell, "Civilizing Pirates"; Müller, "Embodying Piracy."

78 Amirell, "Civilizing Pirates," p. 41; Amirell, *Pirates of Empire*, pp. 229–31; Angster, *Erdbeeren und Piraten*, p. 248; Müller, "Embodying Piracy," pp. 206–7.

79 See Knapman and Müller, "Protector of Aborigines"; Middleton, "Rajah Brooke."

80 For contemporary Spanish accounts see García de Arboleya, *Historia del archipelago*, pp. 30–37, appendice pp. 3–24; Bernaldez, *Reseña historica*, pp. 153–223; Montero y Vidal, *Historia de la piratería*, pp. 404–751. These expeditions and their political contexts are treated in Amirell, *Pirates of Empire*, pp. 48–64; Warren, *Iranun and Balangini*, pp. 343–78.

81 For examples in the period from 1816 to 1845 see Groot, "Piracies." On Dutch naval and amphibious operations, see Teitler, van Dissel and á Campo, *Zeeroof en Zeeroofbestrijding*,

Despite such overt spatial transgressions, the epistemic influence as well as the instrumentalist invocations of the pirate category continued to represent a fundamentally seaborne view; during the nineteenth century this resulted in colonising processes that were firmly embedded in what might be termed the primacy of a seascape episteme.[82] Perhaps, the best way to reconceptualise this dynamic is through the notion of a "maritime frontier" manifested through the imposition of the pirate category upon, first, the coastal zones and the littoral, only later to be pushed further towards the hinterlands.[83] Obviously, this was not a frictionless process; not only did it meet fierce resistance from the communities who found themselves at the receiving end of its thrust, but the change of geographical setting from the smooth seas to the rugged coasts and bended river environments usually tended to water down the legitimacy of the legal regime that followed with the imposition of the piracy category, even if it did not, as already seen, halter the process altogether. The ways in which the category of piracy both served to uphold and at the same time continuously crisscrossed the invisible lines of demarcation between high seas, narrow seas, littoral and terra firma in the nineteenth-century Southeast Asian setting exemplify how an assertive maritime discourse and practice left its indelible mark on the political formations of the period on global, imperial, regional and very local scales.[84] Furthermore, applying a focus on how the usages of the piracy category depended on the kind of spatial framework in which it was invoked turns this thematic in one of those "sites of contestation" that, in the words of Matthew Taylor Rafferty, allow us to "study the overlapping, disputed, and hybridising character of sea law."[85] The nineteenth-century anti-piracy policies in Southeast Asia provide an emblematic example of this. For instance, as pointed out by Lauren Benton and Lisa Ford, the Royal Navy officers on the spot always employed a wide range of registers in order to defend their anti-piracy campaigns; these included, inter alia, appeals to (1) bilateral treaties to police certain waters, (2) the law of war, (3) principles of self-defence, and (4) routine keeping of the king's peace in places adjacent to or within imperial territories.[86] Such a "smörgåsbord" of legitimating strategies resulted in what at times could appear to be a rather cavalier

pp. 191–238, 212–15 on punitive expeditions in the "outlying areas."

82 For a plea for the necessity to stretch, bend and rework the categories of social analysis to accommodate "ocean-centred realities," see Wigen, "Introduction," pp. 16–17.

83 See Maier, "Transformations of Territoriality"; Hopkins, *Ruling the Savage Periphery.*

84 Benton and Perl-Rosenthal, "Introduction," p. 14.

85 Rafferty, "Law Is the Lord of the Sea," p. 54.

86 Benton and Ford, *Rage for Order*, p. 131.

attitude towards who could be considered pirates as well as to how these ought to be dealt with. Instead of taking the alleged pirates prisoners and bringing them back to the cumbersome and uncertain process of a legal trial in a British court of law, there seems to have been a preference among the Royal Navy officers towards exerting a "swift justice"; not infrequently this resulted in outright annihilation of the so-called pirates encountered on the high seas, along the coasts, or in home communities sometimes based up river far away from the open seas.[87] Until the Pirates (Head Money) Repeal Act in 1850 abolished the right established with the Bounty Act in 1825 to collect "head money" for pirates taken, secured or killed (£20 a head) or dispersed (£5 a head) this furthermore carried a considerable economic incentive for the Royal Navy officers on the spot to subsume all acts of undesired indigenous violence on or along the seas in Southeast Asia under one category of piracy that could be extended to include entire communities.[88]

Once branded as "piratical," not just individuals but also entire communities[89] could be subjected to the draconic measures meted out by colonial powers allegedly acting in accordance with an invoked "Law of Nations" and destroyed in a de facto grey area between policing the seas and outright military intervention.[90] Sir James Brooke thus explicitly distinguished between "the general piratical character of a community" and "specific acts of piracy committed by its members," as if it was the first that automatically prompted the latter, rather than the cumulative effects of these acts that resulted in this ascribed "piratical character."[91] Notwithstanding whether primarily defined as *hostis humani generis* (Natural Law) or as *hostis communis omnium* (Positivist Law), the decisive aspect of the figure of the pirate was his inherently un-political, or perhaps better *de-politicised*, nature. As a category, the pirate is, *sui generis,* an entity devoid of political motivation and function—he is discursively placed outside of society and conceptually he lingers beyond the "law of nations"[92]—and as such he represents the exact inversion of Carl Schmitt's foundational figure of the *partisan*.[93]

87 Rubin, *Law of Piracy*, p. 231.

88 On the money paid in 1850 (more than £100,000), and which occasioned the Repeal Law, see "Additional Papers Respecting the Operations Against the Pirates on the North-West Coast of Borneo: Pirates' Head-Money (Borneo), April 15, 1851 and June 27, 1851" in *British Parliamentary Papers. Accounts and Papers: 1852–1853*, vol. 5.

89 See Müller, "Embodying Piracy."

90 Policante, *Pirate Myth*, p. 138.

91 "Letter from Sir James Brooke to Lord Stanley, October 4, 1852" in *British Parliamentary Papers. Accounts and Papers: 1852–1853*, vol. 5.

92 Policante, *Pirate Myth*, pp. 133–39.

93 Schmitt, *Theory of the Partisan*, p. 10.

Schmitt argued in his polemical short tract on "The Concept of Piracy" that, according to the prevalent European views, "action against pirates is unpolitical as well. It is not war, but either criminal justice (according to the English view) or a measure of international maritime police power (according to the Continental construction)."[94] This produced a situation fraught with conceptual tensions and moral ambiguities given that, in the words of Simon Layton, "piracy was politicised at the same time that its politics were denied, and branded with an ideology that stood to challenge the twin projects of commerce and civilisation."[95] In concordance with a naturalist interpretation, British jurisdiction was thus extended to include foreigners who depredated upon other foreigners on the high seas.[96] In addition, this notion of piracy was, as pointed out by Alfred P. Rubin, even extended to "organised political groups operating on land as well as at sea to interfere with the British version of freedom of navigation on the high seas."[97] Following this logic, indigenous maritime violence directed against European shipping—and, by capitalist extrapolation, native shipping too[98]—was systematically deprived of any political meaning and reduced to a criminal activity motivated by greed and, in theory, directed against all mankind.[99] Such delegitimising procedures left little or no space for privateering sanctioned by (internationally recognised!) indigenous states, intertribal warfare or any kind of seaborne violent activities committed by seemingly non-territorially defined ethnic units such as the sea nomads.[100] In his "Memorandum on the Suppression of Piracy" from 1845, Brooke insisted that the right of war could only be conceded to recognised states and not to "chiefs who have seized on territory and arrogate independence making their independence a plea for piracy," given this would imply that "the piratical chiefs, no longer in dread of punishment from European powers, would doubtless declare war against every unwarlike native state which they did not need as a market for the sale of their slaves and plunder."[101] This wide interpretation of what constitutes piracy has by Satsuma Shinsuke been

94 Schmitt, "Concept of Piracy," p. 28.

95 Layton, "Discourses of Piracy," p. 93.

96 Rubin, *Law of Piracy*, p. 227.

97 Ibid., pp. 231, 232, 235.

98 As constituting an integrated and indispensable part of the market network that fed emporia like Singapore with commodities to be shipped on as part of the interregional and increasingly global trade.

99 Policante, *Pirate Myth*, p. 142.

100 See also Rubin, *Law of Piracy*, p. 235.

101 "Memorandum on the Suppression of Piracy and Extension of Commerce in the Eastern Archipelago. Mr. Brooke to the Earl of Aberdeen; Singapore, March 31, 1845, Enclosure. Additional

linked to the contemporary efforts to abolish all privateering as a legal mode of naval warfare, and which resulted in the Declaration of Paris in 1856.[102] Despite obvious discrepancies in political organisation and institutional backing between privately funded European guerre de course and modes of non-European maritime raiding in Southeast Asia, they were both seen as hindrances to the global spread of the kind of aggressively asserted and advocated free trade that went hand in hand with imperial expansion, whether informal or formal.[103] The de- and reterritorialisation involved in these imperial place-making processes exerted a profound influence upon both land and sea—as well as upon the legal, political and indeed epistemological qualities attributed to these spatial entities. The sanctity of the freedom of the seas from predatory incursions from all non-state actors rendered such practices of maritime raiding obsolete within this new world order devised under imperial auspices.

However, we should not be tempted to frame our analysis of colonial conceptualisations of Southeast Asian itinerant communities as "deviants" within a fixed and supposedly unchanging land-sea binary inveterately opposed to each other in a Schmittian sense.[104] As argued, such an approach leaves out the crucial space in between, viz. that of the littoral. Until recently almost always analytically approached from a land perspective, this overlapping between land and sea has been prone to much neglect in the historiography.[105] Yet, as stressed by geographer Philip E. Steinberg, "in fact, just as coasts are exceptionally rich spaces for understanding the linkages that constitute maritime regions, they are useful spaces for unpacking the fundamental binary between land and water (or dry space and wet space) that underpins the modern notion of state territoriality."[106] Rather than constituting a fixed border between two well-demarcated domains, the littoral represented a porous frontier open to influxes from both sides.[107] It was here, twixt land and sea, that notions of territory clashed and coalesced

Papers Respecting the Operations Against the Pirates on the North-West Coast of Borneo" in *British Parliamentary Papers. Accounts and Papers: 1851*, vol. 26, 1.

102 Satsuma, "Plunder and Free Trade," pp. 59–65.

103 Satsuma, "Plunder and Free Trade," p. 64. See also Lapian, "Violence and Armed Robbery," pp. 132–39; Pearson, "Piracy in Asian Waters," pp. 18–20.

104 Schmitt, *Land and Sea*, pp. 11–20.

105 Bentley, Bridenthal, and Wigen, *Seascapes*; Pearson, "Littoral Society"; Steinberg, *Social Construction*.

106 Steinberg, "Of Other Seas," p. 163.

107 Benton and Perl-Rosenthal, *A World at Sea*; Hopkins, *Ruling the Savage Periphery*; Maier, "Transformations of Territoriality."

with sea space, or maritory,[108] and created a both fixed and fluid abode for what Michael N. Pearson has termed "littoral societies."[109] Nowhere was this more pronounced than in the Southeast Asian region.[110] It was precisely in this space that the Southeast Asian sea nomads stayed, moved and lived, and the nature of which they—with their elusive manners—proved instrumental in defining to the bewildered European who found himself struggling with ascribing them an adequate place within his preconceived conceptual frameworks and implemented political strategies. Seemingly moving as effortlessly across the high seas as lurking along the littoral, the "Orang Laut or sea gypsies" encountered by Thomson along the eastern coast of Johore in 1849 had the coastal areas along Java, Sumatra and the Malayan Peninsula within their usual "cruizing ground."[111] The sea connected rather than it separated, such as these seas in particular came to do with the imposition of the modern territoriality-based colonial regimes of the Dutch, British and Spanish. These colonial regimes operated according to the rationale of the "cartographic state"—with its hegemonic notion of territorial exclusivity divided by either well demarcated boundaries between civilised states or by a dynamic frontier between the civilised and the primitive.[112] In contrast, the life-worlds of these sea nomads seemed to be organised as an oceano-centric space, and in which the littoral (approached from seaside) made up the defining periphery. In between these littoral peripheries lay an empty oceanic space that primarily mattered to the sea nomads as a smooth space that allowed mobility between these geographically separated but ecologically similar littoral abodes. It was along these coasts that they eked out their meagre living by fishing, collecting, bartering as well as, whenever deemed possible, pillaging and abducting people to be sold into slavery.[113] The problem was not only their—in the European eyes[114]—illegal activities and the pernicious implications they had upon trade and prosperity in the region; their seemingly elusive existence and the impossibility to pin them down to any fixed place proved to be almost as challenging to

108 To use Tom Gunnar Hoogervorst's terminology. Hoogervorst, "Aquatic Lifestyles," p. 245.

109 Pearson, "Littoral Society."

110 However, as emphasised by late historian Peter Boomgaard, compared to the "strong emphasis on everything terrestrial" Southeast Asian historiography has always been characterised by its "concomitant neglect of aquatic aspects." Boomgaard, *World of Water*, p. 1.

111 Thomson, "Eastern Coast of Johore," p. 141.

112 This term is borrowed from Branch, *Cartographic State*.

113 Thomson, "Eastern Coast of Johore," pp. 141–45.

114 Modern anthropological field-studies suggest, however, that their movements were by no means as erratic and uncontrolled as imagined by these nineteenth-century observers and administrators. Chou, "Tenure of Territoriality," pp. 616–17.

imperial place-making regimes based on notions of territoriality absolutely incompatible with the maritime mobility and seaborne perspective of the sea nomads.[115]

## Settling the Problem of Sea Nomadic Piracy

Although never seen as the worst of the Southeast Asian pirates, the different sea nomadic groups inhabiting maritime Southeast Asia were nonetheless often presented as the most emblematic; their itinerant ways both epitomised the purportedly original state of Malay civilisation, and its non sedentary, maritime foundations constituted the very antithesis to modern territoriality. As such, the solutions proposed to eradicate their alleged piratical habits and the methods to be applied in these processes can be seen as exemplary of the circumscribing and executing modalities through which imperial place-making regimes worked on the ground—as well as on the seas and along the littoral. In dealing with the allegedly inherent piratical nature of the sea nomads three solutions repeatedly came up: making them more manageable through assimilation into sedentary society; annihilation by violent means or a gradual and "natural" extinction; and integration into an imperially controlled global capitalism as gatherers of maritime products.

The first one relies on the assumption that they would gradually be absorbed into the sedentary parts of the local Malay populace, such as it was intimated by the anonymous author of an article entitled "Rajah Brooke and Borneo" (1850). Driven by "the new principles of society introduced from the West," he predicted that these sea nomads "will probably disappear from history, melting away into those populations which, more numerous and powerful than themselves," and who were more prompt to comply these new organising principles of society according to territorial precepts.[116] Rather than constituting an actively pursued and well defined policy, this solution often only appeared implicitly in the discourses on piratical sea nomads; it was presented as the natural—or, indeed, inevitable—result of the dynamics of progress of civilisation. If the (former) sea nomads should stand any chance of survival they must submit to the requirements that followed with the imposition of colonial order and a capitalist economy,

115 The sea nomads also practiced a sort of territoriality, as argued by Cynthia Chou, but this often manisted itself in the "concept of a network of interrelated territorialities." Chou, "Tenure of Territoriality," pp. 616–19.

116 Anon., "Rajah Brooke and Borneo," p. 30.

and this left no space for unregulated roaming on the high seas or along the littoral. Besides, it was universally acknowledged within European historical and ethnological theories that a prerequisite for any substantial progress in terms civilisation implied a sedentarisation and intensified cultivation of the soil.

In contrast to this assumedly "natural" process of civilisational progress the Dutch and the Spanish seemed to favour a policy of enforced sedentarisation of former pirates. The latter began to adopt this policy in earnest in the Sulu archipelago during the latter half of the nineteenth century after gaining a firmer military and naval control in the region as a result of a series of decisive operations between 1848 and 1851. They became inclined to pursue this policy more consistently after other imperial powers, in particular the Dutch and the British, began to show genuine interest in the region, and a more activist enforcing of Spanish sovereignty on the ground hence became imminent.[117] It was most vociferously advocated in a small, but influential tract published in 1872 by Santiago Patero, an experienced naval officer.[118] After having pointed to the insufficiency of merely applying armed means to control and curb piracy,[119] Patero suggested that, in addition, firmer control could in the long run only be obtained through diminishing the power of the local aristocracy, introducing Christianity, severing their connections to the Muslim sphere and, not at least, by destroying their marine economy and maritime culture. This would in time "by force or by necessity" make them, or at least their children, embrace an agricultural life and sedentary habits.[120] With regard to those people of the Sulu zone whom he called the race of the "Sameles," and who would have to be the Samal Bajau Laut (Bajau sea nomads),[121] he recommended a policy that would "encourage the emigration from these islands" given that they were deemed "more difficult to subdue."[122] Their aquatic mode of life probably rendered them incorrigible. Besides, they made up a relatively small part of the population; so, an expulsion of them would not be inconceivable.

117 Amirell, *Pirates of Empire*, pp. 48–61; Manzano, *El Imperio Español*, pp. 137–69; Warren, *Sulu Zone*, pp. 116–25.

118 On the impact of this book upon Spanish attitudes towards the Sulu zone, see Warren, *Sulu Zone*, pp. 118–19.

119 Patero, *Sistema que conviene*, pp. 5–8.

120 Ibid., p. 40. My translation. See also Amirell, *Pirates of Empire*, p. 59.

121 See Warren, *Sulu Zone*, p. 67. Patero's description of their physical features fits well with other nineteenth-century accounts of the sea nomads as being darker skinned and physically fitter than their other indigenous counterparts. Patero, *Sistema que conviene*, p. 42.

122 Patero, *Sistema que conviene*, p. 42. My translation.

Cornets de Groot's "Paper Relating to the Piracies Committed in the Indian Archipelago" offers several examples of Dutch attempts to "pacify" former pirates by forcing them to settle inland, leave their maritime ways behind and turn towards tilling the soil. In addressing piracy along the island of Lingga at the southern entrance to the Straits of Malacca, M. H. W. Mutinghe established a clear link between piracy and not cultivating the soil in a report from 1818.[123] These alleged pirates—who by the description of them seem to belong to the Orang Laut or Rayat[124]—preferred to live upon fish and sago rather than having to occupy themselves with the more cumbersome task of producing rice. Turning former pirates towards an agricultural pursuit with its strictly sedentary life was thus not only meant to draw them away from the sea; it also found legitimacy in the assumed civilising effect inherent in hard labour within a fixed framework organised along hierarchal lines that left no space for unsupervised roving. Rear-Admiral Melvill de Carnbee predicted in 1826 that piracy would continue in the region until the European powers had "convinced the native Princes of the disgraceful character of piracy, and until they have shown their subjects a more honourable means of existence in agriculture and the export of their produce."[125] Groot also refers to at least four attempts between 1833 and 1840 to establish new agricultural colonies composed of vanquished pirates. These followed in the wake of a policy suggested in 1821 by M. J. H. Tobias, then Commissioner of Borneo. In 1833 it was decided "to settle some pirates as [agricultural] labourers" at the island of Tana Tjampea near Celebes [Sulawesi]; 500 pirates were transported to Banjermassin on Borneo to be employed in agriculture in 1835; following a plan devised by M. J. N. Vosmaer, the island of Selayar south of Sulawesi was used to settle 400 former pirates from Tobello in 1837; and three years later, 400 new ones seem to have been transferred to Saleayer.[126] De Groot did not comment upon the rate of success of this policy, except from a reference to an article in the "Gazette of Java" from October 21, 1837, and which left the conclusion entirely open: "It is clear, and experience has shown that too much confidence must not be attached to such submissions, still there are reasons for supposing that the trouble which the Governor has taken for diminishing in this way the number of pirates will not be entirely without benefit."[127] However, according to J. N. F. M. á Campo this

123 Groot, "Piracies," p. 66.
124 Ibid.
125 Ibid., p. 79.
126 Ibid., pp. 92, 99, 103, 108.
127 Ibid., p. 103.

relocation of former pirates did not prove to be a sustainable project, and interest in it faded when Vosmaer was transferred to another position.[128]

Notwithstanding their limited range of application and meagre success, these Spanish and Dutch approaches resembled other contemporary colonial policies of curbing and containing peoples on the move. Other examples include the internment of indigenous peoples in reservations, and the processes of enforced "peasantisation" as happened on a much larger scale in British India during the same period. Not only were these approaches comparable; often they were connected on a conceptual, institutional or personal level too. Contemporary commentators were well aware of these shared situations and referred frequently to them. Referring to the successful campaigns against the so-called Thugs—an alleged murderous itinerant tribe or race who, at least in contemporary British imagination, made the roads of India utterly unsafe[129]—a British commentator in 1849 pleaded for the arrival of a "naval Sleeman" who could deal with the roaming pirates in Southeast Asia in the same draconic and efficient manner as done by Captain William Henry Sleeman on the Indian subcontinent a few years earlier.[130]

The second "solution" consisted in the outright physical annihilation of these sea nomads through a mixture of violent anti-piratical campaigns, establishing a firm colonial control and their subsequent dwindling away as a result of lost hunting grounds and curbed possibilities to roam free. If the sea nomads proved too resistant to or incapable of settling down and becoming assimilated into their larger, more sedentary neighbours, then extinction seemed the only alternative. This constituted a foundational premise in the merciless "progress or perish rationality" implied in so much of nineteenth-century European theorising on the future of those people they considered too primitive to cope with the unstoppable modernity they themselves epitomised. Even if not explicitly evoked, narratives of an ever more marginalising displacement as well as determinist extinction discourses always loomed ominously in the background when addressing the place accorded to non-sedentary peoples in future space.

Unlike the two former "solutions," the last one was not predicated upon the disappearance of the sea nomads as ethnically distinct polities. Instead, this solution aimed at eradicating the sea nomads' piratical activities, but

128 The establishment was dismantled in 1842 after a smallpox epidemic. Teitler, van Dissel and á Campo, *Zeeroof en Zeeroofbestrijding*, p. 93.

129 Wagner, *Thuggee*, esp. pp. 213–28.

130 This appeared in an article entitled "Malay Amok and Piracies: What can We do to Abolish Them?" and which is normally attributed to John Richardson Logan, editor of "The Journal of the Indian Archipelago." Logan, "Malay Amok," p. 465.

they should otherwise continue with their established mode of life and traditional modes of production. Their subsistence economy was, however, supposed to be substituted with integration into the world market organised around the colonial centres from which trading networks emanated out to even the remotest corner of the archipelago. These colonial centres should replace the Sulu Zone as the central hub for the trade in Southeast Asian marine goods, and although Singapore ultimately ended up as the uncontested emporium, an intense colonial competition took place during the first half of the nineteenth century which saw many failed attempts at anticipating, emulating or outdoing Singapore.[131] The Sulu Zone had from the late eighteenth century and onwards become a centre for the collection of marine products, such as pearls, agar-agar, tripang (dried sea-slugs), shark fins and edible swallow nests, all destined for the Chinese market. This turned it into an entrepôt in the growing global trade network as the provider of one of the few foreign products in genuine demand in China (before the boom in the opium trade).[132] The inclusion into the global market proved to be crucial, as James F. Warren argues, in the forging the Sulu zone as a culturally heterogeneous but economically integrated and politically unified region revolving around the Taosug dominated Sulu Sultanate and its centre on Jolo. These processes, furthermore, spurred the surge in piratical and slave raiding expeditions led by the Iranun or Balangini that soon became the scourge of most of maritime Southeast Asia. According to Warren, these expeditions first and foremost served to satisfy a demand for labour to collect these marine goods.[133]

This implied that the sea nomads were not merely a piratical nuisance; they could also be seen as a potential source of capitalist exploitation. This was primarily a British concern, but the Dutch were not blind to the possibilities offered by this scenario. Based on information from George Windsor Earl's publications, James Brooke saw in these people a potential ally even before he shipped out to the East in 1838: "The Bajoos [Bajau] or sea gypsies, are another race on whom some dependence may be placed, particularly if they be freed from the trammels of debt, swindled upon them by the Malays."[134] It was implied that after being relieved them from the fetters

131 See Bastin, *Raffles and Hastings*; Graham, *Indian Ocean*, pp. 329–46, 402–43.

132 See Tagliacozzo, "Necklace of Fins," pp. 32–37; Warren, *Sulu Zone*, pp. 67–125; Warren, "Global Economy," pp. 66–81.

133 Warren, *Iranun and Balangini*, pp. 215–51; Warren, *Sulu Zone*, pp. 309–42; Warren, "Global Economy," pp. 82–102. For a critique of this interpretation see Gaynor, "Piracy in the Offing," pp. 842–44.

134 Templer, *Private Letters*, vol. I, p. 30.

of debt bondage the forces of the free market would automatically attract them to Singapore or to a local port within its extensive network. Likewise, the Dutch commissioner Tobias had in 1822 pointed out that the Rayat (Orang Laut) only undertook piratical activities "from poverty; it would be sufficient to furnish them with better means of subsistence to make useful subjects of them."[135] Echoing this assessment, the anonymous author to an influential article on "The Piracy and Slave Trade of the Indian Archipelago" referred to a report written by the Dutch civil servant Johan Christiaan van Angelbeek in 1825. The focus was on how, with regard to the sea nomads, "Mr. Angelbeck [Angelbeek] advocated the re-establishment of the fishing of agar-agar and of tripang, which the Orang Laut had for some time been forbidden to pursue, subject to a system of checks by means of passes &c. as the only means by which the [piratical] propensity of the Orang Laut could be repressed, short of their total extirpation."[136] Driven by equal measures of incentive and control, they either adapted to the capitalist market system or they perished. In addition to their contributions as gatherers of marine goods, they could also facilitate trade in the region by serving as skilled pilots as emphasised by John Tunbull Thomson.[137]

As argued in this chapter, the sea nomads of Southeast Asia became firmly inscribed into discourses that gravitated around the pirate category throughout the nineteenth century. This, then, prescribed a fixed set of, often rather draconic, means devised to combat, control, and regulate these alleged pirates. What particularly mattered was the ascribed spatiality of the high seas and this notwithstanding whether this spatial association corresponded or not with the actual spaces in which the alleged piratical sea nomads mainly moved. Compared to other ethnic groups upon whom the piratical definition was imposed—like the mighty Illanuns, the feared Balagnini, the more "advanced" Malay societies, and the headhunting Sea Dyaks—the different groups of sea nomadic peoples were chiefly represented as petty, abject thieves whose piratical depredations were of an opportunistic character. This influenced the proposed solutions to the particular problem of piracy they posed; instead of launching large punitive expeditions against them (which would also have been futile given their mobility and lack of fixed abodes) the methods employed to drive them away from the high seas consisted either in enforced sedentarisation

135 Quoted in Groot, "Piracies," p. 74.

136 Anon., "Piracy and Slave Trade," p. 635. For more on Angelbeek's report see Teitler, van Dissel and á Campo, *Zeeroof en Zeeroofbestrijding*, pp. 35–38.

137 Thomson, "Eastern Coast of Johore," p. 142.

and taking up agriculture, or in a strictly supervised integration into the capitalist market economy where their traditional competencies were deemed profitable.[138] Whereas nomadism on the high seas almost by its very existence equalled piracy, the culture, character, and essential nature of the very same peoples could, under the right circumstances, be evaluated according to entirely different criteria when assessed in a context focusing on their coastal affiliation. Then it was rather their capacity and potential as gatherers or producers of maritime products, within a framework of an ever more entangled global economy, that mattered. To the colonial mind the only other conceivable alternative to these two scenarios was marginalisation and gradual extinction as a distinct ethnic group defined by its highly mobile, non-sedentary way of life.

## Bibliography

### Archival Sources

British Library, India Office Library and Records. Private Papers, Mack Private 85. Crawfurd, John. "Review of a Grammar of the Malayan Language with an Introduction & Praxis, by William Marsden, F.R.S."

### Published Sources

Abdullah bin Abdul al Kadir. "The Hikayat Abdullah. An Annotated Translation by A. H. Hill, M. A., B.Sc." *Journal of the Malayan Branch of the Royal Asiatic Society* 28, No. 3 (June 1955): 3–354.

Amirell, Stefan Eklöf. "Civilising Pirates: Nineteenth Century British Ideas about Piracy, Race and Civilization in the Malay Archipelago." *HumaNetten* 41 (Autumn 2018): 25–45.

Amirell, Stefan Eklöf. *Pirates of Empire: Colonisation and Maritime Violence in Southeast Asia*. Cambridge: Cambridge University Press, 2019.

Amirell, Stefan Eklöf. "The Making of the 'Malay Pirate' in Early Modern European Thought." *Humanities* 9, no. 3 (2020): 91.

Anderson, J. L. "Piracy in the Eastern Seas, 1750–1850: Some Economic Implications." In *Pirates and Privateers: New Perspectives on the War on Trade in the Eighteenth and Nineteenth Centuries*, edited by David J. Starkey, E. S. van Eyck van Helsinga and J. A. de Moor, 87–105. Exeter: University of Exeter Press, 1997.

138 Müller, "Embodying Piracy," pp. 210–12.

Anderson, John. *Considerations Relative to the Malayan Peninsula and the British Settlements in the Straits of Malacca*. Prince of Wales Island, 1824.

Andaya, Barbara W. "Recording the Past of 'Peoples Without History': Southeast Asia's Sea Nomads." *Asian Review* 32, No. 1 (2019): 5–33.

Andaya, Leonard Y. *Leaves of the Same Tree: Trade and Ethnicity in the Straits of Melaka*. Honolulu: University of Hawai'i Press, 2008.

Angster, Julia. *Erdbeeren und Piraten: Die Royal Navy und die Ordnung der Welt 1770–1860*. Göttingen: Vandenhoeck & Ruprecht, 2012.

Anonymous. "The Piracy and Slave Trade of the Indian Archipelago." *Journal of the Indian Archipelago and Eastern Asia* 3 (1849): 581–88, 629–39.

Anonymous. "Rajah Brooke and Borneo." *Chambers's Papers for the People* 3 (1850): article 34.

Antony, Robert J. "Turbulent Waters: Sea Raiding in Early Modern South East Asia." *The Mariner's Mirror* 99, no. 1 (February 2013): 23–38.

Barnard, Timothy P. "Celates, Rayat-Laut, Pirates: The Orang Laut and Their Decline in History." *Journal of the Malaysian Branch of the Royal Asiatic Society* 80, No. 2 (December 2007): 33–49.

Barnard, Timothy P. "Siak, Piracy and Early Modern Malay Warfare." In *Piracy and Surreptitious Activities in the Malay Archipelago and Adjacent Seas, 1600–1849*, edited by Y. H. Teddy Sim, 17–34. Singapore: Springer, 2014.

Bastin, John. *Raffles and Hastings: Private Exchanges behind the Founding of Singapore*. Singapore: National Library Board, 2014.

Bastin, John. *Sir Stamford Raffles and Some of His Friends and Contemporaries: A Memoir of the Founder of Singapore*. Singapore: World Scientific Publishing Co., 2019.

Begbie, Peter James. *The Malayan Peninsula, Embracing Its History, Manners and Customs of the Inhabitants, Politics, Natural History &c. from Its Earliest Records*. Madras: Vepery Mission Press, 1834.

Bellina, Bérénice, Roger Blench and Jean-Christophe Galipaud, eds. *Sea Nomads of Southeast Asia: From Past to Present*. Singapore: NUS Press, 2021.

Bellina, Bérénice, Roger Blench and Jean-Christophe Galipaud. "Sea Nomadism from Past to the Present." In *Sea Nomads of Southeast Asia: From Past to Present*, edited by Bérénice Bellina, Roger Blench and Jean-Christophe Galipaud, 1–27. Singapore: NUS Press, 2021.

Bentley, Jerry H., Renate Bridenthal and Kären Wigen, eds. *Seascapes: Maritime Histories, Littoral Cultures, and Transoceanic Exchanges*. Honolulu: University of Hawai'i Press, 2007.

Benton, Lauren and Lisa Ford. *Rage for Order: The British Empire and the Origins of International Law 1800–1850*. Cambridge: Harvard University Press, 2016.

Benton, Lauren and Nathan Perl-Rosenthal. "Introduction: Making Maritime History Global." In *A World at Sea: Maritime Practices and Global History*, edited by Lauren Benton and Nathan Perl-Rosenthal, 1–14. Philadelphia: University of Pennsylvania Press, 2020.

Benton, Lauren and Nathan Perl-Rosenthal, eds. *A World at Sea: Maritime Practices and Global History*. Philadelphia: University of Pennsylvania Press, 2020.

Bernaldez, Emilio. *Reseña historica de la guerra al sur de Filipinas, sostenida por las armas españolas contra los piratas de aquel archipelago, desde la conquista hasta nuestros días*. Madrid: Imprenta del memorial de ingenieros, 1857.

Boomgaard, Peter. *A World of Water: Rain, Rivers and Sea in Southeast Asian Histories*. Leiden: KITLV Press, 2007.

Branch, Jordan. *The Cartographic State: Maps, Territory, and the Origins of Sovereignty*. Cambridge: Cambridge University Press, 2014.

*British Parliamentary Papers. Accounts and Papers: 1851*, vol. 26, 1.

*British Parliamentary Papers. Accounts and Papers: 1852–1853*, vol. 5.

Brown, D. E. "Brunei and the Bajau." *Borneo Research Bulletin* 3, No. 2 (December 1971): 55–58.

Campling, Liam and Alejandro Colás. *Capitalism and the Sea: The Maritime Factor in the Making of the Modern World*. London: Verso, 2021.

Chou, Cynthia. "Contesting the Tenure of Territoriality; the Orang Suku Laut." *Bijdragen tot de Taal-, Land- en Volkenkunde* 153, No. 4 (1997): 605–29.

Chou, Cynthia. "Research Trends on Southeast Asian Sea Nomads." *Kyoto Review of Southeast Asia* 7 (2006). https://kyotoreview.org/book-review/research-trends-on-southeast-asian-sea-nomads.

Crawfurd, John. "History and Languages of the Indian Islands." *Edinburgh Review* 23, No. 45 (1814): 151–89.

Crawfurd, John. *History of the Indian Archipelago*. 3 vols. Edinburgh: Archibald Constable and Co., 1820.

Crawfurd, John. "Malay Pirates." *The Asiatic Journal and Monthly Register for British India and Its Dependencies* 29 (January–June 1825): 243–45.

Crawfurd, John. *Journal of an Embassy from the Governor-General of India to the Courts of Siam and Cochin China Exhibiting a View of the Actual State of Those Kingdoms*. London: Henry Colburn, 1828.

Crawfurd, John. "Piracy in the Eastern Archipelago." *The Examiner* 2125 (October 21, 1848).

Crawfurd, John. *A Descriptive Dictionary of the Indian Islands & Adjacent Countries*. London: Bradbury and Evans, 1856.

Dalrymple, Alexander. *A Plan for Extending the Commerce of This Kingdom, and of the East-India-Company*. London, 1769.

Dampier, William. *A Collection of Voyages, In Four Volumes*. London: James and John Knapton, 1729.

Deleuze, Gilles and Felix Guattari. *A Thousand Plateaus: Capitalism and Schizophrenia*. Minneapolis: University of Minnesota Press, 1987 [1980].

Earl, George W. *The Eastern Seas, or Voyages and Adventures in the Indian Archipelago in 1832–33–34*. London: Wm. H. Allen and Co., 1837.

Finlayson, George. *The Mission to Siam and Hue in the Years 1821–2. From the Journal of the Late George Finlayson, Esq., Assistant Surgeon of His Majesty's 8th Light Dragoons, Surgeon and Naturalist to the Mission. With a Memoir of the Author by Sir Thomas Stamford Raffles, F.R.S.* London: John Murray, 1826.

Fitzmaurice, Andrew. *Sovereignty, Property and Empire, 1500–2000*. Cambridge: Cambridge University Press, 2014.

Forrest, Thomas. *A Voyage to New Guinea and the Moluccas, from Balambangan: An Account of Magindano, Sooloo, and Other Islands; and Illustrated with Thirty Copperplates Performed in the Tartar Galley, Belonging to the Honourable East India Company, During the Years 1774, 1775, and 1776*. London: G. Scott, 1779.

Forrest, Thomas. *A Voyage from Calcutta to the Mergui Archipelago, Lying on the East Side of the Bay of Bengal*. London, 1792.

García de Arboleya, José. *Historia del archipelago y sultania de Joló, y noticia de la expedición española que a órdenes del marques de la Solana acaba de destruir a los piratas joloanos*. Habana: Imprenta de M. Soler y Gelada, 1851.

Gaynor, Jennifer L. "Piracy in the Offing: The Law of Lands and the Limits of Sovereignty at Sea." *Anthropological Quarterly* 85, No. 3 (2012): 817–58.

Gaynor, Jennifer L. *Intertidal History in Island Southeast Asia: Submerged Genealogy & the Legacy of Coastal Capture*. Ithaca: Cornell University Press, 2016.

Graham, Gerald S. *Great Britain in the Indian Ocean: A Study of Maritime Enterprise 1810–1850*. London: Oxford University Press, 1967.

Groot, Jhr. J. P. Cornets de. "Paper Relating to the Piracies Committed in the Indian Archipelago, and to the Measures Adopted by the Netherland Government, in the Year 1815 to 1845, for Their Repression." *British Parliamentary Papers. Accounts and Papers: 1851*, vol. 26, 1.

Hamilton, Alexander. *A New Account of the East-Indies Being the Observations and Remarks of Capt. Hamilton Who Spent His Time There, from the Year 1688 to 1723, Trading and Travelling, by Sea and Land, to Most of the Countries and Islands of Commerce and Navigation, Between the Cape of Good-Hope and the Island of Japan*. 2 Vols. Edinburgh: John Mosman, 1727.

Hamilton, Walter. *The East-India Gazetteer; Containing Particular Descriptions of the Empires, Kingdoms, Principalities, Provinces, Cities, Towns, Districts, Fortresses, Harbours, Rivers, Lakes, &c. of Hindostan and the Adjacent Countries, India Beyond the Ganges, and the Eastern Archipelago*. 2 vols. London: Wm. H. Allen and Co., 1828.

Harrington, Gordon K. "Dr. Horsfield's Report on the Island of Bangka: An Imperialist Proposal for Reform in 1813." *Selected Papers in Asian Studies: Western Conference of the Association for Asian Studies*, no. 72 (2003): 1–14.

Heller-Roazen, Daniel. *The Enemy of All: Piracy and the Law of Nations*. New York: Zone Books, 2009.

Hoogervorst, Tom Gunnar. "Ethnicity and Aquatic Lifestyles: Exploring Southeast Asia's Past and Present Seascapes." *Water History* 4 (2012): 245–65.

Hopkins, Benjamin D. *Ruling the Savage Periphery: Frontier Governance and the Making of the Modern State*. Cambridge: Harvard University Press, 2020.

Horsfield, Thomas. "Report on the Island of Bangka." *Journal of the Indian Archipelago and Eastern Asia* 2 (1848): 299–336, 373–427, 705–25 and 779–824.

Kempe, Michael. "'Even in the Remotest Corners of the World': Globalized Piracy and International Law, 1500–1900." *Journal of Global History* 5, no. 3 (November 2010): 353–72.

Kempe, Michael. *Fluch der Weltmeere: Piraterie, Völkerrecht und internationale Beziehungen 1500–1900*. Frankfurt: Campus Verlag, 2010.

Kent, H. S. K. "The Historical Origins of the Three-Mile Limit." *The American Journal of International Law* 48, No. 4 (October 1954): 537–53.

Keppel, Henry. *The Expedition to Borneo of H.M.S. Dido for the Suppression of Piracy: With Extracts from the Journals of James Brook Esq. of Sarawak*. 2 vols. London: Chapman and Hall, 1846.

Keppel, Henry. *A Visit to the Indian Archipelago in the H.M. Ship Mæander. With Portions of the Private Journal of Sir James Brooke, K.C.B.* 2 vols. London: Richard Bentley, 1853.

Knapman, Gareth. *Race and British Colonialism in Southeast Asia, 1770–1870: John Crawfurd and the Politics of Equality*. New York: Routledge, 2017.

Knapman, Gareth and Martin Müller. "Protector of Aborigines or War Criminal: Two Opposing Liberal Views of James Brooke." In *Liberalism and the British Empire in Southeast Asia*, edited by Gareth Knapman, Anthony Milner and Mary Quilty, 167–91. Abingdon: Routledge, 2019.

Lapian, Adrian B. "Violence and Armed Robbery in Indonesian Seas." In *Pirates, Ports, and Coasts in Asia: Historical and Contemporary Perspectives*, edited by John Kleinen and Manon Osseweijer, 131–46. Singapore: Institute of Southeast Asian Studies, 2010.

Layton, Simon. "Discourses of Piracy in an Age of Revolutions." *Itinerario* 35, no. 2 (2011): 81–97.

Leyden, John. "On the Language and Literature of the Indo-Chinese Nations." *Asiatick Researches* 10 (1808): 158–289.

Logan, John Richardson. "Malay Amok and Piracies: What can We do to Abolish Them?" *The Journal of the Indian Archipelago and Eastern Asia* 3 (1849): 463–67.

Low, Hugh. *Sarawak: Its Inhabitants and Productions: Being Notes during a Residence in That Country with H. H. the Rajah Brooke*. London: Richard Bentley, 1848.

Maier, Charles S. "Transformations of Territoriality, 1600–2000." In *Transnationale Geschichte. Themen, Tendenzen und Theorien*, edited by Gunilla Budde, Sebastian Conrad and Oliver Janz, 32–55. Göttingen, 2006.

Maier, Charles S. *Once Within Borders: Territories of Power, Wealth, and Belonging since 1500*. Cambridge: Harvard University Press, 2016.

Manickam, Sandra Khor. *Taming the Wild: Aborigines and Racial Knowledge in Colonial Malaya*. Copenhagen: NIAS Press, 2015.

Manzano, David. *El Imperio Español en Oceania*. Almuzara, 2020.

Mawani, Renisa and Sebastian Prange. "Unruly Oceans: Law, Violence, and Sovereignty at Sea." *TWAILR: Third World Approaches to International Law Review*, March 12, 2021. https://twailr.com/unruly-oceans-law-violence-and-sovereignty-at-sea.

Middleton, Alex. "Rajah Brooke and the Victorians." *The Historical Journal* 52, No. 2 (2010): 381–400.

Montero y Vidal, José. *Historia de la piratería malayo-mahometana en Mindanao, Jolo y Borneo. Comprende desde el descubremiento de dichas islas hasta junio de 1888*. 2 vols. Madrid: Imprente y fundación de Manuel Tello, 1888.

Müller, Martin. "Civilisation, Culture and Race in John Crawfurd's Discourses on Southeast Asia: Continuities and Changes, c.1814–1868." PhD Diss., European University Institute, 2013.

Müller, Martin. "Embodying Piracy, Textualizing the Piratical Body: Defining, Combating and Punishing Southeast Asian 'Piratical Communities' in British Discourse and Practice, c.1810–1860." In *Entre Traces Mémorielles et Marques Corporelles: Regards sur l'Ennemi de l'Antiquité à Nos Jours*, edited by Jean-Claude Caron, Laurent Lamoine and Natividad Planas, 201–21. Presses Universitaires Blaise Pascal, 2014.

Müller, Martin. "Manufacturing Malayness: British Debates on the Malay Nation, Civilisation, Race and Language in the Early Nineteenth Century." *Indonesia and the Malay World* 42, No. 123 (2014): 170–96.

Newbold, Thomas John. *Political and Statistical Account of the British Settlements of the Straits of Malacca, viz. Pinang, Malacca, and Singapore; with a History of the Malayan States on the Peninsula of Malacca*. 2 vols. London: John Murray, 1839.

Noor, Farish A. *The Long Shadow of the 19th Century: Critical Essays on Colonial Orientalism in Southeast Asia*. Petaling Jaya: Matahari Books, 2021.

Osborn, Sherard. *Quedah; Or Stray Leaves from a Journal in Malayan Waters*. London: Longman, Brown, Green, Longmans, & Roberts, 1857.

Osterhammel, Jürgen. *The Transformation of the World: A Global History of the Nineteenth Century*. Princeton: Princeton University Press, 2014.

Ota, Atsushi. "'Pirates or Entrepreneurs?' The Migration and Trade of Sea People in Southwest Kalimantan, c.1770–1820." *Indonesia* 90 (October 2010): 67–95.

Ota, Atsushi. "Introduction." In *In the Name of the Battle against Piracy: Ideas and Practices in State Monopoly of Maritime Violence in Europe and Asia in the Period of Transition*, edited by Ota Atsushi, 1–15. Leiden: Brill, 2018.

Ota, Atsushi. "Trade, Piracy, and Sovereignty: Changing Perspectives of Piracy and Dutch Colonial State-Building in Malay Waters, c.1780–1830." In *In the Name of the Battle against Piracy: Ideas and Practices in State Monopoly of Maritime Violence in Europe and Asia in the Period of Transition*, edited by Ota Atsushi, 115–42. Leiden: Brill, 2018.

Patero, Santiago. *Sistema que conviene adopter para acabar con la piratería que los mahometanos de la Sultania de Joló ejercen en el Archipélago Filipino.* Madrid: Imprenta de Miguel Ginesta, 1872.

Pearson, Michael. "Littoral Society: The Concept and the Problems." *Journal of World History* 17, No. 4 (2006): 353–72.

Pearson, Michael. "Piracy in Asian Waters: Problems of Definition." In *Pirates, Ports, and Coasts in Asia: Historical and Contemporary Perspectives*, edited by John Kleinen and Manon Osseweijer, 15–28. Singapore: Institute of Southeast Asian Studies, 2010.

Pels, Peter and Oscar Salemink, eds. *Colonial Subjects: Essays on the Practical History of Anthropology.* Ann Arbor: University of Michigan Press, 2000.

Policante, Amadeo. *The Pirate Myth: Genealogies of an Imperial Concept.* London: Routledge, 2015.

Prange, Sebastian. "Asian Piracy." *Oxford Research Encyclopedia of Asian History* (September 2017). https://doi.org/10.1093/acrefore/9780190277727.013.32.

Rafferty, Matthew Taylor. "'The Law Is the Lord of the Sea': Maritime Law and Global Maritime History." In *A World at Sea: Maritime Practices and Global History*, edited by Lauren Benton and Nathan Perl-Rosenthal, 53–71. Philadelphia: University of Pennsylvania Press, 2020.

Raffles, Lady Sophia. *Memoir of the Life and Public Services of Sir Thomas Stamford Raffles, F.R.S. &c. Particularly in the Government of Java, 1811–1816, and of Bencoolen and Its Dependencies, 1817–1824; with Details of the Commerce and Resources of the Eastern Archipelago, and Selections from His Correspondence.* London: John Murray, 1830.

Raffles, Sir Thomas Stamford. "A Discourse." *Transactions of the Batavian Society of Arts and Sciences*, 8 (1816): 1–95.

Raffles, Sir Thomas. "Short Account of the Islands of Timor, Rotti, Ende, or Floris, and Sumba, or Sandal Wood Island, in the Indian Archipelago." *The Investigator; Or Quarterly Magazine* 4 (January and April 1822): 34–45.

Risso, Patricia, "Cross-Cultural Perceptions of Piracy: Maritime Violence in the Western Indian Ocean and Persian Gulf Region during a Long Eighteenth Century." *Journal of World History* 12, No. 2 (Fall 2001): 293–319.

Rubin, Alfred P. *The Law of Piracy.* Newport: Naval College Press, 1988.

Sather, Clifford. "Sulu's Political Jurisdiction over the Bajau Laut." *Borneo Research Bulletin* 3, No. 2 (December 1971): 58–62.

Satsuma, Shinsuke. "Plunder and Free Trade: British Privateering and Its Abolition in 1856 in Global Perspective." In *In the Name of the Battle against Piracy: Ideas and Practices in State Monopoly of Maritime Violence in Europe and Asia in the Period of Transition*, edited by Ota Atsushi, 43–65. Leiden: Brill, 2018.

Schmitt, Carl. *The Theory of the Partisan: A Commentary/Remark on the Concept of the Political.* Michigan State University Press, 2004 [1963].

Schmitt, Carl. "The Concept of Piracy (1937)." *Humanity: An International Journal of Human Rights, Humanitarianism, and Development* 2, No. 1 (Spring 2011): 27–29.

Schmitt, Carl. *Land and Sea: A World Historical Meditation.* Candor: Telos Press Publishing, 2015 [1942].

Scott, David. "Colonial Governmentality." *Social Text* 43 (Autumn 1995): 191–220.

Scott, James C. *The Art of Not Being Governed: An Anarchist History of Upland Southeast Asia.* New Haven: Yale University Press, 2009.

Skeat, Walter W. and Henry N. Ridley. "The Orang Laut of Singapore." *Journal of the Straits Branch of the Royal Asiatic Society* 33 (January 1900): 247–50.

Sopher, David, E. *The Sea Nomads: A Study of the Maritime Boatpeople of Southeast Asia.* Singapore: National Museum Singapore, 1977.

Steinberg, Philip E. *The Social Construction of the Sea.* Cambridge: Cambridge University Press, 2001.

Steinberg, Philip E. "Of Other Seas: Metaphors and Materialities in Maritime Regions." *The Atlantic Studies* 10, No. 2 (2013): 156–69.

Tagliacozzo, Eric. "A Necklace of Fins: Marine Goods Trading in Maritime Southeast Asia, 1780–1860." *International Journal of Asian Studies* 1, No. 1 (2004): 23–48.

Tarling, Nicholas. *Piracy and Politics in the Malay World: A Study of British Imperialism in Nineteenth-Century South-East Asia.* Nendeln: Kraus Reprint, 1978 [1963].

Teitler, Ger, A. M. C. van Dissel and J. N. F. M. á Campo. *Zeeroof en Zeeroofbestrijding in de Indische archipel (19de eeuw).* Amsterdam: De Bataafsche Leeuw, 2005.

Templer, John C. (ed.). *The Private Letters of Sir James Brooke, K.C.B., Rajah of Sarawak: Narrating the Events from 1838 to the Present Time.* 3 vols. London: Richard Bentley, 1853.

Thomson, John Turnbull. "Description of the Eastern Coast of Johore and Pahang, with Adjacent Islands." *The Journal of the Indian Archipelago and Eastern Asia* 5 (1851): 85–92 and 135–54.

Verschuer, F. H. van. "De Badjo's." *Tijdschift van het Aardrijkskundig Genootshap* Zevende deel (7th part) 1883: 1–7.

Wagner, Kim A. *Thuggee: Banditry and the British in Early Nineteenth-Century India.* Delhi: Primus Books, 2014.

Warren, James F. *Iraun and Balangini: Globalization, Maritime Raiding and the Birth of Ethnicity*. Singapore: Singapore University Press, 2002.

Warren, James F. *The Sulu Zone 1768–1898: The Dynamics of External Trade, Slavery, and Ethnicity in the Transformation of a Southeast Asian Maritime State*. Singapore: NUS Press, 2007 [1981].

Warren, James F. "The Global Economy and the Sulu Zone: Connections, Commodities and Culture." *Crossroads* 3 (2011): 53–124.

Watkins, Josh. "Spatial Imaginaries Research in Geography: Synergies, Tensions, and New Directions." *Geography Compass* 9, no. 9 (2015): 508–22.

Whatley, A. T. "Historical Sketch of the Law of Piracy." *The Law Magazine and Review* New Series 3 (1874): 536–55 and 618–39.

Wigen, Kären. "Introduction." In *Seascapes: Maritime Histories, Littoral Cultures, and Transoceanic Exchanges*, edited by Jerry H. Bentley, Renate Bridenthal, and Kären Wigen, 1–18. Honolulu: University of Hawai'i Press, 2007.

## About the Author

**Martin Müller** is an independent scholar whose research interests include imperial formations and their intellectual frameworks, colonial knowledge production, and a global history of territoriality during the long nineteenth century.

# Section II

Practices

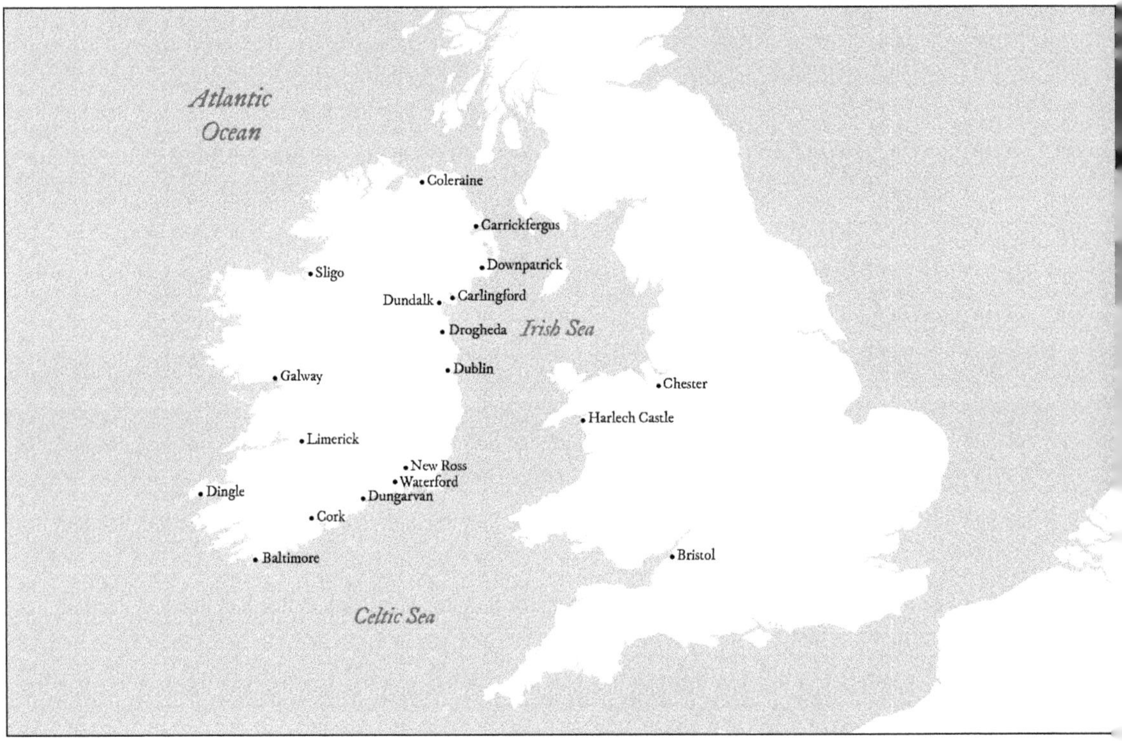

Atlantic Ocean
Coleraine
Carrickfergus
Sligo
Downpatrick
Dundalk
Carlingford
Drogheda
Irish Sea
Dublin
Galway
Chester
Harlech Castle
Limerick
New Ross
Waterford
Dingle
Dungarvan
Cork
Bristol
Baltimore
Celtic Sea

# 3. Scots, Castilians, and Other Enemies: Piracy in the Late Medieval Irish Sea World

*Simon Egan*

**Abstract**

Despite growing interest in Late Medieval Ireland's maritime heritage, little attention has been devoted to exploring the problem of piracy in this period. The fourteenth and fifteenth centuries not only witnessed a rise in attacks by Gaelic Irish marauders on English settlements in Ireland, the period also saw French, Scottish, and Castilian rovers raid English holdings from Ulster to Kinsale. This chapter explores the factors precipitating the rise in marine predation during this period. In particular, it considers the range of different contexts for piracy and explores the economic and plunder-based reasons for piracy as well as the wider context of the Hundred Years War and French and Scottish attempts to open new military fronts against the English in Ireland.

**Keywords:** Gaelic Ireland; Gaelic Scotland; Irish Sea World; Fishing; Hundred Years War

## Introduction

The famous seventeenth-century history of Ireland known as the *Annals of the Four Masters* records a curious incident under the year 1453. The precise time of year is not revealed, but the text notes that a fleet of Breton raiders descended upon Dublin Bay. The English fleet was burned by the marauders and the English archbishop of Dublin, Michael Tregury (d.1471), was captured before being whisked away by the raiders who sailed northwards. The ensuing course of events is not clear, but a rescue mission appears to have been launched by

Coakley, J., C. Nathan Kwan and D. Wilson, *The Problem of Piracy in the Early Modern World: Maritime Predation, Empire, and the Construction of Authority at Sea.*
Taylor & Francis Group, 2024
DOI 10.5117/9789463720960_CH03

the Anglo-Irish citizens of Dublin. A fleet soon departed the colonial capital, intercepted the Bretons, and recovered Tregury.[1] Although the archbishop was rescued, in general he appears to have been plagued by misfortune during his career in Ireland. Two years earlier, while travelling to Ireland to assume his post, Tregury was shipwrecked, losing most of his possessions. Several years later in 1461, Tregury was taken captive once again, this time by the Uí Bhroin dynasty of Wicklow and the Haralds of south Dublin—a lineage who were descended from earlier Viking raiders and settlers.[2]

The series of unfortunate events surrounding Archbishop Tregury's career in Ireland highlights the dangers of both marine predation and travel by sea in the later medieval period. The Breton raid of 1453 was not unusual in itself nor was Dublin the only English town in Ireland to fall victim to maritime raids during the later Middle Ages. Throughout the early fifteenth century, shipping from the nearby ports of Dundalk and Drogheda was frequently preyed upon by Scottish rovers from Ayrshire and Galloway. Scots not only raided the coastlines of Leinster and eastern Ulster, the Isle of Man and northern Wales were each subjected to several large-scale maritime assaults during the fourteenth and fifteenth centuries. Further south and owing to its proximity to the European continent, the bustling entrepôt of Waterford was a prime target for French, Scottish, and Castilian raiders.[3] Moreover, Waterford's relative distance from the colonial heartlands of Leinster made it vulnerable to the predations of neighbouring Gaelic Irish dynasties. It is perhaps a measure of the naval resources at their disposal that in 1368, the Uí hEidirsceoil of Baltimore in West Cork sailed eastwards to Waterford, killing over forty of the townsfolk, including the mayor and sheriff.[4]

The last five decades have witnessed a growing interest in Ireland's medieval maritime heritage and particular attention has been devoted to exploring the island's economic development during the late medieval and early modern periods. From the perspective of the English (also referred to as Anglo-Irish) colony, scholarship has underlined the close political, mercantile, and cultural links between cities in Ireland, such as Dublin and Waterford, and English ports such as Bristol and Chester.[5] Recent

1 J. O'Donovan (ed.), *Annals of the Kingdom of Ireland by the Four Masters* (*AFM*), 1453. All annals references are given as year, rather than page and volume number.

2 For Tregury's career, see Beresford, "Tregury Michael."

3 These attacks are discussed below.

4 J. S. Brewer (ed.) *Calendar of Carew Manuscripts Preserved in the Archiepiscopal Library at Lambeth* (*Cal. Carew MSS*), v, 470–71.

5 Notable works include Childs, "Ireland's Trade"; Childs and O'Neill, "Overseas Trade."; Mac Niocaill, *Na Burgéisí*; O'Neill, *Merchants and Mariners*; Flavin, *Consumption and Culture*; O'Brien,

years have also resulted in an abundance of research on Ireland's western seaboard. Marine archaeology has played a central role in reconstructing the commercialisation of the native Irish economy in the later Middle Ages. Particular attention has been devoted to exploring the development of fisheries in the west, the expansion of Irish maritime lordships, and the western seaboard's position within the international European fishing trade.[6]

Set against the backdrop of these developments, Irish waters clearly experienced a variety of different forms of marine predation in this period. However, scholarship on marine predation and the problem of piracy features largely as a subtheme within the historiography of Ireland's economic development. As noted above, the fourteenth and fifteenth centuries witnessed dramatic rise in difference forms of marine predation within Irish waters as English holdings fell victim to numerous raids by French, Castilian, Scottish, and Irish rovers. James Lydon noted that the outbreak of the Hundred Years War (1337–1453) transformed the Irish Sea "into a battlefield," one that that placed cities such as Waterford at the mercy of France and its allies.[7] Several scholars have alluded to the threat posed by these raiders during key phases of the Anglo-French conflict.[8] However, aside from these notable exceptions, scholars have largely downplayed Ireland's strategic importance within the Hundred Years War.[9] Indeed, Katharine Simms and Simon Kingston have both argued that raids on Ulster and Leinster had little to do with Scottish military strategy and were instead carried out by the Meic Domhnaill of the Western Isles of Scotland, a powerful dynasty which by 1400 had come dominate a huge section of Scotland's western seaboard.[10] There is some evidence to suggest that the Meic Domhnaill may have briefly engaged in some form of maritime raiding during the early 1400s. However, an overwhelming body of primary material indicates that, rather than being inimical to the English interests, the Meic Domhnaill were courted by the English crown as allies in the Irish Sea region.[11] In particular, their extensive naval resources could be bought to bear to protect Ulster

"Politics, Economy, and Society"; Smith, "Late Medieval Ireland."

6 Notable studies include Breen, "Maritime Cultural Landscape"; Breen, *Gaelic Lordship*; Kelleher, "Gaelic O'Driscoll Lords"; McErlean, McConkey, and Forsythe, *Strangford Lough*; Gardiner and McNeill, "Seaborne Trade"; Naessens, "Gaelic Lords of the Sea"; Ní Loinsigh, "Late Medieval North Donegal."

7 Lydon, "Waterford," pp. 5–15, at p. 13.

8 Smith, "Late Medieval Ireland," p. 554; Boardman, "Highland Scots," pp. 231–53, at p. 247, no. 54.

9 For example, see Frame, *English Lordship*, p. 153, no. 102; Sumption, *Divided Houses*, pp. 12–13.

10 Kingston, "Delusions of Dál Riada," pp. 98–114; Simms, "Ulster Revolt," pp. 141–60.

11 See below.

from Scottish attack, be that military operations or opportunist raids on vulnerable sections of the Irish coastline.

The high levels of French and Scottish raiding in Irish waters during the later Middle Ages have yet to be explored in any great detail. In reality, it is only in the later Tudor and Stuart periods that piracy and marine predation in Ireland (as well as the threat of foreign invasion) receives the attention it deserves.[12] Therefore, this chapter seeks to make an intervention in this field of research and investigates the problem of piracy in late medieval Ireland. In particular, it aims to explore and define the different contexts of piracy within Irish waters during this period. Owing to the limitations of surviving sources, it can be challenging to define what exactly constituted piracy in this period. The act of piracy itself is generally understood as a form of unlawful activity at sea, resulting in robbery or violence. Much of the surviving evidence from this period appears to fit this description: Gaelic Irish dynasties raided one another as well as the English colonists as a means of gaining plunder. French and Scottish attacks on the colony served a dual purpose: they undercut England's economic base during the Hundred Years War but they also appear to have been highly opportunistic in nature.

Admittedly, it would be impossible to undertake a complete re-evaluation of this area in such a short piece. Consequently, this chapter takes a "broad brush strokes" approach and explores this topic along the following thematic lines. Firstly, the chapter examines the deeper origins of marine predation in later medieval Ireland. It considers the wider maritime context surrounding the establishment of the English colony in Ireland and how the limitations of English power in Ireland left large areas of the colony (particularly in Connacht and Ulster) vulnerable to raids by Irish and Hebridean dynasts. The section moves on to explore how a combination of factors, both environmental and political, further weakened the English position in Ireland during the fourteenth and fifteenth centuries. In particular, it examines the impact of the Bruce invasions of Ireland and draws upon recent research on the commercialisation of Ireland's western seaboard to contextualise the rapid rise in piracy and raiding during the later Middle Ages. Thirdly, the chapter explores the role of marine predation in the era of the Hundred Years War, examining the importance of piracy as an unofficial yet successful form of warfare within Scottish and French strategy as well as English attempts to

12 For example, see Appleby, "Problem of Piracy"; Chambers, *Granuaile*; Edwards, "Tudor Ireland"; Ekin, *Stolen Village*; Kelleher, *Alliance of Pirates*; Murphy, *Ireland*; Ohlmeyer, "Dunkirk of Ireland."

counter this problem. Ultimately, the chapter argues that piracy presented a very serious threat to English interests in late medieval Ireland. Moreover, it underlines the necessity of locating any examination of piracy with a wide chronological and geographic context and paying close attention to how events within the dynastic world of Ireland and its neighbours shaped the course and problem of piracy in this period.

## The Evolution of Marine Predation in Ireland, c.1100–c.1450

During the medieval period, Ireland was situated at the centre of a "swirling vortex" of political, dynastic, and maritime connections.[13] The sea connected rather than separated the communities of the Irish Sea world and naval power formed a central element in the ambitions of Ireland's kings during the high Middle Ages. By the dawn of the twelfth century, the Uí Bhriain dynasty of Thomond had extended their dominion over Munster and much of Leinster, including the Ostman towns of Limerick, Cork, Waterford, and Dublin. The Uí Bhriain had also set their sights on extending their influence beyond Irish shores: the Isle of Man and Galloway had not only come within the Uí Bhriain sphere of influence, Thomond's kings also meddled in Welsh affairs and backed Welsh princes in their struggles with one another and the encroaching Norman kingdom of England.[14] Similarly, the Uí Chonchobhair kings of Connacht were capable of marshalling significant fleets which they deployed against their rivals on Ireland's western and northern seaboards during the mid-to-late twelfth century. The impressive military resources of the Uí Choncohbhair were further augmented by their ability to recruit fleets of galleys and large consignments of mercenaries from the Isle of Man and the Western Isles of Scotland throughout the later twelfth century.[15] The English invasion of Ireland in 1169 dramatically altered this state of affairs. Irish dynasties which had once been capable of shaping the politics of the wider archipelago were pushed back to the more mountainous and less fertile regions of the island as the invasion gathered momentum and large areas of Leinster, Munster, and eastern Ulster were sub-infeudated.[16] The River Shannon became a

13 McDonald, *Kingdom of the Isles*, p. 87.

14 For the rise of the Uí Bhriain, see Duffy, "Irish Sea Region," pp. 2–47.

15 For the militarisation of Irish society in this period, see Flanagan, "After Brian Bóraime."

16 The most authoritative account of the invasion remains Orpen's *magnus opus*, see Orpen, *Ireland Under the Normans*.

frontier between the Uí Chonchobhair and the expanding English colony on Ireland's eastern seaboard and links between Munster and Wales, if not entirely broken, were drastically reduced.[17]

Despite the relatively rapid expansion of the new colony in Ireland, the English naval presence in the Irish Sea was limited to a handful of galleys which patrolled the strategic lanes between Waterford and Bristol.[18] The limitations of the English navy meant that large areas of the colonial lordship were vulnerable to seaborne assault, particularly as Irish kings sought to recruit troops from the Hebrides in their struggle with the colonial nobility. In attempt to offset this problem, the English crown and its colonial administration negotiated alliances with regional strong players such as the king of Man, Rǫgnvaldr Guðrøðarson (d.1229), and the lords of Galloway, the brothers Thomas (d.1231) and Alan (d.1234). Rǫgnvaldr and the Galloway brothers each commanded large fleets of galleys and their forces were deployed in English service during the early thirteenth century.[19]

From an English point of view, the eruption of internecine feuding on Man in the 1220s, coupled with the partition of Galloway following the deaths of Thomas and Alan in 1231 and 1234 respectively, made it increasingly difficult to police these waters.[20] It is perhaps a measure of the colonial administration's inability to control events in these regions that Magnús Ólafsson (d.1265), a claimant to the kingship of Man, raised a force of mercenaries within Ireland which he used to take Man in 1251.[21] Several years later in 1259, a Hebridean fleet commanded by the lord of Garmoran, Dubghall Mac Ruaidhrí (d.1268) raided the coast of Connemara, plundered a merchant vessel, and killed the local English sheriff before returning unimpeded to the Western Isles.[22] Mac Ruaidhrí was an ally of the Uí Chonchobhair. This Hebridean-Connacht axis posed a serious challenge to the English colonisation project on Ireland's western seaboard, and in the late 1250s, Dubghall and his Uí Chonchobhair allies became embroiled in a scheme to make the king of Tír Eoghain, Briain Ó Néill (d.1260), high-king of Ireland.[23] Although the venture ended in disaster at the battle of Down in the summer of 1260, the limitations of English influence on Ireland's northern and western

17 For the changing relationship between Ireland and Wales, see Duffy, "1169 Invasion."

18 Brown, *Hugh de Lacy*, p. 37.

19 For Rǫgnvaldr's relations with the English crown, see McDonald, *Manx Kingship*, pp. 123–60. For the Galloway brothers, see Stringer, "Periphery and Core."

20 For these developments, see Oram, *Lordship of Galloway*, pp. 140–46.

21 H. S. Sweetman (ed.), *Calendar of Documents Relating to Ireland, 1171–1307* (*CDI*), i, no. 3206.

22 M. Freeman (ed.), *An ála Connacht: The Annals of Connacht* (*AC*), 1258.

23 For the background to these events, see Duffy, "Irish Sea Region," pp. 118–33.

littorals meant that Henry III's Irish colony remained exposed to further threats emanating from the wider North Atlantic world.[24]

The early-to-mid thirteenth century saw England's neighbour, the kingdom of Scotland, drift onto a collision course with the kingdom of Norway. The rise in Scottish-Norse tensions was the direct result of competing claims to sovereignty over the Western Isles of Scotland which had been ceded to Norway by Edgar (d.1107), king of Scots, in 1098.[25] For a century and a half, the Norse monarchy exercised a very loose form of lordship upon these territories. Increasing attempts by the Scottish monarchy to extend its authority over these regions from the 1240s onwards prompted Hákon IV of Norway (d.1263) to lead an expedition westwards in the summer of 1263.[26] Although Hákon's fleet numbered well over a hundred ships, the venture proved a complete failure: the Norwegian king failed to establish a beachhead in the Firth of Clyde and was forced to retreat northwards to Orkney. Despite reaching the Northern Isles, Hákon died from an illness in October 1263. Three years later, his son and successor, Magnus IV (d.1280), decided to cede possession of the Hebrides (as well as the Isle of Man) to Alexander III of Scotland (d.1286), in exchange for a large cash payment and Scottish recognition of Norse rights over Orkney and Shetland.[27] Despite Hákon's death on Orkney, the expedition had inadvertently exposed the vulnerability of the English colony in Ireland. Although Anglo-Norwegian relations had been relatively stable during the mid-thirteenth century, English officials periodically expressed concerns about potential Norse incursions into Ireland.[28] During the Scottish crisis of 1260s, it was rumoured that Hákon's fleet was destined not for Scotland, but Ireland's western seaboard. Seán Duffy has explored this issue in some detail, revealing that the Uí Chonchobhair and their Meic Dubhgaill allies hatched a scheme whereby they offered to make Hákon high-king of Ireland in exchange for Norse aid against the colonists.[29] Although the plan never came to fruition, evidence contained in surviving Irish sources and Icelandic sagas suggest that Hákon gave this Irish option some consideration during the summer of 1263.[30] Had Hákon accepted this offer,

24 S. Ó hInnse (ed.), *Miscellaneous Irish Annals, AD 1114–1437* (*AMisc*), 1261 [recte 1260].

25 For background, see Cowan, "Norwegian Sunset."

26 The expedition is recounted in several contemporary sources, see A. O. Anderson (ed.) *Early Sources of Scottish History* (*ES*), ii, 607–42.

27 *ES*, ii, 653–54.

28 For example, see *CDI*, i, nos. 976, 1179.

29 Duffy, "Prehistory of the Galloglass," pp. 1–27, at pp. 20–23.

30 *ES*, ii, 634.

there was little the colonial administration could have done to prevent a major onslaught in Ireland.

The threat of a Norse intervention in Ireland aside, it was the Scots who posed the first major challenge to the Anglo-Irish colony since its establishment in the late twelfth century. The outbreak and exacerbation of hostilities between England and Scotland during the late thirteenth and early fourteenth centuries had a transformative impact upon the politics of the wider British Isles.[31] Centuries-old ties of cross-border kinship, marriage, and landholding ripped apart while the western section of the archipelago became a hotly contested theatre in the conflict. As a west coast magnate, King Robert I of Scotland (d.1329) enjoyed a close relationship with the Gaelic nobility in both the West Highlands and Islands and Ulster. Gaelic manpower had not only proven essential to the survival of the Bruce case in the early 1300s, Hebridean troops formed the backbone of the Scottish army which invaded Ireland in 1315.[32] The Bruce invasions of Ireland (1315–18) have attracted a considerable body of attention: suffice to say that the venture served to open a second front in Ireland against the English.[33] Although the campaigns ended in disaster, resulting in the deaths of thousands of Scottish veterans as well as Robert's sole surviving brother, Edward (d.1318), the Bruce invasions left a lasting impact upon Irish society. Firstly, the invasions precipitated the fragmentation of colonial Ireland's most powerful dynasties, the de Burghs.

As earls of Ulster and lords of Connacht, the de Burgh dynasty controlled over one third of the island of Ireland during the late thirteenth and early fourteenth centuries. Three years of warfare and occupation by Scottish forces from 1315–18 however, saw the earldom of Ulster battered beyond resuscitation.[34] The death of the last de Burgh earl, William, in 1333 brought an end to the dynasty's ambitions in the northern province and saw the Uí Néill of Tír Eoghain and the Uí Dhomhnaill of Tír Chonnaill re-emerge as the dominant powers therein.[35] Unlike Ulster, where their lordship was destroyed, the de Burgh dynasty retained control of much of Connacht, although control of the port of Sligo was lost to the resurgent Uí Chonchobhair dynasty. Nevertheless, the death of the last earl in 1333 removed the main focal point for the wider dynasty. As a result, the lordship

31 For observations on these profound developments, see Brown, *Disunited Kingdoms*, chs. 1–4.

32 For Bruce's links to the Gaelic world, see Duffy, "Bruce Brothers," pp. 55–86.

33 For Bruce's war aims, see McNamee, *Wars of the Bruces*, ch. 5.

34 For the decline of the earldom, see Orpen, *Ireland Under the Normans*, pp. 539–59.

35 *AFM*, 1333. See also, Simms, "Late Medieval Donegal"; Simms, "Tír Eoghain."

of Connacht split into the two dominant and competing lordships of Mayo (in north-western Connacht) and Clanrickard (in southern Connacht).[36] Moreover, over the course of the fourteenth century, these families became increasingly "Gaelicised": generations of intermarriage and interaction with their Irish neighbours saw the de Burghs become totally assimilated into the socio-cultural world of Gaelic Ireland. Indeed, by the dawn of the sixteenth century the de Búrcaigh (as they had become known) were virtually indistinguishable from their Irish kinsmen.[37]

The fragmentation of the former de Burgh seigneuries coincided with a series of naturally occurring events which contributed to rising levels of maritime activity and piracy in Irish waters. The phenomenon known as the Little Ice Age resulted in a fluctuation of temperatures in Eurasia from the late thirteenth century onwards.[38] The onset of cooler temperatures created immense hardship across much of the European landmass, bringing about famine in human populations as well as panzootics in the animal kingdom. Within an Irish context, wetter weather and a shorter growing, coupled with the advent of the Black Death in 1348, placed severe strain upon an already beleaguered Anglo-Irish colony during the fourteenth century.[39] From the perspective of the Gaelic and Gaelicised aristocracy, these environmental shifts were determining factors leading to the commercialisation and growth of the western seaboard's economy during the later Middle Ages. The lowering of the ocean's temperature saw the vast herring stocks of the North Atlantic migrate to the western shores of Ireland and Scotland.[40] Thousands of fishermen from across western Europe flocked to Ireland's western Atlantic littoral in the hopes of harvesting these fish. Using fleets of galleys, those dynasties controlling the coastlines of Ulster, Connacht, and Munster, were able to police these waters and levy a fee upon foreign fishing vessels.[41] Moreover, Ireland's fisheries were ideally placed to feed Europe's recovering population in the century and a half after the Black Death.[42] The development of these fisheries generated considerable wealth and led to a building boom, seeing new market towns emerge along the western seaboard. The lords of these regions consolidated their control of these settlements by building castles as well as funding the construction

36 For the emergence of these families, see Nicholls, *Gaelicized Ireland*, pp. 170–75.

37 Nicholls, "Worlds Apart?," pp. 22–26, at p. 23.

38 For a recent discussion on these phenomena, see Campbell, *Great Transition*, ch. 1.

39 For the most comprehensive study of the Black Death in Ireland, see Kelly, *Black Death*.

40 Gardiner and McNeill, "Seaborne Trade," pp. 229–62.

41 Kelleher, "Gaelic O'Driscoll Lords," pp. 142–50.

42 For the growing demand for fish in late medieval Europe, see Fagan, *Fish on Friday*.

of new friaries.[43] Luxury goods also became increasingly available: the soubriquet, "an Fhíona" (i.e., "of the wine,") held by lords such as Turlough an Fhíona Ó Domhnaill of Tír Chonnaill (d.1423) and Uilleag an Fhíona de Búrca of Clanrickard (d.1423) perhaps reflects the growing presence of wine on Ireland's western seaboard.[44] The development of these trading links not only provided the Irish lords with access to commodities such as wine or spices, by the dawn of the sixteenth century, Irish dynasties were importing gunpowder weapons from the continent.[45]

By the mid-fifteenth century, the colonial administration in Ireland had become acutely concerned about the Irish fisheries. For instance, in 1465, the Dublin parliament, which was controlled by the English crown, noted that the wealth arising from these initiatives had seen the native Irish become "much advanced and strengthened as well in victuals and harness as divers other necessities" and attempted, with virtually no success, to ban foreign ships from fishing in Irish waters.[46] The commercialisation of the western seaboard's economy aside, the fourteenth and fifteenth centuries also witnessed a rise in piratical activity in these regions. The conflict between the de Búrcaigh of Mayo and Clanrickard offers one such case in point. In 1338, the ruler of Mayo, Edmund Albanach de Búrca (d.1375), murdered his rival, Sir Edmund de Burgh (a brother of the last earl of Ulster), by drowning him in Lough Mask. The deceased Edmund's kinsmen subsequently drove Edmund Albanach from Connacht to the islands off Clew Bay. Here, Edmund Albanach is recorded as raising a fleet which he used to pillage the west coast, before reclaiming his lordship in Mayo.[47] Edmund Albanach's successors in Mayo also appear to have retained the services of a fleet which they used to protect their holdings in northern Connacht and, as evidenced from surviving sources, raid as far afield as Ulster.[48]

The de Búrcaigh were not alone in engaging in piratical activity in these western regions. Their neighbours, the Uí Mháille of the Owles, maintained a fleet of ships which they used to raid the coastline of Connacht as well as raiding territories as far north as Ulster.[49] Indeed, during one expedition to Ulster, a force under Tuathal Ó Máille was blown off course and shipwrecked

43 Gardiner and McNeill, "Seaborne Trade," 26–27.

44 For examples, see *AFM*, 1440, 1488, 1493.

45 For the arrival of gunpowder in Ireland, see Hayes-McCoy, "Guns in Ireland."

46 Berry, *Statute Rolls*, 353–35.

47 *AC*, 1338.

48 *AFM*, *AC*, 1464.

49 *AFM*, 1396, 1427, 1460; *AC*, 1460, 1513.

in Scotland.[50] The lords of western Ulster were also capable of deploying significant naval resources. Unlike the Uí Néill lordship of Tír Eoghain, which was primarily landlocked, the Uí Dhomhnaill lordship of Tír Chonnaill possessed a long coastline running from Foyle Estuary south to mouth of the River Erne. By the dawn of the sixteenth century, the Uí Dhomhnaill had established themselves as one of the most powerful dynasties on Ireland's western seaboard. The commercialisation of the west coast had played a leading role in their development, and the Uí Dhomnaill fisheries were among the most famous in western Europe, with one chieftain Maghnus Ó Domhnaill (d.1565) known on the continent as the "lord of the fish."[51] The desire to the protect these economic resources from seaborne assault saw the Uí Dhomhnaill develop what appears to have been a sizeable fleet of galleys during the later Middle Ages.[52] Combined with their significant land-based forces, these ships allowed the Uí Dhomhnaill to project their power across Ulster and northern Connacht.

While the Uí Dhomhnaill strengthened their hold over these territories, the de Búrcaigh of Clanrickard solidified their position in southern Connacht. The city of Galway was the crown jewel in the western seaboard's economy and by 1400 the port was firmly under the control of the aforementioned Uilleag an Fhíona. An alliance with the Uí Bhriain of Thomond (who held the Aran islands off the coast of modern-day Galway) allowed Uilleag an Fhíona to dominant both the land and sea approaches to the city and levy tribute upon any merchants or traders who travelled there.[53] Such was the challenge posed to English trade in the region that in 1400 a group of Bristol merchants petitioned Henry IV of England (d.1400) to raise an expedition and establish a fortified colony on the Aran Islands as a means to extending English influence over Galway Bay.[54] The venture proved a complete failure and Galway remained under the control of the de Búrcaigh. Southern Connacht was not the only region where prominent English merchants struggled to assert themselves. Shipping from Bristol was equally vulnerable to attack by pirates operating out of Munster. For instance, in 1401 a crew based at Dingle seized and robbed a ship which was travelling from Bristol to Limerick.[55] Dingle was one of the few safe anchorages on the Atlantic coast south of Galway and was under the jurisdiction of the FitzGerald earls of Desmond, a powerful and semi-autonomous comital dynasty that controlled much of

50 *AFM*, *AC*, 1413.

51 *Cal. Carew MSS*, i, 308.

52 Egan, "By Land and By Sea," pp. 16–31, at pp. 22–25.

53 Childs and O'Neill, "Overseas Trade," p. 523.

54 H. C. Maxwell Lyte (ed.), *Calendar of Patent Rolls: Henry IV, 1399–1413* (*CPR*), i, 254.

55 Ibid., 451.

the modern-day counties of Limerick, Kerry, and Cork. However, the earls not only appear to have turned a blind eye to piracy, alongside their kinsmen they appear to have been complicit in these illicit activities.[56] Following their raid on Waterford in 1368, the Uí hEidirsceoill of West Cork continued to threaten English economic interests in Munster. Throughout the fifteenth century, the citizens of Waterford and nearby Tramore were engaged in a long running and low-level war with the Uí hEidirsceoill. Although forces under the mayor of Waterford capture several of the Uí hEidirsceoill in 1413, it did little long-term damage to the Uí hEidirsceoill lordship: Tramore suffered two major raids at the hands of the Uí hEidirsceoill in 1452 and 1461.[57]

The English lordship of Ireland was vulnerable to maritime raids almost from its inception in the late twelfth century. The geographic distance of new colonial lordships in Connacht and Ulster from Dublin meant that large sections of the western seaboard were vulnerable to attack from Irish or Hebridean raiders. The limitations of English naval power in these western regions forced the crown to negotiate alliances with powerful sea-lords such as the rulers of Galloway and Man. Moreover, the vulnerability of the English lordship to external assault by sea became increasingly apparent from the mid-to-late thirteenth century onwards. Although the Norse expedition of 1263 failed to intervene in Ireland, it further exposed the limitations of English power in Ulster and Connacht. The Bruce invasions of the early fourteenth century, followed by the onset of cooler wetter weather, effectively destroyed earldom of Ulster and saw the de Burgh lordship in Connacht disengage from the governance of the colony. The collapse of royal authority across these regions resulted in a rise in piratical activity as lords such as Edmund Albanach used their naval resources to carve out new and autonomous lordships in Connacht. The commercialisation of the western seaboard also contributed to rising levels of maritime predation: Irish dynasties more powerful not only sought to protect their newfound resources, they also sought to undercut the economic base of their rivals through piracy.

## From the Hundred Years War to the Tudors: Piracy and Marine Predation in the Irish Sea World, c.1337–c.1500

The commercialisation of Ireland's western Atlantic seaboard was a key factor in driving the political recovery of the Irish lordships, the decline

56 Childs and O'Neill, "Overseas Trade," p. 523.

57 *Cal. Carew MSS*, v, 471.

of English power, and the growing levels of piracy along the Ireland's coastlines. Although none could have foreseen it, the Bruce Invasions, alongside the aforementioned environmental phenomena, acted as a major catalyst for these political developments. The Scots, however, were unable to exploit this situation to their own advantage until the later fourteenth century. The death of Robert Bruce in 1329 saw Scotland plunged into a series of civil wars which were not resolved until the 1360s.[58] Scotland's growing alignment with France, followed by a series of English reversals in the Hundred Years War during the 1370s, saw the Scots and their French allies becoming increasingly active in the Irish Sea. In June 1372, the Welsh rebel Owain Lawgoch attacked Guernsey with a French fleet.[59] The motivation for this assault remains uncertain, but it has been suggested that Owain and his French patrons hoped to establish a beachhead on Guernsey before attempting a seaborne invasion of Wales.[60] In 1377, another French fleet raided Castle Rushen on the Isle of Man and forced the Manx to buy the French off for one thousand marks; five years later the French targeted the south coast of Ireland, burning the town of Kinsale.[61]

While the French operated primarily in the southern sections of the Irish Sea, their Scottish allies targeted Ulster, Man, and Leinster. The Otterburn campaign of 1388 was the most successful instance of the Scots opening multiple fronts against England since the kingship of Robert Bruce.[62] During the late summer of 1388, the Scots launched a three-pronged assault on northern England and Ulster. A force under Robert (d.1420), earl of Fife, marched through the West Marches, burning large sections of the northern English countryside and defeated an English levy near Carlisle. A second force under James (d.1388), earl of Douglas, raided as far south as Newcastle-upon-Tyne before defeating an English army at Otterburn on August 5. A third force under Earl James' kinsman, Sir William Douglas of Nithsdale (d.1391), sailed westwards to Ulster, burning colonial settlements along Carlingford Lough, followed by an assault on the Isle of Man.[63] French military support, coupled with English entanglement in the continental war, had proven crucial to the survival of an independent Scotland during the fourteenth century. The peace of the 1390s, followed by the deposition

58 For a succinct overview of these developments, see Brown, *Wars of Scotland*, ch. 11.

59 *Sir John Froissart's Chronicles of England, France, Spain and the Adjoining Countries*, iv, p. 165.

60 See Carr, *Owen of Wales*, pp. 28–31.

61 Oliver, *Monumenta*, p. 73. See also Tuck, *Richard II*, p. 16; Preest, *Cronica Maiora*, pp. 108–9.

62 MacDonald, *Border Bloodshed*, pp. 104–5.

63 Amours, Thomas, and Neilson, *Andrew of Wyntoun*, vi, pp. 321–24.

of Richard II (d.1400) and ascension of Henry IV (d.1413), saw this dynamic altered somewhat.

The French monarchy's unwillingness to renew the war with England left Scotland dangerously isolated during the early years of the fifteenth century as the new Lancastrian regime sought to subjugate its northern neighbour. Without a war to distract the English in France, Scotland was vulnerable to English incursions. Several factors helped to steer the kingdom of the Scots through these difficult times. Firstly, the eruption of Owain Glyn Dŵr's rising in Wales (1400–1415) placed a serious strain on English resources. Secondly, the outbreak of the Percy rebellion (1403) in Northumberland undercut Henry's influence within northern England and gave the Scots an added measure of security in the border region.[64] It is from within this context of growing dissent within the Lancastrian kingdom that the Scots unleashed successive waves of piracy within the Irish Sea.

Although the earldom of Ulster had been in a state of decay since the early-to-mid fourteenth century and reduced to thin strip of land hugging the coastlines of the modern-day counties of Antrim and Down, it appears to have remained economically viable through a lucrative fishing trade and strong mercantile connections with the nearby ports of Dundalk and Drogheda.[65] Consequently, the coastlines of eastern Ulster and northern Leinster became prime targets for numerous Scottish raids during the early years of the fifteenth century. The Chronicle of Henry Marleburrough notes that in late May 1400, a Scottish fleet clashed with an English force on Strangford Lough. The outcome of the engagement is not discussed but the chronicler states that many Englishmen were killed.[66] Two years later in the summer of 1403, Carrickfergus Castle is recorded as being burned by unspecified enemies of the king.[67] The following year in March 1404 merchants from Drogheda were attacked by Scottish pirates, and in the summer of the same year a large-scale Scottish foray into Ulster took place. The settlements of Coleraine, Downpatrick, and the Abbey of Inch were burned and a group of English merchants from Somerset were captured by Scottish rovers and taken to Bothwell Castle near Glasgow.[68] This recent Scottish attack caused considerable alarm within the colonial administration and in the late summer of 1404, the Irish council ordered 800 troops be dispatched

64 For Percy revolt, see Lomas, *House of Percy*, ch. 5.

65 O'Neill, *Merchants and Mariners*, pp. 30–33.

66 Ware, *Ancient Irish Histories*, ii, 17.

67 "Patent Roll 4 Henry IV, no. 241" in Crooks, *Circle*, https://chancery.tcd.ie/document/patent/2-henry-iv/23.

68 Smith and Dryburgh, *Handbook*, pp. 170–71; *AMisc*, 1404.

north to aid in the recovery of lands devastated by recent Scottish raids.[69] These efforts to bolster the defence of the region appear to have had little impact and over the course of 1405 and 1406, the coastlines of Ulster and northern Leinster were raided by pirates from Galloway who took several captives from the colonial mercantile community.[70] The northern coast of Wales was also targeted by the Scots and though Owain Glyn Dŵr negotiated an alliance with Charles V of France (d.1422) in 1405, the Welsh rebel also dispatched emissaries to the royal court in Scotland.[71]

The French defeat at Agincourt in October 1415 saw the English momentarily gain the ascendancy in the Hundred Years War. From this point onwards, Scottish military support became increasingly important to the resuscitation of the French war effort. It is a measure of Scotland's contribution to France that thousands of soldiers were recruited for French armies in the late 1410s and early 1420s. The exportation of these fighting men not only underlines the confidence of the kingdom of the Scots during these years but, strikingly, these raids also coincided with rise in further Scottish and French piratical activity in the Irish Sea world.[72] For instance, in early 1417, the English constable of Harlech Castle claimed to have intelligence that Owain Glyn Dŵr's son, Maredudd ab Owain, conspired with the Scots to land troops in Wales during the summer.[73] Ulster was once again raided the following year and in April 1421 (a month after the major Franco-Scottish victory at Baugé) the colonial government complained of repeated attacks by French, Scottish, and Castilian pirates.[74] English concerns with regards to the security of Ireland were only partly motivated by raids on Ireland's eastern and southern seaboards. By this time, the Scottish government had managed to negotiate an alliance with the Uí Dhomhnaill of Tír Chonnaill. Although there is evidence to suggest that this alliance has its origins in the early 1380s, it is not until this period of rising piratical activity in the Irish Sea, coupled with the French resurgence, that the historical record reflects growing English anxieties about Scottish links to the Uí Dhomhnaill.[75] Situated on Ireland's western Atlantic littoral,

69 "Patent Roll 5 Henry IV, no. 84." in Crooks, *Circle,* https://chancery.tcd.ie/document/patent/5-henry-iv/84.

70 *Ancient Irish Histories*, 19; *CPR*, iii, 88, 114.

71 R. R. Davies, *Owain Glyn Dŵr*, pp. 193–95, 169–70; Nicholson, *Scotland*, pp. 226–27.

72 For Scottish recruitment, see Ditcham, "Employment," ch. 2.

73 Smith, "The Last Phase," pp. 250–60, at p. 259.

74 *AFM*, 1418; "Patent Roll 9 Henry V, no. 116" in Crooks, *Circle*, https://chancery.tcd.ie/document/patent/9-henry-v/116.

75 For the origins of this alliance, see Egan, "Richard II," pp. 221–52, at pp. 239–40.

the lordship of Tír Chonnaill was well beyond the effective reach of the colonial administration. Moreover, during the early decades of the fifteenth century, the Uí Dhomhnaill had established themselves as a potent force within Irish politics. The military and maritime resources of the lords of Tír Chonnaill thus provided the Scots with a valuable ally in a region where English influence was minimal.

The threat posed by the Uí Dhomhnaill to the English colony should not be underestimated. By the mid-1420s, the Uí Dhomhnaill had established a hegemony over much of Ireland's northern and western seaboard.[76] In 1423, the head of the dynasty, Niall Garbh Ó Domhnaill (d.1439) led a large-scale raid on the northern boundaries of the colony, moving unopposed from the western Atlantic seaboard to the Irish Sea before attacking and forcing tribute from the town of Dundalk.[77] The Uí Dhomhnaill juggernaut was eventually derailed by their rivals, the Uí Néill of Tír Eoghain, who, with the support of the colonial administration (see below), managed to defeat the Uí Dhomhnaill and re-establish Uí Néill dominance in Ulster from the 1430s through to the 1460s. Nevertheless, under the leadership of Aodh Ruadh Ó Domhnaill (d.1505), the Uí Dhomhnaill gradually reasserted their former power and influence within the northern of Ireland during the later fifteenth century.[78] By the mid-1470s, the English colony was once again vulnerable to raids from the Uí Dhomhnaill. In 1475 Aodh Ruadh brought a force drawn from across Ulster and Connacht and attacked the western boundaries of the colony in Leinster.[79] Twenty years later in 1495, Aodh Ruadh travelled to Glasgow for a conference with the Scottish king, James IV (d.1513).[80] Links between the lordship of Tír Chonnaill were further strengthened in the first half of the sixteenth century. Aodh Ruadh's son and successor, Aodh Dubh (d.1537), travelled to Edinburgh for council of war in the summer of 1513 and appears to have helped the Scots organise a diversionary attack on Belfast Lough and Carrickfergus Castle in the lead-up to the fateful Flodden Campaign of September 1513.[81] Following the disastrous Scottish defeat, the regency of the young James V (d.1542), remained in contact with Aodh Dubh. The lord of Tír Chonnaill was sent artillery in 1516 while in 1521 the colonial government believed that a full-scale Scottish invasion of Ulster

76 For rise of the Uí Dhomhnaill, see Simms, "Niall Garbh O'Donnell."

77 *AFM*, *Annála Uladh (Annals of Ulster): A Chronicle of Irish Affairs, AD 431 to AD 1540*, eds. William Hennessy and Bartholmew MacCarthy (*AU*), 1423.

78 For their recovery, see MacGettigan, "Late Medieval Lordship."

79 *AFM*, 1475.

80 *AU*, 1495.

81 *AFM*, *AU*, 1513.

was imminent.[82] Although no such invasion occurred, the threat of Scottish incursions loomed large in the mindset of Tudor officialdom throughout the early sixteenth century.[83]

The potential for a major Scottish invasion of Ireland, coupled with the ongoing problem of piracy in the Irish Sea, saw the English crown pursue several strategies during the later Middle Ages. The fortification of the eastern and southern seaboards was one such scheme. Kinsale's exposure to the major French raid of 1380 resulted in Irish council's decision to enclose the town with a stone wall in 1382.[84] Throughout the later fourteenth century, the citizens of Waterford and nearby Dungarvan as well the port of New Ross, situated on the River Barrow each petitioned the crown for funds to enwall their respective settlements.[85] Similar plans were mooted for the coastlines north and south of Dublin city. Dundalk was granted funds to repair its wall in 1412, though this did little to spare it from Niall Garbh's raid of 1423.[86] The government considered fortifying Wicklow Head in 1454 and 1460 as well as Lambay Island in 1467 while the administration also discussed building a fortified harbour in the 1490s.[87] However, none of these schemes ever came to fruition and it, in reality, it was not until the reign of Edward VI (d.1553) that the crown invested in the construction of new fortifications along Ireland's eastern and southern seaboards.[88]

Despite the lack of commitment to shoring up the coastal defences, the English crown and its colonial administration were more successful in forging alliances with regional dynasts in Ireland and western Scotland. In Ireland, this task was largely delegated to the Butler earls of Ormond. As acculturated magnates capable of operating across the English and Gaelic interface, the earls of Ormond used their political and dynastic links

82 *AC*, 1516; *Cal. Carew MSS*, i, 18–19.

83 See Cathcart, "James V."

84 "Patent Roll 5 Richard II, no. 193" in Crooks, *Circle*, https://chancery.tcd.ie/document/patent/5-richard-ii/193.

85 "Patent Roll 30 Edward III, no. 42" in Crooks, *Circle*, https://chancery.tcd.ie/document/patent/30-edward-iii/42; "Patent Roll 49 Edward III, no. 59" in Crooks, *Circle*, https://chancery.tcd.ie/document/patent/49-edward-iii/59; "Close Roll 51 Edward III, no. 46" in Crooks, *Circle*, https://chancery.tcd.ie/document/close/51-edward-iii/46; "Patent Roll 12 Richard II, no. 33" in Crooks, *Circle*, https://chancery.tcd.ie/document/patent/12-richard-ii/33; "Patent Roll 12 Richard II, no. 252" in Crooks, *Circle*, https://chancery.tcd.ie/document/patent/12-richard-ii/252.

86 "Patent Roll 13 Henry IV, no. 98" in Crooks, *Circle*, https://chancery.tcd.ie/document/patent/13-henry-iv/98.

87 Berry, *Henry VI*, 311, 671; Berry, *Edward IV*, 467. See also Childs and O'Neill, "Overseas Trade," p. 523.

88 See White, "Edward VI in Ireland," pp. 197–211, at p. 205.

with their Irish neighbours to promote English influence. During his long career in Irish politics, James Butler (d.1452), fourth earl of Ormond, succeeded in drawing the Uí Néill of Tír Eoghain into an alliance against the Uí Dhomhnaill of Tír Chonnaill.[89] Moreover, Ormond's ability to operate across these linguistic boundaries most likely allowed him access to intelligence, particularly in relation to events on Ireland's western seaboard, which would have otherwise been impossible to obtain. For instance, during sessions of the Irish parliament in the 1420s Ormond claimed to have received word that the Uí Dhomhnaill were not only in league with the Scottish government, but that they were actively receiving military support from Scotland.[90] Following Ormond's death in 1452, the Uí Néill continued to cooperate with the colony well into the later fifteenth century.[91]

The English crown sought to safeguard against Scottish incursions by negotiating with the Mac Domhnaill lordship of the Isles. Under the leadership of the so-called first lord of the Isles, Eóghan Mac Domhnaill (d.1387), the Meic Domhnaill lordship came to encompass the entire Hebridean archipelago as well as a sizeable chunk of Scotland's western seaboard.[92] Relations between the Meic Domhnaill and the English crown can be traced to the late 1330s, and throughout the fourteenth century Eóghan moved between English and Scottish allegiance while extending his lordship over the northern Highlands.[93] During the tenure of his son and successor, Domhnall (d.1423), second lord of the Isles, the Meic Domhnaill acquired possession of Antrim in eastern Ulster. As noted in the introduction, certain scholars have argued that the high level of Scottish piracy in the Irish Sea world can attributed to the presence of the Meic Domhnaill in this region. Domhnall's brother and lord of Antrim, Eóghan Mór (d.1428) received a pardon for unspecified activities in 1410.[94] While it is entirely possible that Eóghan Mór engaged in piracy during this period, a growing body of scholarship has revealed that Eóghan Mór, his brother Domhnall of the Isles, and their successors, were courted as allies by successive English monarchs from Edward III through to Henry VIII (d.1547).[95] For instance, throughout the later fourteenth and early fifteenth centuries, Domhnall and

89 For Ormond's career, see Matthew, "Lancastrian Lordship."

90 "Close Roll 6 Henry VI, no. 21" in Crooks, *Circle*, https://chancery.tcd.ie/document/close/6-henry-vi/21.

91 Simms, "Earldom of Ulster."

92 For expansion, see Boardman, *Early Stewart Kings*, pp. 89–91.

93 For which, see Egan, "Early Stewart Kings," pp. 61–78, at pp. 63–65.

94 *CPR*, iv, 190.

95 For main studies, see Cathcart, "The Forgotten '45"; Petre, "Domhnall Balloch."

Eóghan Mór appear as allies of the English crown on Anglo-French truce proclamations.[96] Moreover, it was largely thanks to the agency of Richard II that Eóghan Mór secured possession of Antrim in the 1390s.[97] Perhaps most famously, the first Yorkist king of England, Edward IV (d.1483) sealed the so-called "Treaty of Westminster-Ardtornish" with Domhnall's son and successor, Eóghan (d.1503) in 1462. Under the terms of this treaty, Eóghan and his kinsmen swore allegiance to Edward. In return for their fealty, Eóghan was to be granted all Scotland north of the Forth in the event of a successful English conquest of the kingdom.[98] Collectively, these initiatives served to bolster English influence and frustrate Scottish and French ambitions in the region. Although the lordship of the Isles was eventually dismantled by the resurgent Stewart monarchy in the later fifteenth century, fragmenting into a series of smaller lordships, the Meic Domhnaill retained considerable influence in the northern Irish Sea world.[99] They would remain a potent force in Anglo-Scottish relations well into the seventeenth century.

By opening a second front against the English, in Ireland, the Scots (and their French allies) could draw English forces away from Scotland and keep them occupied in different theatres. This strategy became especially important during the early years of the fifteenth century when the French were reluctant to renew the war with England. Nevertheless, Scottish raids upon English holdings in Ireland, Man, and Wales, should not be viewed solely within this context of military strategy. Assaulting English possessions in these regions undoubtedly weakened England's capacity to wage war against the Scots. However, many of the Scottish rovers appeared to have purely economic motivations for engaging in marine predation. While Otterburn offers an excellent example of a three-pronged military and naval strategy, Scottish raids on Dundalk and Drogheda should also be viewed as an attempt by the Scots to disrupt the activities of their mercantile rival and, indeed, enrich themselves in the process. Scottish raids upon Ireland, whatever their motivation, were a serious concern for the English government and took on an even more worrying dimension following the emergence of the Uí Dhomhnaill as a regional power. The threat of Scottish raids, coupled with an Uí Dhomhnaill-Stewart accord, was a key factor shaping the English crown's relationship with the Meic Domhnaill of Antrim and the Western Isles. Only by harnessing the resources of these dynasts

96 Rymer, *Foedera*, vii, 622–29, 639–40, 714–16, 777, 824; Stamp, *Henry V*, i, pp. 108, 369.

97 For the acquisition of Antrim, see Egan, "Richard II," pp. 232–33.

98 Munro and Munro, *Lords of the Isles, 1336–1493*, pp. 111–16.

99 For collapse, see Boardman, *Campbells*, chs. 9–10.

could the English government hope to exert itself within the wider world of the North Channel.

## Conclusion

Irish waters experience a variety of different forms of marine predation during the later Middle Ages. From its inception in the late twelfth century, the western sections of the English colony were vulnerable to seaborne assault. Not only did Hebridean pirates plunder western lordships in Connacht, the colony was itself very nearly subjected to a Norse invasion in the 1260s. The fourteenth and fifteenth centuries did little to improve this situation. The impact of the Bruce invasions and the onset of cooler wetter weather severely undercut the English colony in Ulster and Connacht. Dynasties which had once formed the backbone of colonial governance in these regions became increasingly Gaelicised and many, such as the de Búrcaigh of Connacht or the FitzGeralds of Desmond, became active participants in piratical ventures. The commercialisation of the western seaboard enabled Irish lineages to expand their power base and recover lands lost during the early conquest. Protecting these newly won marine resources became increasingly important, and Irish lineages such as the Uí hEidirsceoil deployed their ships against their rivals but also in the pursuit of plunder. The outbreak of the Hundred Years War and England's growing entanglement in France made it increasingly difficult to devote the resources necessary to the protection of its western insular satellites. Equally, Scottish vulnerability within this conflict saw the Scots engage in rising levels of piracy within the Irish Sea world. As discussed, these raids enabled the Scots to undercut their opponents in numerous. However, Scottish raids should not necessarily be viewed through this prism alone: many Scots, not unlike the counterparts on Ireland's western seaboard, had economic motivations for engaging in what amounted to little more than robbery.

Throughout the high and later medieval period, successive English governments remained concerned about the activities of pirates, rovers, and raiders in Irish, Welsh, and Manx waters. During the later twelfth and early thirteenth centuries, the English crown had sought to mitigate against the limitations of their own naval resources and harness the fleets of the rulers of Galloway and Man. To all intents and purposes, this strategy continued under the later Plantagenet kings through to the Tudor period: England's monarchs negotiated successive alliances with the Meic Domhnaill of Antrim as a means of safeguarding their interests against maritime raids.

While the support of the Meic Domhnaill went someway to protecting English holdings, in reality it would take several wars of conquest and colonisation before the Tudors and their Stuart successors could finally secure their western satellites from piracy, Gaelic or otherwise.

## Bibliography

### Printed Primary Sources

*Acts of the Lords of the Isles, 1336–1493*, edited by Jean Munro and R. W. Munro. Edinburgh: Scottish History Society, 1986.

*Ancient Irish Histories*, edited by James Ware, 2 vols. Dublin: Hibernia Press, 1809.

*Annals of the Kingdom of Ireland by the Four Masters*, edited by John O'Donovan, 7 vols. Dublin: De Búrca, 1990.

*Annála Connacht: The Annals of Connacht*, edited by Martin Freeman. Dublin: Dublin Institute for Advanced Studies, 1944.

*Annála Uladh (Annals of Ulster): A Chronicle of Irish Affairs, AD 431 to AD 1540*, edited by William Hennessy and Bartholmew MacCarthy. Dublin: H. M. Stationary Office, 1887–1901.

*Calendar of Carew Manuscripts Preserved in the Archiepiscopal Library at Lambeth*, edited by John S. Brewer, 6 vols. London: Stationary Office, 1867–1873.

*Calendar of Close Rolls: Henry V, 1413–1422*, edited by A. E. Stamp, 2 vols. London: Public Record Office, 1929–1932.

*Calendar of Documents Relating to Ireland, 1171–1307*, edited by Henry S. Sweetman, 5 vols. London: Public Record Office, 1875–1886.

*Calendar of Patent Rolls: Henry IV, 1399–1413*, edited by H. C. Maxwell Lyte, 4 vols; London: Public Record Office, 1903–1909.

*Early Sources of Scottish History, A.D. 500–1286*, edited by Alan O. Anderson, 2 vols. Edinburgh: Oliver and Boyd, 1922.

*Foedera*, edited by Thomas Rymer, 20 vols. London: A & J Chruchill, 1727–1735.

*Handbook and Select Calendar of Sources for Medieval Ireland in the National Archives of the United Kingdom*, edited by Brendan Smith and Paul Dryburgh. Dublin: Four Courts Press, 2005.

*Miscellaneous Irish Annals, AD 1114–1437*, edited by Séamus Ó hInnse. Dublin: Dublin Institute for Advanced Studies, 1947.

*Monumenta de Insula Manniae*, edited by J. R. Oliver. Douglas: Manx Society Publications, 1860.

*Sir John Froissart's Chronicles of England, France, Spain and the Adjoining Countries*, 12 vols. London: W. Smith, 1808.

*Statute Rolls of the Parliament of Ireland: Reign of Henry VI*, edited by Henry F. Berry. Dublin: Stationary Office, 1910.

*Statute Rolls of the Parliament of Ireland: First to the Twelfth Year of King Edward IV*, edited by Henry F. Berry. Dublin: Stationary Office, 1914.

*The Chronica Maiora of Thomas Walsingham, 1376–1422*, edited by David Preest. Woodbridge: Boydell and Brewer, 2005.

*The Original Chronicle of Andrew of Wyntoun*, edited by F. J. Amours, John Thomas, and George Neilson, 6 vols. Edinburgh: W. Blackwood and Sons, 1903–1914.

## Printed Secondary Sources

Appleby, John C. "The Problem of Piracy in Ireland, 1570–1630." In *Pirates? The Politics of Plunder, 1550–1650*, edited by Claire Jowitt, 41–55. New York: Palgrave MacMillan, 2007.

Boardman, Stephen. *The Early Stewart Kings: Robert II and Robert III, 1371–1406.* Edinburgh: East Linton, 1996.

Boardman, Stephen. *The Campbells, 1250–1513*. Edinburgh: East Linton, 2005.

Boardman, Stephen. "Highland Scots and Anglo-Scottish Warfare, 1300–1513." In *England and Scotland at War, c.1296–c.1513*, edited by Andy King and David Simpkin, 231–53. Leiden: Brill, 2012.

Breen, Colin. "The Maritime Cultural Landscape in Medieval Gaelic Ireland." In *Gaelic Ireland, c1250– c.1650: Land, Lordship, and Settlement*, edited by Patrick J. Duffy, David Edwards, and Elizabeth FitzPatrick, 418–36. Dublin: Four Courts Press, 2001.

Breen, Colin. *The Gaelic Lordship of the O'Sullivan Beare: A Landscape Cultural History*. Dublin: Four Courts Press, 2005.

Brown, Daniel. *Hugh de Lacy, First Earl of Ulster: Rising and Falling in Angevin Ireland*. Woodbridge: Boydell and Brewer, 2016.

Brown, Michael. *The Wars of Scotland, 1214–1371*. Edinburgh: Edinburgh University Press, 2004.

Brown, Michael. *Disunited Kingdoms: Peoples and Politics in the British Isles, 1280–1460*. Harlow: Pearson, 2013.

Campbell, Bruce M. S. *The Great Transition: Climate, Disease and Society in the Late-Medieval World*. Cambridge: Cambridge University Press, 2016.

Carr, Anthony D. *Owen of Wales: The End of the House of Gwynedd*. Cardiff: University of Wales Press, 1991.

Cathcart, Alison. "James V, King of Scotland—and Ireland?" In *The World of the Galloglass: Kings, Warlords, and Warriors in Ireland and Scotland, 1200–1600*, edited by Séan Duffy, 124–43. Dublin: Four Courts Press, 2007.

Cathcart, Alison. "'The Forgotten' 45: Donald Dubh's Rebellion in an Archipelagic Context." *Scottish Historical Review* 91, no. 2 (October 2012): 239–64.

Chambers, Anne. *Granuaile: The Life and Times of Grace O'Malley.* Dublin: Wolfhound Press, 1983.

Childs, Wendy. "Ireland's Trade with England in the Later Middle Ages." *Irish Economic and Social History*, 9 (1982): 5–33.

Childs, Wendy and Timothy O'Neill. "Overseas Trade." In *A New History of Ireland II: Medieval Ireland, 1169–1534*, edited by Art Cosgrove, 492–524. Oxford: Oxford University Press, 1987.

Cowan, Edward J.. "Norwegian Sunset—Scottish Dawn: Hákon IV and Alexander III." In *Scotland in the Reign of Alexander III, 1249–1286*, edited Norman H. Reid, 103–31. Edinburgh: East Linto, 1990.

Davies, Rees R. *The Revolt of Owain Glyn Dŵr.* Oxford: Oxford University Press, 1995.

Duffy, Seán. "The Bruce Brothers and the Irish Sea World, 1306–29." *Cambridge Medieval Celtic Studies* 21 (1991): 55–86.

Duffy, Seán. "The 1169 Invasion as a Turning Point in Irish-Welsh Invasions." In *Britain and Ireland, 900–1300: Insular Responses to Medieval European Change*, edited by Brendan Smith, 98–113. Cambridge: Cambridge University Press, 1999.

Duffy, Seán. "The Prehistory of the Galloglass." In *The World of the Galloglass: Kings, Warlords, and Warriors in Ireland and Scotland, 1200–1600*, edited by Séan Duffy, 1–27. Dublin: Four Courts Press, 2007.

Edwards, David. "Tudor Ireland: Anglicization, Mass Killing, and Security." In *The Routledge History of Genocide*, edited by Cathie Carmichael and Richard C. Maguire, 23–37. New York: Routledge, 2015.

Egan, Simon. "Richard II and the Wider Gaelic World: A Re-Assessment." *Journal of British Studies* 57, no. 2 (April 2018): 221–52.

Egan, Simon. "By Land and By Sea: The Role of the Maritime Sphere in the Expansion of O'Donnell Power, ca.1380–1500." *Journal of the North Atlantic* 12 (2019): 16–31.

Ekin, Des. *The Stolen Village: Baltimore and the Barbary Pirates.* Dublin: O'Brien, 2006.

Hayes-McCoy, Gerald A. "The Early History of Guns in Ireland." *Journal of the Galway Archaeological and Historical Society* 18, no. 1–2 (1938): 43–65.

Flanagan, Marie T. "After Brian Bóraime: The High-Kingship and the Kings of Connacht." In *Medieval Dublin XVI*, edited by Séan Duffy, 218–59. Dublin, 2017.

Flavin, Susan. *Consumption and Culture in Sixteenth-Century Ireland: Saffron, Stockings, and Silk.* Woodbridge: Boydell and Brewer, 2014.

Frame, Robin. *English Lordship in Ireland, 1318–1361.* Oxford: Oxford University Press, 1982.

Gardiner, Mark, and T. E. McNeill. "Seaborne Trade and the Commercialisation of Fifteenth- and Sixteenth-Century Ulster." *Proceedings of the Royal Irish Academy* 116 (2016): 229–62.

Kelleher, Connie. "The Gaelic O'Driscoll Lords of Baltimore, Co. Cork: Settlement, Economy, and Conflict in a Maritime Cultural Landscape." In *Lordship in*

*Medieval Ireland: Image and Reality*, edited by Linda Doran and James Lyttleton, 130–59. Dublin: Four Courts Press, 2007.

Kelleher, Connie. *The Alliance of Pirates: Ireland and Atlantic Piracy in the Early Seventeenth Century*. Cork: Cork University Press 2020.

Kelly, Maria. *A History of the Black Death in Ireland*. Dublin: Tempus, 2001.

Kingston, Simon. "Delusions of Dál Riada: The Coordinates of Mac Domhnaill Power, 1461–1550." In *Gaelic Ireland, c.1250–c.1650: Land, Lordship, and Settlement*, edited by Patrick J. Duffy, David Edwards, and Elizabeth FitzPatrick, 98–114. Dublin: Four Courts Press, 2001.

Lomas, Richard. *The Fall of the House of Percy, 1368–1408*. Edinburgh: East Linton, 2007.

Lydon, James F. "The City of Waterford in the Later Middle Ages." *Decies* 12 (1979): 5–15.

MacDonald, Alastair J. *Border Bloodshed: Scotland, England and France at War, 1369–1403*. Edinburgh: East Linton, 2000.

MacGettigan, Darren. "The Renaissance and the Late Medieval Lordship of Tír Chonaill, 1461–1555." In *Donegal: History and Society*, edited by William Nolan, Liam Ronayne, Mairead Dunleavy, 203–28. Dublin: Geography Society, 1995.

Mac Niocaill, Gearóid. *Na Burgéisí XII–XV Aois*. 2 vols. Dublin: Cló Morainn, 1964.

McErlean, Thomas, Rosemary McConkey, and Wes Forsythe. *Strangford Lough: An Archaeological Survey of the Maritime Cultural Landscape*. Belfast: Blackstaff Press, 2002.

McDonald, R. Andrew. *The Kingdom of the Isles: Scotland's Western Seaboard, c.1100–c.1336*. Edinburgh: East Linton, 1997.

McDonald, R. Andrew. *Manx Kingship in Its Irish Sea Setting, 1187–1229: King Rognvaldr and the Crovan Dynasty*. Dublin: Four Courts Press, 2007.

McNamee, Colm. *The Wars of the Bruces: Scotland, England and Ireland, 1306–1328*. Edinburgh: East Linton, 2006.

Murphy, Elaine. *Ireland and the War at Sea, 1641–1653*. Woodbridge: Boydell and Brewer, 2012.

Naessens, Paul. "Gaelic Lords of the Sea: The Coastal Tower Houses of South Connemara." In *Lordship in Medieval Ireland: Image and Reality*, edited by Linda Doran and James Lyttleton, 217–35. Dublin: Four Courts Press, 2007.

Ní Loinsigh, Máire. "An Assessment of Castles and Landownership in Late Medieval North Donegal." *Ulster Journal of Archaeology* 57 (1994): 145–58.

Nicholls, Kenneth. "Worlds Apart? The Ellis Two-Nation Theory on Late Medieval Ireland." *History Ireland* 7, no. 2 (March/April 1999), 22–26.

Nicholls, Kenneth. *Gaelic and Gaelicized Ireland in the Middle Ages*. Dublin: Lilliput, 2003.

Nicholson, Ranald. *Scotland: The Later Middle Ages*. Edinburgh: Oliver and Boyd, 1974.

O'Brien, A. F. "Politics, Economy, and Society: The Development of Cork and the Irish South-East Region, c.1170–c.1583." In *Cork: History and Society*, edited by Patrick O'Flanagan and Cornelius G. Buttimer, 82–155. Dublin: Geography Society, 1993.

Ohlmeyer, Jane. "The Dunkirk of Ireland: Wexford Privateers During the 1640s." *Journal of the Wexford Historical Society* 12 (1988–89): 23–49.

O'Neill, Timothy. *Merchants and Mariners in Medieval Ireland*. Dublin: Irish Academic Press, 1987.

Oram, Richard. *The Lordship of Galloway*. Edinburgh: East Linton, 2000.

Orpen, Goddard H. *Ireland Under the Normans, 1169–1333*, edited by Séan Duffy, 2nd Edition. Dublin: Four Courts Press, 2005.

Petre, James. "Domhnall Balloch, the Treaty of Ardtornish-Westminster and the MacDonald Raids of 1461–3." *Historical Research* 88, no. 242 (August 2015), 599–628.

Simms, Katharine. "Niall Garbh O'Donnell, King of Tír Conaill, 1422–39." *Donegal Annual* 12 (1977): 7–21.

Simms, Katharine. "The 'King's Friend': O'Neill, the Crown, and the Earldom of Ulster." In *England and Ireland in the Later Middle Ages*, edited by James F. Lydon, 214–36. Dublin: Irish Academic Press, 1981.

Simms, Katharine. "Late Medieval Donegal." In *Donegal: History and Society*, edited by William Nolan, Liam Ronayne, Mairead Dunleavy, 183–202. Dublin: Geography Society, 1995.

Simms, Katharine. "Late Medieval Tír Eoghain: The Kingdom of the Great O'Neill." In *Tyrone: History and Society*, edited by Charles Dillon and Henry A. Jefferies, 127–62. Dublin: Geography Society, 2000.

Simms, Katharine. "The Ulster Revolt 1404—An Anti-Lancastrian Dimension?" In *Ireland and the English World in the Late Middle Ages*, edited by Brendan Smith, 141–60. Basingstoke: Palgrave MacMillan, 2009.

Smith, Brendan. "Late Medieval Ireland and the English Connection: Waterford and Bristol, ca.1360–1460." *Journal of British Studies* 50, no. 3 (July 2011): 546–65.

Smith, J. Beverly. "The Last Phase of the Glyn Dŵr Rebellion." *Bulletin of the Board of Celtic Studies* 12 (1966–67): 250–60.

Stringer, Keith J. "Periphery and Core in Thirteenth-Century Scotland: Alan son of Roland, Lord of Galloway and Constable of Scotland." In *Medieval Scotland: Crown, Lordship and Community*, edited by Alexander Grant and Keith J. Stringer, 82–113. Edinburgh: Edinburgh University Press, 1998.

Sumption, Jonathan. *Divided Houses: The Hundred Years War*. London: Faber and Faber, 2012.

Tuck, Anthony. *Richard II and the English Nobility*. London: Edward Arnold, 1973.

White, Dean G. "The Reign of Edward VI in Ireland: Some Political, Social, and Economic Aspects." *Irish Historical Studies* 14, no. 55 (1965): 197–211.

## Unpublished PhD Theses

Ditcham, B. "The Employment of Foreign Mercenary Troops in French Royal Armies, 1415–1470." PhD Diss., University of Edinburgh, 1978.
Duffy, Séan. "Ireland and the Irish Sea Region, 1014–1318." PhD Diss., Trinity College Dublin, 1993.
Matthew, E. "The Governing of the Lancastrian Lordship of Ireland in the Time of James Butler, Fourth Earl of Ormond." PhD Diss., Durham University, 1994.

## Online Resources

Beresford, David. "Tregury Michael, c.1399–1471." In *Dictionary of Irish Biography*, edited by James McGuire and James Quinn. Cambridge: Cambridge University Press, 2009. http://dib.cambridge.org/viewReadPage.do?articleId=a8630.
*Circle: A Calendar of Irish Chancery Letters, c.1244–1509*, edited by Peter Crooks. Trinity College Dublin. https://chancery.tcd.ie.

## About the Author

**Simon Egan** is a Lecturer in Medieval Irish History at Queen's University Belfast. His research interests include lordship, aristocratic culture, warfare, and piracy in late medieval/early modern Ireland and Britain.

# 4. Boston, Logwood, and the Rise and Decline of the Pirates, 1713 to 1728

*Steven J. Pitt*

**Abstract**
After the end of Queen Anne's War (1702–1713), the economic fate of Atlantic seafarers and colonial Boston hinged on an unlikely commodity—logwood. This dyewood from the Spanish-claimed Yucatan Peninsula provided a haven from depressed wages and unemployment for seafarers and bankruptcy for Boston merchants. The growing British presence in the Bay of Campeche angered Spanish authorities who considered them pirates. In late 1716, the Spanish attacked, permanently ending that branch of the logwood trade. This act unleashed a decade of piracy, as the hundreds of seafarers and loggers engaged in the trade sought new avenues to riches. Reprieve from this onslaught only came when highly motivated Bostonians re-established the logwood trade in the Bay of Honduras in the mid-1720s.

**Keywords:** New England; London; Belize; Baymen; Edward Low; Blackbeard; Sam Bellamy

Historians point to 1716 as the pivotal year when pirates irrupted from the belly of the Atlantic to devour the shipping of "civilised" nations. This irruption has often been attributed to the end of Queen Anne's War (1702–1713) and the subsequent unemployment, wage reductions, and mistreatment of seafarers.[1] Recently, historians have identified two other important reasons for this rise in piracy. Mark Hanna argues that colonial emphasis on cash crops such as sugar and tobacco reduced the need for

1 Rankin, *Golden Age*; Earle, *Pirate Wars*; Rediker, *Villains of All Nations.*

Coakley, J., C. Nathan Kwan and D. Wilson, *The Problem of Piracy in the Early Modern World: Maritime Predation, Empire, and the Construction of Authority at Sea.*
Taylor & Francis Group, 2024
DOI 10.5117/9789463720960_CH04

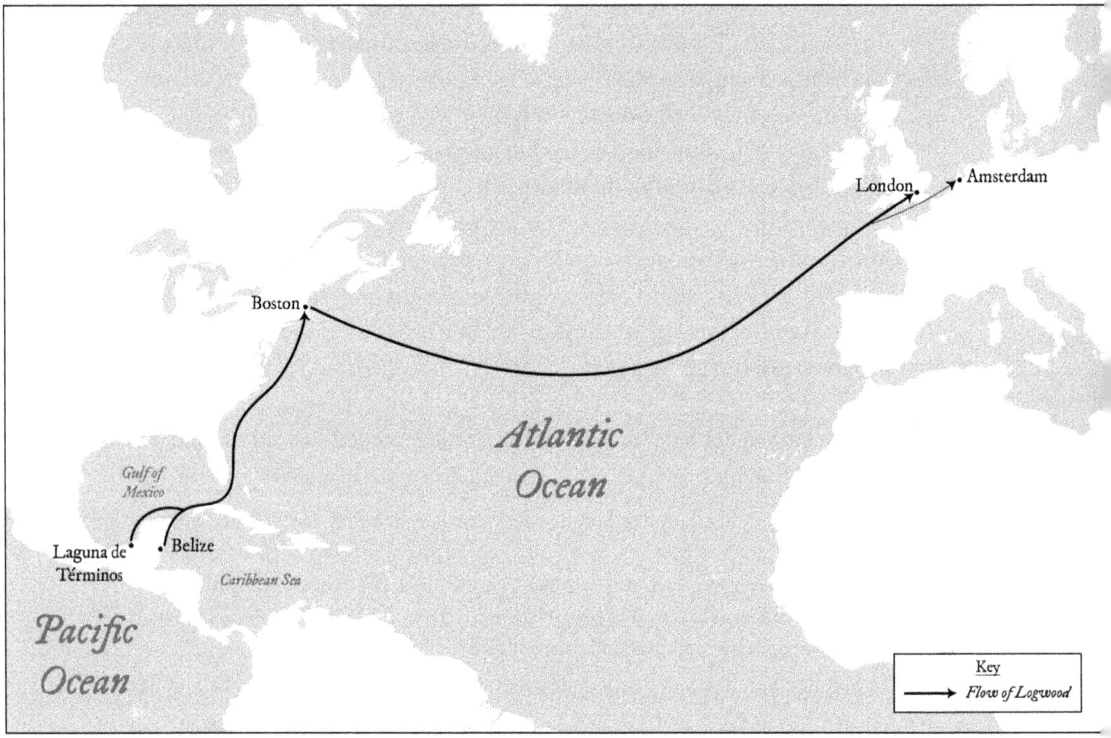
Boston
London
Amsterdam
Atlantic
Ocean
Gulf of
Mexico
Laguna de
Términos
Belize
Caribbean Sea
Pacific
Ocean
Key
Flow of Logwood

pirate silver and gold, as well as required stable, predictable trade routes. New imperial laws and institutions turned "Pirate Nests" in the West Indies and North America away from harbouring and trading with pirates.[2] David Wilson, meanwhile, contends that historians have neglected to examine the impact of continuing conflict with Spanish guardacostas in disputed territories like the coast of Florida and the Laguna de Términos. According to Wilson, Spanish depredations in these locations greatly impacted Jamaica's shipping and economy, which "displaced a sizeable seafaring population and encouraged their turn to piracy."[3] These important arguments add more nuance to our understanding of the rise in piracy in the aftermath of Queen Anne's War, and in the case of Wilson, hints at a major underlying cause. Historians, however, have failed to recognise the importance of colonial Boston and the Atlantic logwood trade in producing the conditions necessary for the dramatic upsurge of piracy in 1716, as well as its decline in the mid-1720s.

Boston had a complex, evolving, and intertwined history with pirates, privateers, and logwood cutters dating back to the 1670s. During much of the seventeenth century, pirates and privateers confidently sailed into Boston rich with Spanish or Portuguese gold and silver and found a warm welcome. They outfitted their ships, enlisted additional crew members, sold stolen goods, and rubbed elbows with Boston's political and economic elites. Boston, like many North American and West Indian ports, relied heavily on the pirates' gold and silver to overcome hard currency shortages. Beginning in the mid-1690s, political and religious authorities turned against pirates. Bostonians prosecuted pirates and their benefactors, even as other colonial ports like New York, Philadelphia, and Newport continued to be enriched by pirate exploits in the Indian Ocean. By the early 1700s, Boston elites sought to capture and hang pirates such as John Quelch and his crew as a warning to the entire seafaring community. In 1717, Boston again led the charge to suppress piracy and hang pirates. Pirates retaliated by singling out Boston and its vessels for special vengeance. Indeed, the infamous pirate Ned Low, a direct product of Boston's shipbuilding industry and Bay of Honduras logwood trade, terrorised the port for two years, hindering trade and greatly damaging the economy.

Boston's persecutorial attitude towards pirates in the 1690s coincided with a significant shift in its relationship with logwood cutters in the Laguna de Términos. During the 1670s, Boston merchants and ship captains

2 Hanna, *Pirate Nests.*

3 Wilson, *Suppressing Piracy*, p. 29.

contracted with resident logwood crews, like that of William Dampier, the noted privateer and explorer, in 1675–76. They paid well and the relationship proved mutually beneficial. In 1690s, Boston merchants began to wring greater profits out of the logwood by employing their sailors, as well as enslaved African and Indigenous peoples, as logwood cutters rather than purchasing it from locals. In the process, Bostonians working the lagoons and shipping logwood became pirates in the eyes of the Spanish who claimed those lands and waters. Like tobacco for Virginia and sugar for Jamaica, logwood became Boston's "cash crop" and helped propel the port's economy forward by funding shipbuilding and the importation of European goods. Boston's lust for logwood after Queen Anne's War resulted in the destruction of the Laguna de Términos trade, unleashing a vengeful tide of unemployed seafarers willing and able to wreak havoc on the Atlantic system of trade through piracy.

Boston's role in the rise of piracy in 1716 has been little explored by historians for two reasons. First, Boston's economic dominance in the Americas, especially in maritime and financial pursuits, has not been fully appreciated. Mark Peterson's recent study, *The City-State of Boston*, provides an important intellectual approach to Boston's unique status in the Atlantic World, but the watery world of seafarers, pirates, and logwood cutters barely makes an appearance.[4] Second, the logwood trade has been a sorely neglected area of study by Atlantic historians, despite the fact it deserves recognition alongside sugar, tobacco, coffee, fish, and cacao as one of the most important Atlantic commodities.[5] The secretive and illicit nature of the trade undoubtedly contributes to our dearth of knowledge, but the evidentiary traces are available in colonial newspapers, shipping and merchant papers, travel narratives, and court records. Drawing from these sources, this chapter argues that we cannot fully grasp the rise of piracy in 1716 and its fall in 1726 without understanding Boston's place in the Atlantic economy, dominance of the logwood trade, and relationship with, and suppression of, pirates after 1716.

4 Peterson, *City-State of Boston*.

5 The most significant studies on the logwood trade are Wilson, "Logwood Trade"; Thornton, "English at Campeachy"; Craig, "Logwood"; Joseph, "British Loggers"; Joseph, "Part II"; Bolland, *Colonialism and Resistance*, chs. 1, 2; Finamore, "A Mariner's Utopia"; Cromwell, "Life on the Margins"; Jarvis, *Eye of All Trade*, pp. 218–25; McDonald "Sailors from the Woods." These studies neglect to identify the importance of Boston in the Atlantic logwood trade. Newton highlights the Boston connection with the logwood trade during the late 1720s to the 1740s but does not explore the economic importance of logwood for Boston or its connection to piracy. Newton, "Good and Kind Benefactors."

## Logwood: Boston's "Cash Crop"

Boston hummed in 1713. Shipping and shipbuilding had nearly doubled over the course of Queen Anne's War (1702–1713), creating a remarkable economic transformation.[6] The town ranked third in the British Empire in shipping, behind only London and Bristol. It also established itself as the primary location for British manufactured goods that Boston vessels reexported to frontier communities up and down the North American coast. This economic success led to the highest town population in British America, as well as an increased standard of living for merchants, shipbuilders, waterfront labourers, and seafarers. Boston escaped the problem of specie-scarcity that plagued European colonial possessions by using creative book debts and bills of exchange. The war's end, however, threatened much of this economic success. Privateer crews disbanded, wages for sailors plummeted, and unemployment rose among the maritime and shipbuilding community. Shipping contracted across the British Empire and powerful London merchants began clamouring for debts owed. Boston merchants had a solution for these economic woes. They expanded their trade in logwood—a relatively short, crooked, and prickly tree from the Yucatan Peninsula that produced red, purple, and black dyes for Europe's textile and furniture industries.

Boston merchants had relied on logwood to help balance their accounts with London merchants since the 1670s when English buccaneers settled the Laguna de Términos after the Treaty of Madrid. Boston appears to have been the most dominant actor in shipping logwood even in this early period. Twelve of the seventeen vessels William Dampier identified by their homeport during his time in the Laguna de Términos came from Boston and New England. Indeed, Dampier's crew had been contracted by a Boston ship captain to harvest one hundred tons of logwood. These "Baymen,"[7]

6 From 1699–1702, Boston built forty-five vessels (2,807 tons) while from 1711–1714, the town constructed ninety vessels (5,102 tons), see Bailyn and Bailyn, *Massachusetts Shipping*, Table XXa, 116.

7 The term "baymen" or "bayman" is likely more complicated than historians have typically applied it. Historians have used this term to indicate a person residing in the Bays of Campeche and Honduras. Anecdotal evidence from my research, however, suggests that anyone involved in trading to the Bays of Campeche or Honduras could be labelled "bayman." This included merchants, ship captains, and sailors that resided in Boston. For example, the Boston ship captain Thomas Porter was referred to as a "bayman" in The National Archives, UK (TNA) Colonial Office Papers (CO) 152/10. Masters of vessels in the Bay of Campeachy to John Cample, December 10, 1716. I have also examined the careers of some of the "baymen" in Newton, "Kind Benefactors," and found evidence that they resided in New England with their families, including Andrew

as they were called, earned at least four times more, but as high as ten times more, than a typical common seaman for their efforts. As Dampier remarked, working the lagoons represented "a Place where a Man might have gotten an Estate," no matter their background or occupation.[8] Bostonians were content to leave the dangerous and hard work of harvesting logwood to the ex-buccaneers. On the few occasions when Boston ship captains and sailors braved the logwood cutters' treacherous environment to go hunting, they lacked the necessary experience, knowledge, and skill. Some of them never returned from those jaunts while others suffered from severe dehydration and would have died without aid. The harsh environment of the logwood trade empowered local logwood cutters in negotiations with Boston merchants and ship captains, creating common interests, cooperation, and economic prosperity for all parties.[9]

In the late 1690s, Boston's more ambitious merchants decided to forgo typical business arrangements in the logwood trade by employing sailors and enslaved African and Indigenous peoples as logwood cutters to increase their profits.[10] This important transition decreased the per ton of logwood labour costs for Boston merchants by roughly 75 per cent. The competition with local cutters increased production and drove down the price of logwood heading into the eighteenth century. Colonial merchants could expect roughly £16 per ton in Europe from 1700 to 1717 instead of the typical £30 to £40 per ton in the seventeenth century. Volume more than made up for the price drop, as fleets of ships employing hundreds of seafarers descended upon the Laguna de Términos.[11]

The new labour arrangements benefited Boston merchants but created significant tensions and hardships between ship captains, mates, and sailors. Injury, disease, warfare with the Spanish, and death plagued sailors and logwood cutters in the lagoons. Some sailors found themselves tricked or "forced into the Laggoons" to cut wood due to the vagueness of contracts that

Woodbury of Boston, see *Early History of Boston*, vol. 28, p. 117. Nicholas Winkley who resided in Portsmouth with his wife Sarah, see New Hampshire, *Probate Records*, 1635–1771, vol. 32, p. 759. For the purposes of this chapter, I plan to use the term "resident" or "local" and "cutters" or "loggers" to refer to people living in the two bays.

8 Masefield, *Dampier's Voyages*, p. 157.

9 Dampier, *Voyages*, vol. 2, pp. 114, 125, 132, 142–43, 177, 183, 189–90, 225.

10 Michael Jarvis argues that Bermudans were among the first to employ enslaved people in the Bay of Honduras logwood trade during the 1720s, but Boston-based crews brought enslaved people to work in the Laguna de Términos logwood trade in the late seventeenth century. Jarvis, *Eye of All Trade*, p. 226.

11 For more on this important transitional period in the logwood trade, see Pitt, "City upon the Atlantic Tides," pp. 170–82.

included the Bays of Campeche and Honduras in the "West Indies."[12] Those willing to work in the trade came from difficult and violent backgrounds, oftentimes having been pirates or privateers. They had a fierce independent streak and resented ship captains overbearing authority.[13] It did not help that most captains remained on their vessels, safe from the rampant mosquitos, biting flies, and alligators. As his crew toiled, the captain spent his time drinking with friends, making deals with other captains, and occasionally hunting turtles, fowl, and wild cows. The chief mate had the unpleasant task of overseeing the construction of the "barkudeer," or barracks, and harvesting the logwood, which took at least two months for a 100-to-150-ton ship.[14] Given such wretched conditions on board, violence, desertion, and mutiny lurked behind every logwood voyage, even for those that did not employ their sailors as logwood cutters.

These competing forms of labour coexisted in the lagoons surrounding Triste, modern-day Carmen de Isla, through Queen Anne's War (1702–1713). English, West Indies, and some North American vessels continued to trade with the roughly 200 to 250 self-employed loggers rather than subject their sailors to cutting wood. In late 1713 and into early 1714, however, Boston merchants launched a veritable armada of thirty-seven vessels to the Laguna de Términos to alleviate the difficult post-war economic circumstances. The lagoons swarmed with Boston woodcutting crews and their "barkudeers" occupied many of the prime locations for easily gathering logwood. Merchants from London, Bristol, Amsterdam, Jamaica, Barbados, Rhode Island, and New York also saw opportunity to profit from the Yucatan's forests, swelling the number of vessels in the Bay of Campeche to "about 70 or 80 Sayle [sic]" in March of 1714 alone.[15] With an average crew size of fifteen, roughly 1,200 total sailors, 550 from Boston vessels, along with 250 local cutters, resided in the Laguna de Términos. Spanish officials in nearby Campeche Town could only look on in horror as roughly 1,500 pirates robbed their territory.

12 Suffolk County Massachusetts Supreme Judicial Court Files (SCSJCF), Massachusetts State Archives (MSA), case no. 6112. Depositions of Richard James and Benjamin Girdler.

13 For a good example of this behaviour, see Uring, *Voyages*, pp. 243–63.

14 Massachusetts Archives Collection (Mass. Arch.) 62: 368–69. Depositions of John Viall, March 18, 1701, and Peter Cutler, March 20, 1701; SCSJCF, MSA, case no. 4992. Deposition of Ephraim Nichols.

15 The total number of Boston vessels in the Bay of Campeche during the 1714 logwood cutting season is probably higher. Many Boston vessels voyaged directly from the Bay of Campeche to Europe, which meant upon their arrival in Boston their last port of departure would not have reflected their time in the Bay of Campeche. Fortunately, the March 1714 report covered some of those captains. *Boston News-letter*, March 21, 1714; "Entrances and Clearances," *Boston News-letter*, January 3, 1714–December 27, 1714.

Large numbers of vessels and sailors continued to operate in the Laguna de Términos into 1715 and, to a lesser extent 1716, providing employment for seafarers and wealth for merchants, especially from Boston.

Logwood provided Boston with a lucrative "cash crop" to draw in British-owned vessels, support its ship outfitting and building industry, and dominate the reexport trade of European manufactured goods to other North American colonies. The numbers were impressive. From June 1714 till the end of the year, Britain imported 1,129 tons of this logwood directly from Boston. This represented 30 per cent of Great Britain's total trade in logwood and amounted to £23,000 sterling or 45 per cent of the value of all New England's, not just Boston's, exports to Britain. Incredibly, these numbers only reflect half of Boston's trade in logwood in 1714. Fifteen Boston-owned vessels, mostly larger ships with higher tonnages, sailed directly for Europe with their cargoes of logwood. Many of these sailed for Britain while some broke the Navigation Acts and smuggled their cargoes into Amsterdam, Lisbon, Leghorn, or Hamburg where high demand increased prices. Boston merchants, however, often had their foreign contacts remit their profits to London merchants. If we estimate ninety tons per vessel for those sailing directly to Europe, Bostonians extracted 2,784 tons of logwood from the Laguna de Términos worth a total of roughly £44,500, or 86 per cent of the value of New England's exports to Great Britain. In the unlikely event all this logwood made it to Great Britain, it equalled 57 per cent of all logwood imports into the mother country. In 1715, Europe's direct imports of logwood from Boston increased to 1,875 tons. Forty-three different vessels, employing 403 sailors, carried this wood that comprised little over half the tonnage. Importantly, British merchants owned and operated 75 per cent of the vessels.[16] This enriched Boston and its merchants because every vessel required repairs and provisions that supported the local economy and provided a way to move book debts at the expense of outsiders.[17]

Boston merchants and ship captains emerged from the aftermath of Queen Anne's War as the leading employer of seafarers in the British colonies, including those in the Laguna de Términos logwood trade. They bet heavily on the commodity, and the labour of seafarers, to balance accounts with creditors in London. It worked brilliantly, at least in 1714 and into 1715, but

16 TNA, CO 5/848. Massachusetts Shipping Returns, 1688 to 1719; *Boston Newsletter*, 1714 to 1716; Wilson, "Logwood Trade," pp. 1–15. For New England's export and import numbers, see Whitworth, *State of the Trade*, pp. 63–64.

17 For a more detailed discussion of how Boston merchants used outsiders' vessels to move debts see Pitt, "Building and Outfitting Ships," pp. 900–904.

trouble brewed on the horizon. The scale of operations in the Laguna de Términos strained available local resources. Crews had to venture further inland to find wood. Food became scarce because Boston vessels intending to cut their own wood sailed in ballast and only carried provisions for the voyage. The price of logwood plummeted reflecting Boston's saturation of the market. Finally, ominous reports began to circulate that the Spanish intended to punish the logwood cutters as pirates.

## The End of the Laguna de Términos Logwood Trade and the Rise of Pirates

While historians identify 1716 as the decisive year when pirates irrupted upon Atlantic waters, the Spanish first experienced this epidemic in late 1713 and into 1714 with 1,500 pirates pillaging their forests near the Laguna de Términos. The British merchants, ship captains, sailors, and loggers engaged in the trade believed they had a customary right to the lagoons derived from the Treaty of Madrid of 1670. The Spanish disagreed and even had allies in Britain that considered anyone engaged in the trade as pirates. In the summer of 1714, *Mercator*, a London newspaper, declared "The Trade of Jamaica is known to be principally depending upon their Robbing the Spaniards: On one hand their Logwood and Nicoraga [sic]-wood. Cutting! What is it but cutting down their Neighbour's Woods in the Gulph of Honduras, and the Bay of Campeachy?" Although unfairly laying blame for the trade at the feet of Jamaica, the author sympathised with Spain's hostile position towards loggers.[18] To destroy the pirate nest at Laguna de Términos, however, would require significant cooperation among Spanish officials and the naval and military commanders in the provinces surrounding the Bay of Campeche. Bostonians' intense exploitation of the trade provided the nudge to provoke the Spanish to act cooperatively.

Throughout 1714, the sheer number of vessels and men in the lagoons overwhelmed the possibility of a Spanish attack. Despite constant complaints from Jamaican merchants about Spanish depredations in the greater Caribbean, there was not a direct attack on vessels in or near Laguna de Términos in 1714. Nor did the Spanish threat intimidate Boston merchants. In mid to late 1714, they prepared twenty-eight vessels to depart for the Bay

18 *Mercator*, July 1, 1714. These debates over the legality of the logwood would continue into the late 1710s, 1720s, and '30s, contributing to three wars between Britain and Spain, including the long War of Jenkins' Ear (1739–1748).

of Campeche. Of these, we know the cargoes of eighteen. Only four listed provisions as part of their cargo, nine sailed in ballast, indicating they planned to cut their own wood, and another four had alcohol and European goods. Bartholomew Cheever in the 130-ton *Indian Emperor* only declared a cargo of prostitutes, which again indicated an intent to stay a few months cutting his own wood.[19] According to one observer, "above 30 English Ships" were in the Bay in early 1715, which would have included many of these Boston vessels. The fleet of British vessels continued to intimidate Spanish officials but the guardacostas began to target fully loaded vessels sailing for home.[20] The main threat to local cutters, however, came not from the Spanish but rather those empty Boston vessels with crews designing to cut their own wood and hunt fresh meat.

The diligent Boston crews successfully chopped, chipped, and loaded their cargoes of logwood by late winter and some of the smaller sloops returned to Boston by late April and early May. They arrived only to learn that the price of logwood had plummeted to £12 per ton in London. Merchants in the London trade considered this a boon. As Jeremiah Dummer proudly trumpeted a few years later, the "Industry of the people of New England" had managed to reduce the cost of logwood for the mother country from "30 and 40£ per Ton" to "12£ per Ton."[21] The merchants and ship captains bringing in logwood from the Bay of Campeche, however, believed the price of logwood "must be dearer for which end they are going to cellar what they have" in Boston and wait for higher prices.[22]

Back in the Laguna de Términos, Boston's larger ships, like Cheever's *Indian Emperor*, continued to labour at filling their holds but instead of supplying provisions for the local cutters, they, along with other crews cutting their own wood, drained local resources. Compounding the problem, the low price of logwood made it difficult to buy food. As the Boston merchant Thomas Moffat related, "all the cutters are going a pirating or to leave the bay the price [of logwood] is so low."[23] By the late summer of 1715, resident cutters also faced starvation, "Provisions being very scarce and extraordinary dear." Circumstances deteriorated to the point where several enslaved Africans "dyed there for want of Provisions." Desperate and left with few good options, "near Two Hundred Men" turned pirate, causing widespread

19 TNA, CO 5/848. Massachusetts Shipping Returns, 1688 to 1719.

20 *Weekly Journal*, August 13, 1715.

21 Dummer, *New England Charters*, p. 11.

22 New York Public Library (NYPL), MssCol 2026. Thomas Moffat to John Angier, August 10, 1715. Special thanks to David Wilson for sharing this source.

23 Ibid.

panic among the remaining Boston crews. One Boston ship captain reported "that all Persons that can get away, carry what they have with them, and desert the Place, there was but one ship in the Bay when he came away." Boston ship captains, including Cheever in the *Indian Emperor*, scurried home, far more frightened of possible retribution from the resident cutters than the Spanish threat.[24]

The famished loggers turned pirates threatened all British shipping in the Bay of Campeche. According to one report, "the Bristol Galley, called the *Royal George*, of 18 Guns, being in the Bay cutting Logwood," was presumed lost.[25] Bostonians clearly did not monopolise using their own crews for cutting wood and, unsurprisingly, local cutters targeted these vessels. The primary target for the ex-loggers, however, was the nearby Spanish town of Campeche, undoubtedly chosen for its proximity and provisions. Roughly two hundred and fifty starving men on two sloops, likely captured from the British, and periaguas, a small boat, assaulted the town. The Spanish successfully repelled the attack and the survivors scattered. Radicalised by their experience in the Laguna de Términos logwood trade, some of these men joined pirate crews operating out of the Bahamas.

Boston's intense exploitation of the Bay of Campeche logwood trade and its collapse in 1715 provided pirates with radical recruits willing to attack British shipping. Governor Alexander Spotswood of Virginia immediately recognised the potential threat the men posed, writing to England in July of 1716, "A nest of pirates are endeavouring to establish themselves at Providence and by the addition they expect of loose disorderly people from the Bay of Campeachy, Jamaica and other parts, may prove dangerous to British commerce, if not timely suppressed." Spotswood's concerns became a reality when many seafarers liberated from Spanish prisons and former logwood cutters settled in the Bahamas where, as he related from depositions from men fleeing the islands, they helped populate the pirate base at Nassau. Their arrival in the Bahamas coincided with a major ideological shift among the freebooters.

Ex-privateers like Benjamin Hornigold, John Jennings, and Philip Cockram refrained from attacking British and Dutch shipping and spent their energies on looting Spanish and French vessels. In May of 1716, however, Sam Bellamy overthrew the command of Hornigold with promises to attack vessels from all nations. Desperate seafarers previously engaged in the

24 TNA, CO 5/848. Massachusetts Shipping Returns, 1688 to 1719; "Entrances and Clearances," *Boston News-letter*, January 3, 1714–December 27, 1714; *Boston News-letter*, September 19, 1715.
25 *Evening Post* (London, 1709), January 21, 1716.

exploitative logwood trade and loggers had reason to spurn national and imperial loyalties. Their presence in the pirate crews may have been a deciding factor in Bellamy's success. John Brown, previously employed in the logwood trade, verified that vessels engaged in that trade provided pirates with many recruits. Shortly after departing Hornigold, Bellamy seized two ships heading to the Bay of Honduras. The two logwood ships provided twenty-seven new volunteers, bringing the total size of his crew to 120. In comparison, a Bristol ship trading with Jamaica only provided Bellamy with three new recruits.[26]

The attack on Campeche in 1715 foretold the end of the Laguna de Términos logwood trade. The grim news of the resident loggers turning pirate and the failed attempt on Campeche arrived in Boston just as they would have started to prepare their logwood fleet in September of 1715. Boston merchants fatefully decided against sending their vessels and crews. Without hundreds of Boston sailors, and a reduced population due to famine and piracy, the Laguna de Términos was ripe for an attack by the Spanish. In nearby Tabasco, as anticipated, they mustered their naval and military strength. They launched a successful attack in late February 1716 and "entirely cut off the Western Lagoons and the Island of Treas, took 150 Whites and Blacks, burnt all the logwood, destroy'd the Craft, carry'd away a Pink and a Sloop" and destroyed a few others. The victors then carried their prisoners to Campeche where they "erected a large Pair of Gallows, and threaten to hang them all," just as they would any pirate. The Spanish undoubtedly felt confident that the British would discontinue pillaging their lands with such a forceful message.[27]

In Boston, meanwhile, an economic crisis brewed due to their failure to send a logwood fleet in late 1715. Ships from London, Bristol, and other British ports began arriving in January. The town had enough stored logwood for the first few vessels but ran dry by late April.[28] As one Boston merchant, Samuel Keeling, explained to his partners in London, "For please to observe that

26 *Calendar of State Papers, Colonial: America and the West Indies* (CSPC), vol. 29 (1716–1717), no. 240. Lt. Governor Spotswood to the Council of Trade and Plantations, July 3, 1716; "London, March 19," *Boston News-letter*, July 22, 1717; "Examination of John Brown, May 6, 1717" in Jameson, *Privateering and Piracy*, pp. 293–95. For more on Bellamy's crew and recruitment of logwood cutters, see Woodard, *Republic of Pirates*, pp. 124–25.

27 *Weekly Journal or British Gazette* (London), June 30, 1716.

28 Logwood accounted for roughly half the cargoes of the first ships Britain. Logwood would start arriving in Boston during the late summer and once again began to make up a larger amount of the cargo of vessels leaving Boston for Britain. TNA, CO 5/848. Massachusetts Shipping Returns, 1688 to 1719.

here was soe many Ships that there was noe Freight to be had for London." In desperation to pay their creditors, Keeling and other Boston merchants prepared a fleet in early summer to send to the Bay of Campeche despite the dangerous circumstances with the Spanish and the hurricane season. As Keeling rationalised this "Charter Party for ye Bay of Campeche ... was at that time look'd upon to be the best improvement that could be made for shipping from this Place [Boston]."[29] Thomas Moffat also considered logwood a commodity on the rise due to the pirates and Spanish aggression. Keeling and Moffat, like all Boston merchants and seafarers, understood that the Laguna de Términos and its logwood were extensions of the town's economy and success. Boston had essentially created a floating colony in the Bay of Campeche with the purpose of extracting resources and wealth.

The staggered arrival of Boston's fleet at the Laguna de Términos, along with several British and Dutch vessels, triggered Spanish authorities to act, especially the Spanish General and Admiral in chief, Don Alonso Philipo de Andrades, in Campeche. Exasperated with the resettlement of the logwood cutters, they sent word to Marquis de Monteleon, "Embassador [sic] Extraordinary" in Europe. Monteleon then delivered an ultimatum to the British Secretary of State in mid-November, stop cutting logwood and leave the Laguna de Términos within eight months or be forcibly dislodged. Furthermore, the Spanish considered "the carrying on the Trade to the Laguna de Terminos" to be "interpreted an Invasion, and the Logwood cutters accounted as Pirates."[30] In Campeche, Andrades planned a swifter resolution for the pirate problem than Monteleon anticipated.

On the morning of November 29, 1716, twenty-three vessels, nearly half from Boston and by far the largest contingent of British vessels, quietly rode at anchor near the Island of Triste. John Alden Jr., captain of the *Prudent Betty*, had loaded his vessel with over one hundred tons of logwood and was nearly ready to sail back to Boston. His employer, Samuel Keeling, "look't on my Adventure soe good that I did not make one farthing insurance for myself."[31] It would prove a critical mistake. Besides Alden's crew of thirteen, the other vessels had skeleton crews with a total of sixty-seven men, primarily made up of ship captains and their servants. The rest of their

29 Mass. Arch., 62: 267. Samuel Keeling to Atkins Windham and Dodd, Boston, August 29, 1718; NYPL, MssCol 2026. Thomas Moffat to John Angier, June 21, 1716.

30 TNA, CO 5/383. Report Relating to the Right of the Subjects of Great Britain to the Cutting of Logwood in the Bay of Campeachy, September 25, 1717.

31 Mass. Arch. 62: 267. Keeling to Atkins Windham and Dodd, Boston, August 29, 1718.

sailors, estimated at two to three hundred, were in the lagoons chopping logwood. The Spanish had already destroyed Triste and attacked and seized vessels at the Laguna de Términos three times during 1716. Nevertheless, the British and Dutch ship captains prioritised profits over safety by leaving their vessels defenceless. It proved a fateful decision that morning of November 29.

As the officers prepared for the day's activities, Andrades arrived with three ships of war, one fire ship, three sloops of war, and a host of periaguas for transport, along with 854 sailors and soldiers. Andrades had one purpose—to permanently end the Bay of Campeche logwood trade in retribution for the attack on Campeche by the loggers the previous year. He slyly deployed a large force of periaguas to block the creeks heading into the Laguna de Términos, trapping the sailors working the lagoons. He then approached the outgunned and outmanned British and Dutch vessels. He demanded an unconditional surrender or else they would "be deemed as Pirates and suffer the Law of Arms for you to live in a Place, which belongs unto me." After some deliberation, the British and Dutch captains responded in the affirmative but pleaded with Andrades to allow them keep their vessels and goods, including the logwood already loaded. He denied them such generous terms but agreed to give them one sloop with provisions enough to reach the nearest British or Dutch port, as well as a pass to protect them from further Spanish harassment. The captains surrendered with their own safety assured but abandoned their employees in the lagoons. In short order, the Spanish killed or captured and imprisoned those men. Andrades then set about fortifying and settling the Island of Triste, ensuring future Spanish control of the Bay of Campeche and its lucrative logwood.[32]

Contemporaries recognised the end of the Laguna de Términos logwood trade as the most important factor in the rise of piracy. One official noted that "most of them that first turn'd pirates" had "always sailed in these parts [American colonies] in privateers and lived in the Bay of Campechia." Two other colonial officials recommended swift negotiation with the Spanish to reopen the Bay of Campeche to the English because it was "the most likely thing to divert them [the pirates] from their piratical courses" and it would help "to bring the pyrates to become good subjects." In the trial of Stede Bonnet in 1718, the Attorney General understood the reason for why pirates hunted American waters:

32 TNA, CO 152/10. Don Alonso Philipo de Andrades to Masters of Vessels in the Bay of Campeche, November 29, 1716 and Masters of Vessels in the Bay of Campeachy to John Cample, December 10, 1716.

> It is not my business to call in question the conduct of the Spaniards, in breaking up the Bay of Campeachy. They could not but think the turning away such a number of profligate wretches, as were got together, must put them on a worse course of life. They done them more harm since than cutting their logwood: for nine parts in ten of them turned pirates, and have lived upon robbing and plundering them and us ever since that time.

Some contemporaries even knew of Boston's role in the fall of the logwood trade and subsequent rise in piracy. Jeremiah Dummer, discussing the economic impact of the closure of the Bay of Campeche on Boston, remarked, "the Spaniards have at several times fallen upon our people [Bostonians] whom they found cutting wood in the Bays beforementioned, and seiz'd their ships, whereby we have lost that trade; and the mariners who were employ'd in it to the number of 3000, have since turn'd pirates and infested all our seas." Dummer's noteworthy assessment came from a position of authority, having lived in the major centres of the logwood trade, Boston, the Netherlands, and London. Interestingly, Dummer's report on the 3,000 seamen who lost employment due to the end of the logwood trade closely matches the historians' best estimates for how many pirates operated in the Atlantic during the first six years of the Golden Age.[33] This is not to suggest all pirates were ex-logwood cutters or involved in the logwood trade—they were not—but rather that contemporaries understood how important unemployment was to the rise of Atlantic piracy and believed the logwood trade would solve the crisis. Ultimately, their analysis proved prophetic.

## The Bay of Honduras Logwood Trade and Boston's Pirate: Ned Low

The end of the Laguna de Términos logwood trade certainly reverberated around the British Atlantic World but for Boston the economic and human consequences were catastrophic. More than half the vessels seized by Andrades came from the port, including five large ships like the *Prudent*

33 CSPC, vol. 30 (1717–1718), no. 660. Extracts of several letters from Carolina, August 19, 1718; CSPC, vol. 31 (1719–1720), no. 479, Nicholas Lawes to Council of Trade and Plantations, December 6, 1719; CSPC, vol. 31 (1719–1720), no. 548. Thomas Bernard, a Councillor of Jamaica, to John Chetwynd, a Commissioner for Trade and Plantations, February 4, 1720; CSPC, vol. 31 (1719–1720), no. 578. Jeremiah Dummer to the Council of Trade and Plantations, February 25, 1720; "The Trials of Major Stede Bonnet and Thirty-three Others for Piracy. Charleston, South Carolina, 1718" in Lawson, *American State Trials*, p. 658; Dummer, *New England Charters*, p. 11. For number of pirates see Rediker, *Villains of All Nations*, pp. 29–30.

*Betty*. Boston lost an additional three vessels that arrived shortly after Andrades fortified Triste. As the *Boston Newsletter* despondently reported, "most of the English Vessels taken in the Bay belong'd to Boston." Although Boston had established itself as the most important British shipping centre in the Americas, it could not bear the weight of losing fourteen vessels. One merchant estimated the losses to be over £20,000, not including the lost revenue from the logwood. The town descended into a flurry of panic, litigation, and recriminations. Merchants and ship captains involved in the trade lost fortunes and judges confined a few in debtors' prison. The downward economic spiral continued for years, leading to unemployment and hardship for many, especially in the maritime sector.[34]

Boston merchants nevertheless still believed logwood would solve their economic woes. They, along with British merchants, petitioned the king to force Spain to reopen the Laguna de Términos trade and compensate merchants for their losses of November 29, 1716. These petitions resulted in a declaration of war against Spain on December 16, 1718, a short-lived engagement that provided Boston with temporary relief from pirates in 1719. The Bay of Campeche remained closed to Boston and the British. Despite the disappointment, Boston merchants plotted a new course of action to extract logwood from the Yucatan Peninsula. They would attempt to establish a base of operations in Belize and the Bay of Honduras.

Boston's transition to the Bay of Honduras proved difficult. Very few New Englanders, or British seafarers more generally, had experience there since, as one official put it, "the whole [logwood] Trade" had centred in the easily accessible Laguna de Términos.[35] In comparison, the Bay of Honduras had an abundance of dangerous shoals, reefs, and rocks that required a far higher level of expertise, skill, and luck. Nathaniel Uring complained that he had to pay "extraordinary Wages and Privileges" to a first mate and pilot due to the dangerous waters. Despite the man's high compensation and supposed experience, Uring's ship ended up floundering on the shoals and sinking.[36] The lawless nature of the Bay of Honduras also constituted a significant threat to Bostonians attempting to re-establish their place in the logwood

34 *Boston News-letter*, April 8, 1717; Mass. Arch. v. 62 no. 267. Samuel Keeling to Atkins Windham and Dodd, Boston, August 29, 1718. For more on the economic downturn in 1717, see Pitt, "Building and Outfitting Ships," pp. 890–95.

35 Prior to the closure of the Bay of Campeche, Boston had eight voyages to the Bay of Honduras between 1700 and 1716 versus 130 voyages to the Bay of Campeche during that same period. TNA, CO 5/383. Report Relating to the Right of the Subjects of Great Britain to the Cutting of Logwood in the Bay of Campeachy, September 25, 1717.

36 Uring, *Voyages*, p. 342.

trade. An unofficial war raged between Spaniards, pirates, logwood cutters, the British Royal Navy, and local natives during most of the 1720s. The chaos emboldened common seamen from Boston to mutiny, turn pirate, or both when treated poorly. Importantly, Boston had also alienated pirates by hanging members of Sam Bellamy's crew on November 15, 1717, and some, like the infamous Blackbeard, sought revenge. The Bay of Honduras provided a perfect location for pirates to terrorise Boston's shipping.

Boston's first attempt to re-establish its logwood trade in Belize departed with pomp and ceremony and ended in disaster. William Wyer, captain of the *Protestant Caesar*, a 400-ton behemoth,[37] set sail from Boston in October 1717 as the trial of Sam Bellamy's crew gripped the town.[38] Since July, rumours swirled in Boston that "if the Prisoners [the pirates] Suffered" then pirates "would Kill every Body they took belonging to New England." Boston learned that Edward Teach, also known as Blackbeard, made a similar threat just as Wyer departed for Belize: "One of our Pilates that was on board Teach the Pirates says that they very much threaten New-England men, in case any of their fellow Pirates suffer there, that they will revenge it on them." Town authorities disregarded such threats and proceeded confidently in their trail and execution of the pirates. Boston's logwood trade suffered the consequences.[39]

Wyer arrived on the outskirts of the Bay of Honduras on March 28, 1718, after a stop in Jamaica. The pirate Stede Bonnet greeted Wyer with an ill-advised assault. In a drawn-out engagement, the massive *Protestant Caesar* outgunned and outduelled Bonnet's *Revenge*. Bonnet scurried away to the Turneffe Island off the coast of Belize where Blackbeard reprovisioned *Queen Anne's Revenge* and continued harassing passing logwood vessels. Upon learning of Boston's great logwood ship, Blackbeard sailed to exact revenge for the hanging of Sam Bellamy's crew. Wyer, meanwhile, believed his bravery against Bonnet had ended the threat to his voyage and continued into the Bay of Honduras to cut his logwood. Roughly a week later, a pirate fleet of one large ship and four sloops sped towards the *Protestant Caesar*, which was anchored and loading logwood. Wyer "call'd his Officers and Men up on Deck asking them if they would stand by him and defend his

37 This may be the largest ship built in Boston during the entire colonial period. I have found no evidence of a larger one.

38 "Entrances and Clearances," *Boston News-letter*, October 14, 1717.

39 "Trial of Simon Van Vorst and Others, October 1717" in Jameson, *Privateering and Piracy*, pp. 303–6; Some historians have speculated that Teach left his long-time partner Benjamin Hornigold in Nassau to sail for North America with the sole purpose to attack Boston and rescue Bellamy's crew from prison, see Woodard, *Republic of Pirates*, p. 197.

Ship, they answered, if they were Spaniards they would stand by him as long as they had Life, but if they were Pirates they would not Fight." The crew deserted the ship without a fight upon recognising Stede Bonnet's sloop and Blackbeard's *Queen Anne's Revenge*. After a few days hiding in mosquito-infested lagoons, Wyer and his crew surrendered to Blackbeard only to learn the pirate intended to "burn his Ship because she belonged to Boston, adding he would burn all Vessels belonging to New-England for Executing the six Pirates at Boston." Blackbeard understood that the loss of a 400-ton ship and its cargo of logwood hurt Boston and its merchant elites far more than killing or terrorising the seafaring men employed by those merchants. In the end, the massive ship, despite its testament to Boston's wealth and power in the Atlantic World, could not escape pirates' retribution.[40]

The years 1716–1719 had been tough on Boston's shipping. The Spanish took over a dozen vessels in 1716. Pirates plundered, commandeered, and burned at least seventeen Boston-owned vessels, including seven large ships, brigs, and snows. The 400-ton *Protestant Caesar* alone was worth roughly £4000 sterling not including its cargo. Boston suffered more than other colonial ports because pirates often burned the port's vessels in revenge for hanging Bellamy's crew in 1717. Pirates also disrupted Boston's most lucrative trade routes, including their attempts to re-establish their logwood trade in the Bay of Honduras.[41] The port was inundated with terrifying reports from all over the Atlantic, some of which described pirates' burning hatred of Boston. In 1718 alone, two of every three issues of the *Boston News-letter* referenced the terror sown by Atlantic pirates and the ineffective steps the colonies and London took to suppress them.[42]

The port's fear of pirates peaked in 1723 when its own pirate—Edward Low—haunted New England's coast. "Ned" Low was born in Westminster,

40 *Boston News-letter*, June 18, 1718; Woodard, *Republic of Pirates*, pp. 240–43.

41 Boston lost at least fifty-four vessels to pirates from 1716 to 1726. See *Boston News-letter*, May 13, 1717, July 22, 1717, August 19, 1717, October 28, 1717, March 10, 1718, June 16, 1718, August 18, 1718, November 17, 1718, December 29, 1718, August 1, 1720, August 22, 1720, November 21, 1720, December 5, 1720, April 30, 1722, May 7, 1722, June 18, 1722, March 21, 1723, August 8, 1723, September 19, 1723, October 18, 1723, July 16, 1724, April 15, 1725, February 11, 1725; *Boston Gazette*, June 27, 1726; *New England Courant*, June 17, 1723; *American Weekly Mercury*, August 8, 1723, October 4, 1723; CSPC, vol. 30 (1717–1718), no. 660. Extracts of several letters from Carolina, December 17, 1717; CSPC, vol. 30 (1717–1718), no. 797 iv. Deposition of Jonathan Bull, December 19, 1718; "Deposition of Ralph Merry and Samuel Roberts May 11 and 16, 1717" and "Trial of Thomas Davis, October 28, 1717" in Jameson, *Privateering and Piracy*, pp. 301–7; Defoe, *General History*, pp. 326, 338, 355, 587; Earle, *Pirate Wars*, p. 170.

42 *Boston News-letter*, January 1, 1718–December 31, 1718.

London. He lived there in impoverished conditions and spent much of his youth picking pockets and gambling. When Low reached an appropriate age, his elder brother enticed him to make a living at sea. Low's employment eventually took him to Boston where he settled in 1711. Interestingly, Low turns up in Boston-based portledge bills preparing for a logwood voyage to the Bay of Campeche with Captain Thomas Porter, who eventually became the unfortunate captain taken at the Laguna de Términos in 1716 and then by Stede Bonnet a year later, on the ship *Content*. While waiting for the *Content* to set sail, Low must have encountered his future wife, Elizabeth Marble of Boston. Like Ramblin' Jack Cremer who fell for a "Red headed," "well-Riged" Boston girl and Cremer's Chief Mate who "got married" in Boston "god help him," Elizabeth transformed Ned Low's watery world.[43]

Fortunately for Low, the *Content* required major repairs and although he was hired on October 4, 1711, the ship did not sail until May 2, 1712. He decided the voyage with Porter, which would have taken close to two years, was too long to be away from his newfound love. He also knew that seafarers involved in the logwood trade had poor reputations and faced many dangers. He received permission from the *Content*'s supervising merchant to ship out on board the *Francis* instead for a voyage to the West Indies and back to Boston. In a little less than a year, Low returned with £16.02 in his pocket after an uneventful voyage. Afterwards, Low joined a growing multitude of waterfront labourers in Boston as a rigger. Low's new profession allowed him to benefit from stable employment based on the exploitation of foreign resources and maritime labour. He also would have developed local credit relations with Boston's merchant elite. In other words, Low was now settled and ready to marry a local girl.[44]

On August 12, 1714, Low's hard work and commitment finally paid off when he married his sweetheart, Elizabeth. In late 1719, the couple joyfully welcomed a baby girl into their lives. Low's happiness, however, abruptly ended when Elizabeth died shortly after delivering their daughter due to post-pregnancy complications. Elizabeth's death could not have come at a worse time for Low both emotionally and economically. As historian Elaine Crane has demonstrated, Boston's waterfront women contributed significantly to the financial success of the household. Low's family almost certainly relied heavily on Elizabeth and her family for financial resources

43 Massachusetts Historical Society (MHS), Jeffries Family Papers, box 17. Portledge Bills for *Content*; Boston, *Early History of Boston*, vol. 28, p. 53; Bellamy, *Ramblin' Jack*, pp. 112–14.

44 MHS, Jeffries Family Papers, box 17. Ships' Papers, *Content* and *Francis*; Dow and Edmonds, *New England*, pp. 141–42.

and connections. The closure of the Bay of Campeche logwood trade and economic downturn of 1717 also triggered new laws targeting the credit economy on which so many merchants, waterfront tradesmen, and labourers relied. Low, like many other waterfront tradesmen and labourers in Boston, likely suffered from sporadic employment, reduced wages, and limited access to credit. Together, these local and international factors compelled Low to return to seafaring in 1721 to support his motherless daughter. Hitherto, Low had the opportunity to participate in Boston's logwood trade but declined in favour of settling down. Ten years later, he fatalistically returned to his previous path and signed a contract in the brutal Bay of Honduras trade.[45]

Low's experience there mirrored many other Boston mates and seamen in the logwood trade. The trade was especially dangerous with both the Spanish and pirates preying on British logwood vessels. Low's captain placed him in charge of the woodcutting crew. He desired to load his logwood quickly and depart post-haste. The captain consequently pushed Low's crew hard. One evening just before dinner, after Low and his crew had finished loading a bundle of logwood, Low "desired that they might stay and Dine; but the Captain being in a Hurry for his Lading, ordered them a Bottle of Rum, and to take t'other Trip." Famished after a hard day of labour, Low became enraged, firing his musket "at the Captain, but missing him, shot another poor Fellow thro' the Head." Although we cannot possibly know what happened after this accidental killing, it seems likely that the captain commanded the crew to seize Low and imprison him aboard the sloop. The crew, who had spent weeks labouring with Low, refused and joined Low in mutiny.[46]

Low acted impulsively and in response to the captain's provocation.[47] After the crime, Low surely grasped that his only chance of returning to Boston and seeing his daughter was to turn pirate. From living and working on the Boston waterfront, Low would have been well aware of King George's multiple pardons for pirates, which the *Boston News-letter* and other proclamations broadcasted. Low's turn to piracy and subsequent actions makes far more sense in this context, as he hoped that another pardon that included murder on the high seas would be issued. The first-person account of impressed Marblehead fisherman, Philip Ashton, supports the notion that Low believed or hoped he may eventually be reunited with his

45 For impact of new laws, see Pitt, "Building and Outfitting Ships," pp. 890–95; Defoe, *General History*, p. 319; Crane, *Ebb Tide*, pp. 17–18.

46 Defoe, *General History*, p. 319.

47 Dow and Edmonds, *New England*, p. 143; Some historians have speculated that Low and his men had planned to become pirate prior to this incident but that theory has no supporting evidence. Flemming, *Point of a Cutlass*, pp. 10–11.

daughter in Boston, as he frequently told the forced Marblehead fisherman that Ashton "should go home when he did and not before."[48]

Historians have demonised Ned Low as the most bloodthirsty pirate of the Golden Age. As Hugh Rankin remarked of Low, "no one mourned his passing—except, maybe, the Devil."[49] No historian, however, has thoroughly examined his experiences prior to becoming pirate and motivations during his time as a pirate. His story is incomplete without understanding his Boston and logwood connections. Low's sanity may have been compromised after the death of his wife, separation from his daughter in Boston, poor treatment in the Bay of Honduras, and realisation that he may never see his daughter again. Ashton's narrative reflects a man tortured by his past, but it also humanised the beleaguered pirate. According to Ashton, Low refused to take any married sailor despite relying heavily on impressment to fill his crew. His past played an important role in that decision:

> It seems his design was to take no Married Man away with him, how young soever he might be, which I often wondred at; till after I had been with him some considerable time, and could observe in him an uneasiness in the sentiments of his Mind, and the workings of his passions towards a young Child he had at Boston (his Wife being dead, as I learned some small time before he turned Pirate) which upon every lucid interval from Revelling and Drink he would express a great tenderness for, insomuch that I have seen him sit down and weep plentifully upon the mentioning of it.

Low's humanity asserted itself in other ways. He willingly risked his own life to save an impressed ship's doctor, John Kencate, when his vessel, the pink *Rose*, suffered a careening accident near Guiana. Low had already escaped the sinking vessel when he noticed Kencate trapped by the water rushing in through the portholes. He managed to grab the doctor by the shoulder and drag him to safety.[50]

Low also recognised the paradox of impressing sailors and treating them poorly to serve his interests. At one point, Ashton and a few other impressed men designed to murder the small pirate crew under the command of Low's quartermaster, Francis Spriggs, and escape to New England on board a stolen schooner. Spriggs learned of their plot and told Low who, in response, humorously "said he did not know, but if it had been his own case, as it was

48 Mather, *Sad Effects of Sin*, pp. i–x; Barnard, *Ashton's Memorial*, p. 16.

49 Rankin, *Golden Age*, p. 151.

50 Barnard, *Ashton's Memorial*, pp. 3, 10.

ours, he should have done so himself." Low did not punish Ashton and his compatriots even though most ship captains in the merchant marine or Royal Navy would have certainly beaten mutinous crew members bloody. Upon uncovering a potential mutiny on board his vessel, Nathaniel Uring, another veteran of the logwood trade, "chastiz'd" crew members "very handsomely" by thrashing them with a cane. He gave one sailor "two or three such Strokes with a Stick I had prepared for that purpose, that he soon changed his Note, the Blood running about his Ears, he pray'd for God's sake that I would not kill him." Pirates did not hold a monopoly on torture, violence, and murder on the high seas. Nevertheless, Low became increasingly violent as his crew suffered losses at the hands of the Royal Navy and colonial port cities and the days and months went by without a new pardon issued by the king.[51]

Low and his crew employed terror far more frequently and with greater violence than many other pirates. Predominantly Catholic, Portuguese, Spanish, and French seafarers received Low's most ruthless treatment. In one instance, he and his crew slaughtered dozens of Spaniards in the Bay of Honduras. This massacre, however, likely had its roots in the bloody frontier warfare between British logwood cutters and the Spanish in the Bay of Honduras logwood trade. Like so many pirate crews, many of Low's crew had been employed in that trade and probably felt justified considering the atrocities committed by the Spanish on British cutters and seafarers. Low also took revenge upon New England ship captains and merchants through torture and destruction of property because, as the *General History* relates, he had an "irreconcileable Aversion to *New-England* Men."[52]

Boston and the logwood trade shaped Low's tragic circumstances, and he intended to exact revenge on the port. Low and his compatriots soon joined another mutineer pirate, George Lowther, who had worked in conditions far worse than the logwood trade on board a slave ship. Low was appointed a lieutenant, and the two men began cruising around the Bay of Honduras for potential prizes. Boston's resuscitating logwood trade suffered disaster at the hands of these two determined pirates. In January 1722, they seized a 200-ton Boston-owned ship called the *Greyhound*, commanded by Benjamin Edwards. After a brief resistance, Edwards struck his colours but the pirates, according to the *Boston Newsletter*, savagely beat the men on board the ship and impressed others. Some of the impressed, however, like the second mate Charles Harris, soon recognised that their own interests coincided with the pirates and willingly cooperated. The pirate crew set the *Greyhound* ablaze at Low's insistence. Low

51 Ibid., 15; Uring, *Voyages*, pp. 260–61.

52 *American Weekly Mercury*, May 2, 1723; Defoe, *General History*, p. 326.

and Lowther then seized two more Boston-based brigantines in the Bay of Honduras and like the *Greyhound* burned them with their cargoes of logwood.[53]

Low continued to damage Boston's shipping throughout his career as a pirate. He seized at least eleven Boston-based vessels, which more than doubles the total of any other single pirate. Low, like Edward Teach before him, burned or sank most of these vessels in revenge. Unlike Teach, Low tortured Boston ship captains and crews, including cutting the ears off Captain John Welland of the *Amsterdam Galley*. The pirates then sank the *Amsterdam Galley*, placing Welland and his crew on board another captured vessel from Piscataqua. Welland's earless appearance undoubtedly horrified Boston's merchants and ship captains.[54] The terror induced by Ned Low in 1723 reduced Boston's entrances and departures by over 50 per cent from the previous year from May through September. In June 1723 only 32 cleared out, typically one of the most active months for shipping.[55] Boston's departures returned to normal in October 1723 only after word arrived that Low had abandoned the New England and Newfoundland coasts to sail for the Azores. Unlike other pirates, Low had terrified Boston's merchants and ship captains and his presence on the coast decreased Boston's clearances by almost 20 per cent from the entire previous year. Many merchants and ship captains appear to have waited till the subsequent year to set sail because clearances went from 485 in 1723 to 731 in 1724, an increase of 34 per cent and the greatest increase recorded in Boston during the colonial period. Bostonians would remain terrified of Low, a product of their economic woes and logwood trade, until the end of the Golden Age.[56]

## Conclusion: Where did the Pirates Go? The Return of Boston's Logwood Trade and the Decline of Pirates

On July 12, 1726, the pirate captain William Fly and two of his crew met a grisly end in Boston. Fly's body "was hung up in Irons, as a Spectacle for the

53 *Boston News-letter*, May 7, 1722.

54 Court of Vice-Admiralty for Rhode Island, *Tryals of Thirty-Six Persons*.

55 Even when Boston issued a seven-day embargo in June 1717 due to the presence of pirate Paul Williams, departures in June and July equalled 71 and 78 respectively. *Boston News-letter*, May 27, 1717; "Entrances and Clearances," *Boston News-letter* January 7, 1717–December 29, 1718.

56 "Entrances and Clearances," *Boston News-letter*, January 1, 1722–December 31, 1724; The *Boston News-letter* regularly reported any news of Low, see *Boston Newsletter*, October 18, 1723, October 31, 1723, March 19, 1724, March 27, 1724, May 7, 1724, October 1, 1724, October 8, 1724, October 15, 1724, February 11, 1725; *Boston Gazette*, April 25, 1726, June 27, 1726. See also Dow and Edmonds, *New England*, pp. 216–17.

Warning of others, especially Seafaring Men" on an island in Boston harbour while the other two received a burial. A fourth pirate, George Condick, had received a reprieve from the gallows in front of the large crowd.[57] Fly's death and warning foreshadowed the end of the Golden Age of Piracy. It also foreshadowed the fate of ex-pirates. Over the decade, Boston had zealously pursued pirates at sea, whether by fitting out privateers or sending out Royal Navy vessels and hanging them if captured. The town's political and religious leaders condemned the actions of pirates although only a few understood Boston's role in facilitating their rise through destroying the Laguna de Términos logwood trade in the first place. Boston, paradoxically, also held the key to the decline of piracy—employment in the logwood trade.

Boston's economy remained shellshocked by the loss of the logwood trade through the much of the early 1720s. Boston merchants, ship captains, and sailors worked diligently to re-establish the trade in the Bay of Honduras, but at every turn pirates, or the Spanish, disrupted those efforts. In 1723, Boston sent fifteen vessels to the Bay but lost three to Low and Lowther. Pirate captain Shipton captured a couple more in 1724 but by then Boston's trade had revived to about half its size prior to 1716.[58] Not surprisingly, the number of pirates plaguing the Atlantic also declined by 50 per cent. Good news for Boston arrived in April of 1725 that the HMS *Diamond* patrolled the Bay of Honduras for pirates, both British and Spanish.[59] Feeling confident, Boston merchants sent a logwood fleet of twenty-eight over the year, their largest since 1715. Pirates again declined by more than half. In 1726, roughly 400 sailors found employment in Boston's and New York's logwood trade. Finally, in 1727 eighty-three North American vessels, fifty-four from Boston, with roughly 830 men sailed for the Bay of Honduras for logwood. The trade now equalled or exceeded the numbers involved in 1714. These numbers also did not include vessels from Great Britain or Jamaica that likely made up another dozen or two vessels and hundreds of sailors. Men who would be pirates once again had opportunity for employment in their favoured trade, fulfilling contemporaries' prescient vision for the end of Atlantic piracy. Once again, the Spanish encountered close to a thousand "pirate-loggers" pillaging their land and waged war against them.[60]

57 *Boston News-letter*, July 14, 1726.

58 See sources in footnote 41 for captured Boston vessels.

59 *Boston News-letter*, April 22, 1725. For more on the HMS *Diamond* and suppression of piracy in the Bay of Honduras, see Wilson, *Suppressing Piracy*, pp. 217–18.

60 For these numbers see Table 1 below.

**Table 1: North American Ports, the Logwood Trade, and Employment, 1714 to 1727**

| | 1714 | 1715 | 1716 | 1717 | 1718 | 1719 | 1720 | 1721 | 1722 | 1723 | 1724 | 1725 | 1726 | 1727 |
|---|---|---|---|---|---|---|---|---|---|---|---|---|---|---|
| **Boston Vessels** | 37 | 29 | 20 | 2 | 2 | 0 | 4 | 9 | 15 | 18 | 19 | 28 | 26 | 54 |
| **New York and Newport Vessels** | 4 | 6 | 2 | 0 | 2 | 0 | 0 | 4 | 2 | 10 | 6 | 9 | 15 | 29 |
| **Estimated Employed Sailors** | 615 | 525 | 330 | 20 | 40 | 0 | 40 | 130 | 170 | 280 | 250 | 370 | 410 | 830 |
| **Estimated Pirates** | ? | ? | 1,500 to 2,000 | | | ? | 1,800 to 2,400 | | | 1000 | 500 | >200 | >200 | ? |

**Note on Method:** I have compiled a database of over 2,000 voyages from North American ports to the Bays of Campeche and Honduras from 1688 to 1776. The numbers primarily come from entrances and clearances in colonial American newspapers or written accounts from the two bays in those papers. I have also drawn from Massachusetts Shipping Returns, merchant papers, court records, and published travel accounts. I have estimated fifteen sailors per vessel for voyages to the Bay of Campeche (1714 to 1716) and ten sailors per vessel for voyages to the Bay of Honduras (1717 to 1727). I found in my research that the Bay of Campeche logwood trade used larger ships and crews whereas the Bay of Honduras trade relied heavily on smaller sloops and crews, see Pitt, "City upon the Atlantic Tides," chap. 5 "The Atlantic Logwood Trade."

**Sources:** TNA, CO 5/848. Massachusetts Shipping Returns, 1688–1719; "Entrances and Clearances," *Boston News-letter*, January 1, 1714–December 31, 1727; "Entrances and Clearances," *Boston Gazette*, January 1, 1721–December 31, 1727; "Entrances and Clearances," *New England Chronicle*, January 1, 1722–January 31, 1727; "Entrances and Clearances," *New England Weekly Journal*, January 1, 1727–December 31, 1727; "Entrances and Clearances," *American Weekly Mercury*, January 1, 1721–December 31, 1727; MSA, Suffolk County Court of Common Pleas. Cunningham & Co. v. Foxwell Curtis Captain of the Sloop Fortune (1721); Uring, *Voyages and Travels*, 342–52; For estimated number of pirates, see Rediker, *Villains of All Nations*, 29–30. I have excluded the year 1719 when Britain warred with Spain. Most historians and contemporaries agree that many pirates became privateers for either the British or Spanish during the short-lived war.

Most historians have understood the importance of seafaring unemployment in the rise of Atlantic piracy in 1716 but most have argued that state-sponsored executions or Royal Navy suppression was the key factor in its decline. Only a fraction of pirates, however, died on the gallows or were killed in combat. The rest dispersed, but without Boston's revitalised logwood trade, would they have? We cannot know exactly how many ex-pirates became loggers in the Bay of Honduras. Contemporary observers certainly understood the correlation between the end of logwood cutting and the rise of pirates so if "nine parts in ten of them [loggers] turned pirates" than predictably "nine parts in ten" returned to the trade once Boston re-established it, as illustrated in the table above.[61]

We do know of two ex-pirates that returned to the trade—the reprieved George Condick and an escapee from William Fly's crew. In April of 1728, Condick shipped out on Thomas Belcher's sloop for the Bay of Honduras as part another large Boston fleet of forty-seven vessels.[62] While in the Bay, Condick and Belcher encountered Alexander Mitchel working in the logwood trade. Condick identified Mitchel as a "chief actor in the Murder on board the Snow where Fly and others turn'd Pirates." Boston authorities maintained a warrant for his arrest and Belcher clapped Mitchel in irons to face justice. The ex-pirate had hoped to hide in the supposedly lawless Belize, but Bostonians had, for all intents and purposes, colonised the lagoons, just as they had done with the Bay of Campeche. Two years after hanging Fly, the long arm of the town stretched out to seize the hapless logwood cutter. Mitchel's luck, however, held out. After loading their logwood, Belcher and Condick set sail for Boston only to fall afoul of a prowling Spanish guardacosta. Belcher abandoned his sloop and Mitchel with it. The same guardacosta seized three more Boston vessels in their continuing war against the pirates, whisking those vessels and crews away to Havana where long prison sentences awaited the Bostonians. Belcher's crew fortunately escaped, making their way home to Boston to share their harrowing tale.[63]

The fates of Boston and pirates intertwined with that of the logwood trade from 1713 to 1727. Most contemporaries knew the importance of the trade in employing a distinctive, difficult to control, group of seafarers. Bostonians, meanwhile, knew the importance of the logwood to their pocketbooks, ability to pay for English manufactured goods, support their shipbuilding industry, and employ their seafaring community. In 1715, Boston's over exploitation of its cash crop led to a severe drop in prices and famine in the Laguna de Términos. Resident logwood cutters emerged as the first willing to pirate British shipping and fatefully attacked the Spanish town of Campeche. The following year, Spain's retribution permanently closed Britain's Bay of Campeche logwood

61 "The Trials of Major Stede Bonnet" in Lawson, *American State Trials*, p. 658.

62 "Entrances and Clearances," *Boston News-letter*, January 1, 1728–December 31, 1728; "Entrances and Clearances," *Boston Gazette*, January 1, 1728–December 31, 1728; "Entrances and Clearances," *New England Weekly Journal*, January 1, 1728–December 31, 1728.

63 *New England Weekly Journal*, December 9, 1728.

trade, leading to hundreds, if not thousands, of unemployed seafarers joining the ranks of the pirates. These two pivotal events and the collapse of the logwood trade explain the rise in piracy against all nations in late 1716 and early 1717.

Boston had an inverse relationship with pirates. As piracy skyrocketed, Boston's economy plummeted due to the loss of its logwood trade and subsequent decline in ship outfitting and building, along with pirate attacks on the town's efforts to re-establish the logwood trade in the Bay of Honduras. The combination of these factors led to the creation of Boston's own pirate, Ned Low, who terrorised Boston both in the Bay of Honduras and on the New England coast. He impacted the town's ability to regain its place in the Atlantic economy from 1722 through early 1724. Bostonians, however, persisted in attempting to regain their logwood trade. By 1725, the combination of suppression by the British navy, hangings in colonial ports, and employment in Boston's revived logwood trade in Belize led to a significant decline in pirates and an economic boom in Boston. For would-be pirates, the logwood trade represented a viable occupation with just enough independence, increased earnings, and elements of danger to keep them content. For the Spanish, Belize and its logwood cutters represented just another pirate haven in need of eradication.

## Bibliography

### Printed Primary Sources

Barnard, John. *Ashton's Memorial: An History of the Strange Adventures and Signal Deliverances of Mr. Philip Ashton*. Boston: Samuel Gerrish, 1725.

Bellamy, R. Reynell, ed. *Ramblin' Jack: The Journal of Captain John Cremer, 1700–1774*. London: Jonathan Cape LTD., 1936.

Boston. *Records Relating to the Early History of Boston*, vol. 28. Boston: Municipal Printing Office, 1894.

Court of Vice-Admiralty for Rhode Island. *Tryals of Thirty-Six Persons for Piracy*. Boston: Kneeland, 1723.

Defoe, Daniel. *A General History of the Pyrates*. Edited by Manuel Schonhorn. New York: Dover Publications, 1999.

Dummer, Jeremiah. *A Defence of the New England Charters*. London: W. Wilkins, 1721.

Jameson, John Franklin, ed. *Privateering and Piracy in the Colonial Period: Illustrative Documents*. New York, New York: Augustus M. Kelly, 1970.

Masefield, John, ed. *Captain William Dampier: Dampier's Voyages*, 2 vols. Edinburgh and New York, 1906.

Mather, Cotton. *The Sad Effects of Sin: A True Relation of the Murder Committed by David Wallace, on his Companion Benjamin Stolwood*. Boston, 1713.
New Hampshire. *Probate Records of the Province of New Hampshire, 1635–1771*, vol. 32. New Hampshire: Rumford Printing Company, 1914.
Public Record Office. *Calendar of State Papers, Colonial: America and the West Indies*, 45
volumes. London: Her/His Majesty's Stationery Office, 1860–1994.
Uring, Nathaniel. *A History of the Voyages and Travels of Captain Nathaniel Uring*. London, 1726.
Whitworth, Sir Charles. *State of the Trade of Great Britain in its Imports and Exports, Progressively from the Year 1697*. London, 1776.

## Secondary Sources

Bailyn, Bernard and Lotte Bailyn. *Massachusetts Shipping 1697–1714: A Statistical Study*. Cambridge, Massachusetts: The Belknap Press of Harvard University, 1959.
Bolland, O. Nigel. *Colonialism and Resistance in Belize: Essays in Historical Sociology*. Barbados: University of West Indies Press, 2003.
Craig, Alan. "Logwood as a Factor in the Settlement of British Honduras." *Caribbean Studies* 9, no. 1 (1969): 53–62.
Crane, Elaine Forman. *Ebb Tide in New England: Women, Seaports and Social Change 1630–1800*. Boston: Northeastern University Press, 1998.
Cromwell, Jesse. "Life on the Margins: (Ex) Buccaneers and Spanish Subjects on the Campeche Logwood Periphery, 1660–1716." *Itinerario* 33 (2009): 43–71.
Dow, George Francis and John Henry Edmonds. *The Pirates of the New England Coast*, reprint. New York: Dover Publications, 1996 [1923].
Earle, Peter. *The Pirate Wars*. New York: Thomas Dunne Books, 2003.
Finamore, Daniel. "A Mariner's Utopia: Pirates and Logwood in the Bay of Honduras." In *X Marks the Spot: The Archaeology of Piracy*, edited by Russell K. Skowronek and Charles R. Ewen, 64–78. Gainesville, Florida: University Press of Florida, 2006.
Hanna, Mark G. *Pirate Nests and the Rise of the British Empire, 1570–1740*. Chapel Hill, NC: University of North Carolina Press, 2015.
Head, David, ed., *The Golden Age of Piracy: The Rise, Fall, and Enduring Popularity of Pirates*. Athens, GA: University of Georgia Press, 2018.
Jarvis, Michael. *In the Eye of All Trade: Bermuda, Bermudians, and the Maritime Atlantic World, 1660–1783*. Chapel Hill, NC: University of North Carolina Press, 2010.
Joseph, Gilbert M. "British Loggers and Spanish Governors: The Logwood Trade and Its Settlements in the Yucatan Peninsula: Part I." *Caribbean Studies* 14, no. 2 (1974): 7–37.

Joseph, Gilbert M. "British Loggers and Spanish Governors: The Logwood Trade and Its Settlements in the Yucatan Peninsula: Part II." *Caribbean Studies* 15, no. 4 (1976): 43–52.

Linebaugh, Peter and Marcus Rediker. *The Many-Headed Hydra: Sailors, Slaves, Commoners, and the Hidden History of the Revolutionary Atlantic.* Boston: Beacon Press, 2000.

McDonald, Kevin P. *Pirates, Merchants, Settlers, and Slaves: Colonial America and the Indo-Atlantic World.* Oakland: University of California Press, 2015.

McDonald, Kevin P. "'Sailors from the Woods': Logwood Cutting and the Spectrum of Piracy." In *The Golden Age of Piracy: The Rise, Fall, and Enduring Popularity of Pirates*, edited by David Head, 50–72. Athens, GA: The University of Georgia Press, 2018.

Newton, Ross A. "'Good and Kind Benefactors': British Logwood Merchants and Boston's Christ Church." *Early American Studies* 11, no. 1: 15–36.

Peterson, Mark. *The City-State of Boston: The Rise and Fall of an Atlantic Power, 1630–1865.* Princeton and Oxford: Princeton University Press, 2019.

Pitt, Steven J. J. "Building and Outfitting Ships in Colonial Boston." *Early American Studies: An Interdisciplinary Journal* 13, no. 4 (Fall 2015): 900–904.

Pitt, Steven J. J. "City upon the Atlantic Tides: Merchants, Pirates, and the Seafaring Community of Boston, 1689–1748." PhD Diss., University of Pittsburgh, 2015.

Rankin, Hugh. *The Golden Age of Piracy.* New York: Holt, Rhinehart and Winston, 1969.

Rediker, Marcus. "'Under the Banner of King Death': The Social World of Anglo-American Pirates, 1716 to 1726." *William and Mary Quarterly* 38, no. 2 (1981): 203–27.

Rediker, Marcus. *Between the Devil and the Deep Blue Sea: Merchant Seamen, Pirates, and the Anglo-American Maritime World, 1700–1750.* Cambridge: Cambridge University Press, 1993.

Rediker, Marcus. *Villains of All Nations: Atlantic Pirates in the Golden Age.* Boston: Beacon Press, 2004.

Thornton, A. P. "The English at Campeachy." *The Jamaican Historical Review* 2 (1953): 27–38.

Wilson, Arthur M. "The Logwood Trade in the Seventeenth and Eighteenth Centuries." In *Essays in the History of Modern Europe*, edited by Douglas McKay, 1–15. New York: Harper & Brothers, 1936.

Wilson, David. *Suppressing Piracy in the Early Eighteenth Century: Pirates, Merchants and British Imperial Authority in the Atlantic and Indian Oceans.* Woodbridge: Boydell Press, 2021.

Woodard, Colin. *The Republic of Pirates.* New York: Harcourt, 2007.

## About the Author

**Steven Pitt** is an Assistant Professor of History at St. Bonaventure University in New York. He is the author of "Cotton Mather and Boston's 'Seafaring Tribe,'" *New England Quarterly*, vol. 85 no. 2, (June 2012): 222–52 and "Building and Outfitting Ships in Colonial Boston," *Early American Studies: An Interdisciplinary Journal*, vol. 13 no. 4 (Fall 2015): 900–904. He is currently working on a book manuscript entitled "Mutiny, Revolution, and Piracy: Boston's War on the Waterfront, 1689–1726."

# 5. Pirate Encounters and Perceptions of Southern-Netherlandish Sailors on the North Sea and the Indian Ocean, 1704–1781

*Wim de Winter**

**Abstract**
This chapter investigates practices of piracy through the vantage point of previously unexplored shipboard documents, notarial archives, and testimonies from eighteenth-century Southern Netherlandish sailors. These sources reveal how Southern Netherlandish expeditions from the port of Ostend to the Indian Ocean perceived and encountered Angrian, Chinese, and Madagascar pirates. Surviving crewmembers reported on pirates' tactics and their enabling conditions. However, Southern Netherlandish sailors would also engage in maritime predation themselves, during wartime on the North Sea and as renegade traders in the Indian Ocean. One may therefore contrast Southern Netherlandish sailors' own shipboard perceptions and encounters with Indian Ocean piracy against their own practices of maritime predation, questioning the ambivalence between piracy as a practice and a label for condemnation.

**Keywords:** piracy; maritime history; Indian Ocean; Angria; Madagascar; China

* This research was supported by, and contributes to the ERC AdG project TRANSPACIFIC which has received funding from the European Research Council (ERC) under the European Union's Horizon 2020 Research and Innovation Programme (Grant agreement No. 833143). Part of the research was carried out during the author's employment as head of the "maritime history & society" research unit at the Flanders Marine Institute (VLIZ).

Coakley, J., C. Nathan Kwan and D. Wilson, *The Problem of Piracy in the Early Modern World: Maritime Predation, Empire, and the Construction of Authority at Sea.*
Taylor & Francis Group, 2024
DOI 10.5117/9789463720960_CH05

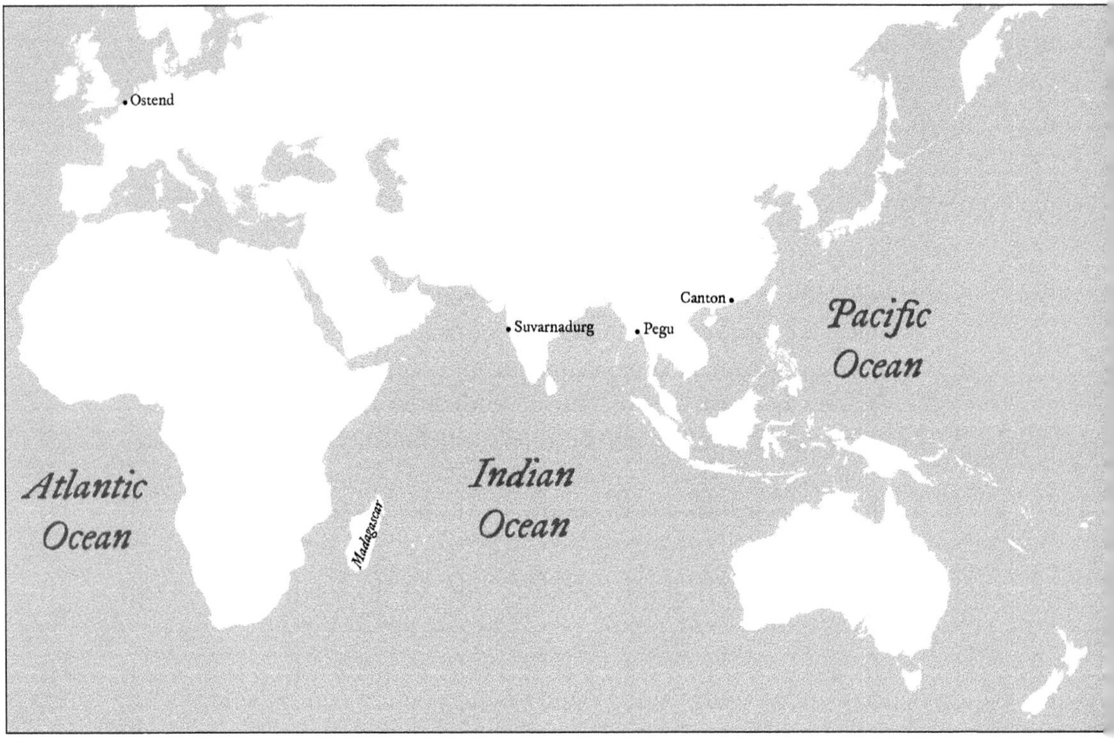
Ostend
Canton
Suvarnadurg
Pegu
Pacific Ocean
Atlantic Ocean
Indian Ocean
Madagascar

## Introduction: 'Illegitimate' Maritime Robbery as Piratical Practice

The 1696 *Tryals of Joseph Dawson et al. for several Piracies and Robberies* served as the British Admiralty court's attempt to define piracy in the Indian Ocean as "a sea term for Robbery [...] the Mariners of any Ship that shall violently dispossess the Master, and afterwards carry away the Ship itself, or any of the Goods [...] with a felonious Intention."[1] This definition was immediately linked to a notion of state jurisdiction, "because the King of England has an undoubted Jurisdiction, and Power, in concurrency with other Princes, and States for the punishment of all Piracies and Robberies at Sea, in the most remote parts of the World."[2] This claim invoked a statist perception and discourse which aimed to extend itself globally across the seas. As Lauren Benton has pointed out, defining piracy itself concerned differing understandings of sovereignty according to European legal traditions.[3] As part of this legal discourse, Derek Elliott and Sebastian Prange have also stated that the indictment of piracy accompanied claims to legal and political sovereignty, which mostly resided in the eye of the beholder.[4] Perceptions of piracy and sovereignty were steeped in specific categories of legal and political thought which historians have inadvertently replicated, whereby they affirmed the model of the European nation state as an implicit reference point for political legitimacy.[5] For the Indian Ocean, David C. Kang would decry such discourse as founded upon a "Westphalian paradigm" neglecting different cultural or legal logics,[6] while the "new thalassologists" studying history on the sea would emphasise the land-based component of such thinking.[7] Instead, they would note that the view from the sea gives the opportunity to transcend conventional Eurocentric historical approaches and biases.[8]

Following Hugo Grotius' seventeenth-century juridical claim that there was no such thing as lawful capture outside a state of war, Benton points out that the term piracy could be applied to multiple violent or transgressive shipboard actions, with a "thin line between privateering and piracy."[9]

1 *Tryals of Joseph Dawson*, p. 6.
2 *Tryals of Joseph Dawson*, p. 6.
3 Benton, "Legal Spaces of Empire," pp. 704–5.
4 Elliott and Prange, "Beyond Piracy," p. 96.
5 Elliott and Prange, "Beyond Piracy," p. 97.
6 Kang, *East Asia*, pp. 3–4.
7 Harlaftis, "Maritime History," pp. 211–12.
8 Harlaftis, "Maritime History," p. 223.
9 Benton, "Legal Spaces of Empire," pp. 705–6.

A fluid "grey zone" existed between piracy and privateering: in the 1680s, English buccaneer William Dampier mentioned how forged, blank, or flawed commissions were given out in order to legitimise French and English freebooters' raiding voyages on Spanish ports in the Pacific. These commissions offered them a margin of what Benton calls "creative interpretation," especially in violent maritime actions towards Spanish possessions in Central and South America.[10]

These remarks call for a need to re-assess both the perception and practice of piracy towards a shipboard perspective, including the "bottom-up" voices of sailors themselves. This chapter will do so from the vantage point of sailors and merchants from the Southern Netherlandish port of Ostend, by showing three eighteenth-century occasions when these sailors testified to their encounters with piracy during their commercial voyages to Asia. These testimonies reveal how Southern Netherlandish sailors perceived pirates, how the pirates they encountered practically and strategically operated, and how similar enabling conditions were used by Ostenders themselves to engage in maritime robbery, even when they refused to be considered as pirates. Ostend sailors occasionally transgressed the "thin line" between piracy and privateering, in the Indian Ocean as well as on the North Sea, yet would afterwards claim their actions to be legitimised or backed by state authority.

Eighteenth-century Ostend sailors were already familiar with privateering from their own context on the North Sea, where Ostend and Dunkirk served as privateering ports. As we shall see, several maritime strategies employed by pirates were similar to those employed by Ostend and Dunkirk privateers during the War of the Spanish Succession (1701–1713) or during the later Seven Years War (1756–1763). The so-called "Prize Papers" of the British High Court of Admiralty, juridical documents concerning the legitimacy of British privateering, testify to Southern Netherlandish privateers' use of false flags, with which they strategically changed from Dutch to English or French colours.[11] These sources also mention how large privateering crews served on small manoeuvrable ships,[12] and how the specificity of waters and islets required the skilled navigation of fishermen-privateers in order to trap their prospective prizes on the North Sea sand banks. An account kept on board the ship *Hope* in 1756 mentioned how a crew of twelve privateers

10 Benton, "Legal Spaces of Empire," pp. 710–11.

11 The National Archives, UK (TNA), High Court of Admiralty Papers (HCA) 32/199/1. "Interrogation De Hope—cpt. Antoine Lux," 1756.

12 TNA, HCA 32/48/69. "Interrogation L'Aurora of Dunkirk—cpt. Francis Demitter," 1703.

armed with cutlasses vied for a prize yielding a cargo of lead and consumable commodities, such as salmon and pickled herrings.[13] Although operating within the confines of wartime privateering, interrogations reveal that officers' personal belongings, such as jackets or gilders of "Flemish money," could be taken by force during privateering.[14] A 1702 letter from Dunkirk ship *St. Antony* reveals how even allied privateers could come on board and indiscriminately rob the captain's cane and hat, clothes, and some guns or goods.[15] This shows us how the distinction between piracy and privateering was practically transgressed.

As Johanna Theresia Verhees–van Meer reported, eighteenth-century literature already designated this "misconduct" of North Sea privateers as piracy, transgressing the rules of privateer engagement. It follows that the maritime behaviour of sailors formed the actual distinguishing criterium, which depended on the measure of state control over actual conduct at sea.[16] Starkey also puts this sharply, as he states that privateers' aims and means held close affinities with those of pirates, as their authority was only valid against enemy property during wartime. As he points out, those aims and means nevertheless "carried the taint of piracy despite their potential utility to the state," as they were ultimately aimed at the destruction of enemy commerce.[17] David J. Starkey noted how the extent in which privateersmen plundered vessels, mistreated prisoners, distorted evidence, or otherwise ignored their instructions remained uncertain.[18] However, research into the Southern-Netherlandish "Prize Papers" reveals more concrete evidence, as a patchwork of sources concerning individual ships shows a micro-level at which plunder and mistreatment took place. Therefore, maritime practices from the North Sea region already confirmed the "grey zone" between piracy and privateering and how certain practices of maritime predation would have already seemed familiar to Southern Netherlandish sailors venturing elsewhere. Common sailors' concrete encounters on more distant seas will reveal how they perceived the threat of violence and robbery at sea.

13 TNA, HCA 32/199/1. De Hoop (Lux), "Translations of the papers seized on board the ship called De Hope, Antoine Lux [...] Condemnation and Confiscation of the ship the Nancy of London," 1756.

14 TNA, HCA 32/247/2. "De Twee Gebroeders" (Feliers)—"Interrogation captain Pieter Feliers"

15 TNA, HCA32/48/49. "The ship St Anthony of Dunkirk—captain John Tiller"—"Letter of Rene Sinou to his wife," 1702–3.

16 Verhees–van Meer, *De Zeeuwse Kaapvaart*, pp. 70–71.

17 Starkey, "Eighteenth-Century British Privateering," p. 76.

18 Starkey, "Eighteenth-Century British Privateering," pp. 76–77.

## A Southern-Netherlandish ship chaplain's perception of Angria and Barbary pirates (1721)

Michael de Febure sailed on the merchant ship *Sint-Pieter* in 1721, venturing from the Southern-Netherlandish port of Ostend to the Indian Malabar coast, as chaplain and chronicler on a commercial expedition fitted out by Ghent merchant Jacobus Maelcamp. One of his tasks was keeping the ship's logbook, which he did in extensive narrative detail.[19] On the September 25, 1721, de Febure's log shows how the crew was acutely aware of the risks posed by pirates. As the ship passed southward by Madagascar, his logbook mentioned how "said Island is very serviceable for sea-robbers as a refuge, as there is a government without police there, and they have none or little correspondence with other Kings or princes, most of them are Heathens."[20] In his perception, the chaplain's consideration of piracy or sea robbery was closely linked to the perceived absence of state or government, and to Madagascar's moral connotation as a "heathen" environment.

Next to De Febure's mention of Madagascar pirates as a potential risk for the Ostend-Asia ventures, he also mentioned another Asian threat: that of the so-called "Angria pirates." On December 20, 1721, de Febure's logbook mentions that "small pirates" from Angria formed a danger on South Indian seas:

> between here and Suratte, caution is advised due to some small pirates from Angria. It is said that captain Larmes, who sailed from Goa to Europe 14 days ago, had some skirmishes which caused a number of wounded. Here at Goa, the ship of Ostend captain Vlaenderlynck had been taken from the wharf by pirates. Here some purport that Vlaenderlynck had an encounter with the pirates of Madagascar, in which he would have broken his neck, others claim he did not.[21]

This citation reveals how the perception of pirate encounters also depended on the exchange of oral information in the maritime environment, where sailors' rumours concerning possible pirate attacks and victims directly influenced the navigational strategies of the *Sint-Pieter*. Due to the risks posed by the "Angria pirates," the crew kept their course ten to twenty

19 De Winter, "Southern-Netherlandish observations," pp. 267–68.

20 Ghent University Library (UBG), HS 0929. Michael de Febure, *Logboek van het schip Sint-Pieter, kapitein Jan Frans Janssens, op zijn reis van Oostende naar Oost-Indië en terug*, 1721–1722, fo. 9.

21 Ibid., fo. 17 vo.

miles off shore in order to remain out of sight. They also foresaw an extra possibility for defence by mounting two cannons in the captain's cabin.[22] On December 22, de Febure noted how the Agria pirates reportedly roamed the seas towards Bombay and up to the docks of Surat.[23] During the return journey, on February 14, 1722, de Febure noted how the ship took a wider berth at the so-called "Height of Angria," in order to get rid of the Angria sea-robbers, "of which it was said they had at least 12 to 13 vessels cruising up to Surat."[24] This led to a panic on the following day, when the ship entered a calm with no wind, as de Febure wrote in his journal how "the silence here is very harmful to us if it would be the case that the Angria Robbers would come for us, while we are near their coasts here, for if we had winds they could do us little damage."[25]

On February 18, the crew spotted five ships, which "made them appear as if they were Algerian [sic.] Robbers, which they then saw sailing towards Surat, so that they believed them to be Bengali ships heading for Surat."[26] Here, de Febure's journal presents us with an interesting slip: unless de Febure really meant that Barbary corsairs would have sailed all the way across the Atlantic into the Indian Ocean, this may instead reveal how Algerian and Angrian corsairs were perceived within a same category as pirates. Margarette Lincoln has also pointed out that commentators at the time did draw parallels between "Angria pirates" and what they knew of North African Muslim corsairs, who took European crewmen as slaves under their own form of state-controlled piracy or privateering.[27] Perhaps de Febure also did this, while confusing them in his manuscript.

De Febure's logbook also mentions the reputed island hideout or base of the Angria pirates. As he approached the island on February 26, 1722, de Febure revealed more perceptions and prejudices concerning the Angria pirates, drawing on possible tropes concerning heathenism, idolatry, and human sacrifice, which formed part of the early-modern European ethnographic imaginary:[28]

> At present at noontime we were close by a small island, which is named, l'île du sacrifice, or the Island of sacrifice by reason that the Robbers, who live around those parts on the Malabar coasts, if they have acquired any

22 Ibid., fo. 17 vo.
23 Ibid., fo. 18.
24 Ibid., fo. 22 vo.
25 Ibid., fo. 22 vo.
26 Ibid., fo. 23.
27 Lincoln, *British Pirates*, pp. 101–3.
28 Rubiés, *Travellers and Cosmographers*, pp. 274–77.

> Robbery over some small vessels, in gratitude of their Robbery that they transport 2 or 3 of the taken persons to said Island, and there sacrifice them to their heathen idols by taking their lives. It is only a small Island, located 4 or 5 miles off Calicut.[29]

This island was already depicted as a place for human sacrifice by John Fryer in 1674, who reported how a rock off the coast of Mangalore was known as Sacrifice Island, "in remembrance of bloody butchery of some English by Malabar pirates."[30] These pirates were identified as Vingorla pirates, operating under the Sawunt Waree chief Kempsant.[31] By the 1720s, this notion had apparently made its way into a legend which was also known and shared among Southern Netherlandish sailors.

The Angria fleet definitely formed a threat to the Ostend-Asia shipping, but not necessarily one to be defined as piracy. Maratha leader Chhatrapati Sambhaji granted Kanhoji Angria the title of *sarkhel* in 1688, which gave him control of the island fortress of Survarnadurg, and put him in charge of the Maratha naval force. His *dastak* system was similar to the earlier *cartaz* used by Portuguese raiders, by which all ships that sailed his waters were required to purchase a pass granting immunity from maritime attacks.[32] Lincoln notes how Kanhoji Angre or Angria had a fleet of numerous small vessels, which he essentially hired out as a "Maratha navy." From 1710 onwards, his fleet would attack ships sailing through the Western Indian waters which had not paid for the immunity granted by his pass or *dastak*, which he used as a source of tribute revenue.[33] Janice E. Thomson also notes how the Angria system of passes or *dastur* was not unlike the passes used by Barbary corsairs in the Mediterranean.[34]

As authors such as Thomson or Patricia Risso point out, Angria was not an "illegitimate" pirate since he held a valid office in his own state, although he did engage in what the British considered as piratical acts, leading to a globalised "unclear norm" of piracy.[35] Risso mentions that Angria was perceived as a pirate to the British, as he opportunistically attacked their shipping, yet operated on a basis of personal non-aggression agreements.[36] He

29 De Febure, *Logboek*, fo. 23 vo.
30 Mentioned in Biddulph, *Pirates of Malabar*, pp. 71–72. Biddulph refers to Fryer, *A new account*.
31 Biddulph, *Pirates of Malabar*, p. 72.
32 Elliott and Prange, "Beyond Piracy," p. 107
33 Lincoln, *British Pirates*, pp. 101–2; Thomson, *Mercenaries*, p. 113.
34 Thomson, *Mercenaries*, p. 117.
35 Thomson, *Mercenaries*, p. 117.
36 Risso, "Cross-Cultural Perceptions," p. 309.

held legitimate power as a privateer-like ally of the Maratha confederation against the Mughal authorities, and the local English and Dutch ships the Mughals occasionally enlisted.[37] Elliott and Prange claim that the British did not, or refused to, understand the Maratha as a legitimate territorial and political power.[38] As a result, "Angria" became an eighteenth-century British synonym for piracy, and spread in newspapers and journals which frequently reported on predations against British shipping.[39] De Febure's log shows that this awareness was not limited to the British context, and that Southern-Netherlandish sailors also shared these views and anxieties concerning the potential risk formed by the Angria.

De Febure's logbook shows how piracy remained a potential risk for the *Sint-Pieter*'s crew throughout the return journey across the Atlantic. On the August 16, 1722, his logbook reports that the captain had received information concerning three Barbary sailors cruising near the Canary Islands:

> The captain came on board with tidings that advice had come by barge from St Michael's Island, as that there were three Robbers in sight, which had already taken about five ships there, to know three Portuguese and two English: which tidings worried us to quite an extent [...] Near St Michael's Island they took a Portuguese ship laden with salt, but after most of its folk had taken up service with the pirate, and the captain of said ship refusing to go with them, they cut off his arm, and allowed him to carry on with his ship.[40]

In this testimony, we see another aspect of pirates' operational procedures: the possibility of their victims' crewmembers taking up service on board the pirate ship, and the use of violence against those refusing to do so, and against authority figures such as captains and officers. This observation seems to fit with Marcus Rediker's remark that desertion to piracy was especially common during the period 1716–1722, as pirate crews were usually recruited through collective mutiny or by victims volunteering to join a pirate crew once their vessel was taken.[41] Rediker considers their desertion as a form of protest and social justice against "the near-dictatorial arrangement of command in the merchant service," as it built on "lower-class/lower-deck

37 Risso, "Cross-Cultural Perceptions," pp. 303–5.
38 Elliot and Prange, "Beyond Piracy," p. 107.
39 Elliott and Prange, "Beyond Piracy," pp. 105–6.
40 De Febure, *Logboek*, fo. 38
41 Rediker, *Villains of All Nations*, pp. 46–48.

values of collectivism, anti-authoritarianism, and egalitarianism."[42] Despite Rediker's emphasis on these values, de Febure's account also shows lower-class sailors' acute anxieties at the prospect of encountering pirates and undergoing their violent tactics. Despite these anxieties, the *Sint-Pieter* did not directly encounter pirates and safely made it back to the harbour of Ostend in September 1722. Although he did not report a direct pirate encounter, de Febure's account has given us a shipboard look at the perceptions and imaginations sailors held towards the potential presence of pirates during their journey. In contrast to the perceptions of the *Sint-Pieter*'s crew, several other Ostend-Asia ships did have the misfortune of encountering pirates, from which we may learn more about actual piratical practices as reported in an Ostend sailor's own voice.

## Navigator Joannes De Vos' 1720 Encounter with "the Dragon Pirate" at Madagascar

Sources such as notarial declarations allow us to learn how sailors directly experienced pirate encounters. In the case of Joannes De Vos' 1720 encounter, this will reveal how violence was used in maritime robbery. He testified which tactics were used by pirates, and how the pirates were perceived as one group or entity belonging to a specific ship, the *Dragon*. As Jan Parmentier has mentioned, the trajectory between the southern Indian Ocean and the Table Bay was especially insecure for the Ostenders during the period 1718–1722, due to roaming pirates. In 1721, pirates Olivier La Buze and John Taylor hijacked the Ostend vessel *Stadt Oostende* near La Réunion.[43] Its captain reported how the pirate ship was more heavily armed, and carried over 700 crewmembers. They plundered the Ostender ship and forced fifty of its crewmembers to join the pirate crew, while an additional fourteen voluntarily decided to join the pirates.[44] This testimony on forced recruitment somewhat nuances Rediker's recruitment thesis, although some Ostend sailors did join voluntarily.

A similar drama befell the ship *Prins Eugenius* in February 1720: Captain Joannes de Clerck spotted three ships flying Dutch and English colours near Cape Good Hope. As they approached, these ships immediately fired their guns and forced de Clerck to surrender, upon which pirate Edward Congdon forced the captain to reveal if he hid diamonds, silver or gold in his vessel. The

42 Ibid., p. 155.

43 Parmentier, *Oostende & Co.*, p. 12.

44 Ibid., p. 105.

consequent plundering lasted five days, after which the kidnapped crew was put on another Southern Netherlandish ship, the *Huys van Oostenryck*, which had only been hijacked two days previously.[45] This ship's pirate encounter is portrayed in a notary testimony by the *Huys van Oostenryck*'s navigator Joannes De Vos. On his arrival from the ship's China expedition, navigator De Vos immediately testified before Ostendian notary Philips Rycx concerning his encounter with "the Dragon Pirate" Edward Congdon.[46] This source grants us a rare glimpse at how this pirate raid took place, and what its consequences were for the crewmembers. It reveals some of the tactics commonly used by pirates or privateers, and how shipboard tensions among English and Southern Netherlandish crewmembers served as a base for pirate recruitment.

De Vos related that when the ship *Huys van Oostenryck*, commanded by British captain James Naish, had reached the Cape of Good Hope, an unknown vessel carrying an English flag passed by and suddenly engaged them in a fight by shooting cannons and muskets, while striking the red flag.[47] Rediker mentions how the pirates' sign of raising the red or "bloody" flag indicated that they would engage in a fight to the death, as part of their customary practice to teach their opponents a lesson by terror, especially when blood had been spilt in battle.[48] The Ostenders immediately surrendered to the "pirates or sea-robbers," as they did not have enough cannons on board, since Captain Naish had sold some of these at Canton in order to stow excess smuggling goods.[49]

As the pirates planned to bring the ship into Madagascar, they started opening all trunks, chests and cupboards for spoils. Common sailor Pieter Joannes De Reuse was brutally questioned by the pirates, as he was accused of hiding guns while planning to take revenge for the ship's capture. The pirates threatened him to show the hiding place of the alleged weapons, under threat of breaking his neck. In addition, some of the *Huys*' British sailors falsely accused De Reuse of hiding gold, which De Vos attributed to their apparent grudge towards Flemish sailors. Due to this accusation, the pirates grabbed De Reuse, tore his clothes from his body and hung him to the mainmast, where they hit him with sticks on the body and head until he swore not to know of either guns or gold. They then aimed a gun at him,

45 Ibid., p. 107.

46 Fijksarchief Brugge (RB), TBO 132/8–12 "1941. Depot notaris F. J. Van Caillie te Oostende—Rijcx, Philips senior Oostende—band 2 januari 1720–29 december 1720—'declaratie Joannis de Vos aan Philips Rycx not. 28 Juny, 30 Juny, eersten July 1720 Joannis De Vos schipper en onderstuurman van den fregat schepe genaemt het Huys van Oestenrycke."

47 RB, TBO 132/8–12, "declaratie Joannis de Vos," fo. 10–11

48 Rediker, *Villains of All Nations*, pp. 83–86.

49 Ibid., fo. 11–12.

and threatened to hang a noose round his neck. Eventually, De Vos reports how the pirates took sixteen pieces of damask, a piece of yellow tafta, twelve pounds of green tea, two pairs of silk stockings, the main part of his clothes and underwear, a quantity of porcelain, and parts of his sailors' tools from his own chest. These goods were brought to the pirate ship and divided among the entire pirate crew or "*zeeroversequipage*."[50] Subsequently, the main part of the supplies, powder, lead, bullets, guns, ropes and other goods were confiscated, and the ship's accompanying sloop or longboat was sunk. Meanwhile, De Vos stressed how the English crewmembers of the *Huys van Oostenryck* voluntarily decided to join the pirate crew in order to "sail for robbery" or "*ten roove vaeren*," which he and several others refused to do. On February 28, 1720, Captains Naish, De Clerck, and the remaining crewmembers were then brought back on board the *Huys*. The ship's surgeons were threatened with violence and forcefully enlisted on board the pirate ship. In this condition, the ship returned to Ostend with a skeleton crew on the June 27, 1720.[51] A second source corroborating De Vos' testimony may be found in a letter that Captain Naish personally wrote and sent to England "Purposely With Desire to be forwarded to My Family, that they might not be in Pain, by my being an absent Ship."[52] In the letter, Naish testified on his capture, indicating the size of the pirate crew and their ship:

> Seiz'd on by the Officers of the Dragon Pirate on Feb:ry 9/20, after as Good a Resistance as we were capable of making, we fell into their hands with The Loss of Chief Third mate & Two Seamen, & many others Slightly wounded; She [Dragon] Sails Princely Well, Mounts forty Guns, & Twenty Brass Pieces upon Swiffles on their Gunnil, Two of them also, with one Chorn, in each Top & Small Arms for Twice her Complement, being Three hundred & Twenty Men when I left them on Feb:ry 16/27; they were compleately Stored with Ammunition, & Provision at full Allowance for fiveteen Months. We found them a Generous, tho' Common, Ennemy, for Our Loss inconsiderable in the Ships Large & Valuable Cargoe.[53]

Naish was remarkably courteous in his judgment of the "officers of the Dragon Pirate," as he found them a "generous enemy." Angus Konstam mentions how

50 Ibid., fo. 14–16.

51 Ibid., fo. 17–18.

52 (British Library) BL, India Office Records (IOR)/E/1/11, fo. 226–27v. June 15, 1720—"Letter 139 James Naish on board the House of Austria off of the coast of Dover to the Court reporting that their ship was taken by the prates after some resistance."

53 BL, IOR/E/1/11, fo. 226–27.

captain Congdon, who some sources erroneously mention as Christopher Condent, originally fled New Providence with pirate captain Charles Vane in 1718. He captured a small Dutch sloop-of-war off the Cape Verde islands, which he renamed the *Flying Dragon*. By the summer of 1719, Congdon had reached Madagascar and spent the next year cruising the Red Sea and the northern Indian Ocean for targets.[54] This was undoubtedly the ship captain Naish and navigator De Vos fell prey to. Contrary to Konstam's work, the Southern-Netherlandish shipboard sources did not focus on the identity of pirate captains. Instead, they indicated the pirate ship, its sailing strategies, and the general behaviour of the pirate crew as a group or collective.

Despite his violent encounter with Congdon and the "Dragon pirate," Glyndwr Williams mentions how nineteen years later, Naish himself proposed an enterprise of maritime robbery against the Spanish in the Pacific. After his brief Ostendian foray, Naish had again taken service with the British East-India Company until he was accused of smuggling and fraud in 1731.[55] In 1739, Naish reportedly talked to "the First Lord of the Admiralty about the possibility of capturing Manila, the great Spanish entrepôt in the North Pacific and the destination of the Acapulco treasure galleon."[56] Williams mentions how Naish argued that a British conquest of the Philippines would give a decisive advantage to the British trade in the East, connecting the "commerce to Cochin-China, Cambodia, Siam, Ichore, the coasts of China, an in any islands upon them; and to Japan, and even to the coasts of Mexico, Peru, and Chile; and to every island and place in the vast Pacific Ocean."[57] Here too, the "grey zone" between maritime robbery and privateering was invoked. Having seen how Ostend sailors perceived and encountered European pirates on the Indian Ocean and beyond, we may compare these views with another Ostend source reporting on local encounters with Chinese pirates in the Pearl River delta.

## 1727 GIC-Merchant Robert Hewer's Testimonies on Chinese Pirates

Robert Hewer's testimony on Chinese pirate encounters reveals how a specific geographical environment was conducive to piracy: the Pearl River Delta, consisting of creeks and inlets, allowed for swift raiding, hiding, and smuggling.

54 Konstam, *Piracy*, p. 266.

55 His involvement in the *Huys'* expedition, and the fate of the ship's cannons seem to confirm his involvement in such practices.

56 Williams, *South Sea*, p. 216.

57 Personal notes by Naish cited in Williams, *South Sea*, p. 217.

It also shows how specific economic circumstances allowed pirates to prey on staple goods during times of famine or poverty. The commercial maritime presence of the GIC during the 1720s helped to create a commodity flow to be preyed on. Since 1722, the GIC or "Ostend Company" had been officially founded in the Austrian-Netherlands, in order to systematically continue the Ostend-Asia trade to Mughal India and China as a joint-stock trading company.[58] The company annually sent a ship and several merchants to Canton for its China trade. Once arrived, the merchants had to enter Canton by sloop or barge, in order to arrange permits for further trade. In his 1726 diary, GIC-merchant Robert Hewer noted how his barge got lost at night, and its crew encountered pirates in the Chinese Pearl River Delta:

> When we were passing the boat by the boac or mouth of the Canton River [...] lost our way in the Night, and came among a parcel of Rogues up a deep River, and was in great danger of having all our throats cut, being forced to stand to our Arms all night, however after abundance of troubles the next day we found out our way again, and after much Difficulty and trouble, and our People almost half dead, we arrived at Canton after having spent three days and nights in the Boat, and in much perplexity.[59]

In his diary, Hewer further mentioned the modus operandi of such "Rogues or thieves," who would also rob the ship's supply boats or sampans:

> the Sampan dispatched yesterday with the Company goods had been robbed, and the Rogues were detected along with the Ship being nine or ten Chinese men that had hidd themselves under the Boards, up which the Chest of Silk was stowed in the Sampan [...] thing we not discovered till all the Merchdize as discharges (opend and made so fast that impossible to discover for the Chief Writer and 13 sailors, by which I observe what dextrous Rogues these are [...] The Chinese on their discovery immediately all jumped over Board into the water, but most of them taken up and made Prison by our People (some bundles of Silk taken in the struggle).[60]

Hewer mentions how these robbers were subsequently punished by the Chinese authorities, as they were carried away in chains and severely

58 De Winter, "Ostend Company's Worlds," p. 157.

59 Stadsarchief Antwerpen (SAA), GIC 5757 (MF 002 K). "Chine. Résolutions et Transactions par Robert Hewer 1726."—"A diary of transactions by Mr Robert Hewer," p. 9.

60 SAA, GIC 5757 (MF 002 K). "A diary of transactions by Mr Robert Hewer," p. 94.

punished for stealing the company's staple goods of tea and silk.[61] This kind of robbery was endemic near Canton, as mentioned throughout GIC sources. Capt. Reijngoet of the GIC ship *Concordia* mentioned how a "Case of tea was stolen by the Bankers [suppliers], and that he caught him with a Sack of tea with which he jumped overboard."[62] It appears that staple trading goods such as tea or silk were frequently robbed or stolen by way of the rivers in the delta. Presumably these would be stocked and resold elsewhere. Dian Murray describes how eighteenth-century Chinese pirate communities operated as "a few struggling bandits who sheltered in the offshore islands and carried out nocturnal depredations in the shallow waters."[63] GIC sources show that these "struggling bandits" formed a continuous presence, even if they were merely considered as a "coastal nuisance" by Chinese authorities. Robert Antony provides several points of comparison between these pirates and those found elsewhere throughout the Indian Ocean world: the use of passes, kidnapping of specialists, and piracy as an occupational background for the labouring poor, as an escape from poverty and hardships in making a living.[64] Calanca also points out that most Chinese pirates in this period are to be situated in marginalised groups of sailors or fishermen, who took to crime on a part-time basis as a response to increasing food prices or fraught economic conditions.[65]

The GIC's pirate encounters in China seem to fit the pattern concerning small-scale, ad hoc, and petty piratical collaborations, rather than large, well-organised bands of marauders. Nevertheless, Hewer's encounter with pirates near the mouth of the Canton River may not have been entirely coincidental. Paola Calanca notes how Chinese maritime security reports mentioned how two pirate fleets from Guangdong and Fujian cooperated in 1726, during a period of famine and widespread poverty. Their banners displayed the slogan "rob the rich and succor the poor" (*jiefu jipin*), their pirate leaders wore long red robes "like Buddha's disciples" (*luohan*), and the crew dressed "ninja-like in close fitting red and green garments."[66] There is a possibility that Hewer encountered an offshoot of this pirate presence, operating in an economic and geographic context which enabled such piratical practices.

61 SAA, GIC 5757 (MF 002 K). "A diary of transactions by Mr Robert Hewer," p. 94.

62 SAA, GIC 5700. "Journal of Groot Boeck van de supercargos op de schepen de Concordia en de Marquis de Prié (Canton 1727)," fo. 21

63 Murray, "Chinese Pirate Communities," p. 47

64 Antony, *Like Froth*, pp. 83–84.

65 Calanca, "Piracy and Coastal Security," p. 91.

66 Ibid., p. 92.

## Piratical Practices of the Austrian Netherlandish Presence in Bengal (1723–1744)

We noted several common elements to the piratical practices encountered by Southern Netherlandish sailors in Asia: an advantageous geographical area harbouring islands or creeks, swift tactics in manoeuvrable ships, the theft of trade goods and staple commodities, and the use of ransoming or a system of passes. Although Southern Netherlandish sailors and merchants in Bengal would not have considered their own activities in the Ganges delta as piracy, they employed some of these same elements as they engaged in maritime predation of their own. This predation did not occur on explicit orders by a state or company, but on local agents' own initiative, resulting from their situation in the intra-Asian trade, fuelled by the profitability of private trade and smuggling.

In 1744, former GIC employees under the leadership of merchant-director François de Schonamille engaged in piracy, as they no longer received financial support from their homeland due to the Company's abolishment in 1731 and had to find a way to maintain their subsistence in Bengal. As some GIC agents had previously done in 1724,[67] they initially decided to hijack Ganges-shipping and gathered a small army of hirelings in order to conduct a so-called "holy war," which they used as an excuse for plundering. This venture escalated as the hirelings started drinking and robbing indiscriminately, and their band had to escape. While making their escape, they robbed a Mughal ship on the Ganges, which they discovered to be full of treasure. They buried the treasure on an island in the mouth of the river, hoping to return and collect it when the situation calmed down. They then proceeded to Burma or Arakan, near an area which had previously been a hotbed of piracy.[68] In the *Records of Old Fort William*, the English at Calcutta described de Schonamille's conflict as follows:

> Having just now received Information that the Ostenders have quitted their Factory at Banquibazar and gone a board Two Ships they have taken from the Moors [...] and that Several Moors Forces are upon the Banks of the River at those Places.[69]

67 SAA, GIC 5927. "Stukken uit de Economisch-historische tentoonstelling 1930: 7. Brief van kapitein Adriani aan Cobbé 1724, juni 18."

68 Parmentier, *De holle compagnie*, pp. 80–90.

69 Wilson, *Records of Old Fort William*, Range 1, vol. XVII, p. 178.

The British account thus seems to confirm the Ostenders' piracy on the Ganges, and continued to describe the situation which led up to the 1744 conflict:

> Schonamille was accused of helping the Marathas, and a large indemnity was demanded [...] hostilities broke out. On 22 September 1744 he attempted in vain to storm the fort of Hughli. The town, however, was set on fire, and two ships with rich cargoes were captured. [...] Schonamille and a few of his companions escaped, sailed out of the Ganges and took refuge in Pegu.[70]

Only two members of de Schonamille's band of marauders survived the ordeal and were taken prisoner by the Dutch East-India Company at Batavia. They were accused of piracy as they had robbed a Mughal ship and caused damage to Armenian or Bengali merchants engaged in business with the Dutch United East India Company (VOC). This in turn sparked a diplomatic discussion in Europe, due to which the prisoners' testimonies have been kept in the Belgian state archives. In the testimony directed towards his VOC captors, officer Bernard Maccaffry replied: "What is called waging a sort of War? I know nothing of it, but that I pushed back the Moor attacks, and sought to destroy as many of them as I could by the ordinary means of War."[71] He claimed how the small vessel taken by his marauders was declared by Schonamille to hold an expired *dastur* or passport, and had been destined towards the Malacca straights. This indicates it to have been a ship destined for intra-Asian commerce, which indirectly made it a rival ship to the local Southern Netherlandish commerce. It held a cargo of 400 swordblades, which Maccaffry claims made it valid "as a rightful prize," as if the Southern Netherlandish band engaged in privateering in Bengal.[72]

This declaration raises many questions as to which authority could have justified privateering in this case, as the plundering happened on their own initiative. Perhaps prize-taking was invoked as a practice from the Ostenders' familiar North Sea wartime context. The passports or *dastur* claimed or handed out by Austrian-Netherlandish officers had a questionable legal basis,

70 Ibid., vol. XVII, p. 179.

71 Algemeen Rijksarchief Brussel (ARA), Departement voor de Nederlanden van de Hof- en Staatskanselarij te Wenen, nr. 645, document 52 – "Réponse faite et exhibée à Monsieur Reyier Stapel president et à messieurs du Conseil de Justice de Batavia par Bernard de Maccaffry [...] prisonnier depuis le 19 Decembre 1745 jusques à ce jour, sur une accusation de Piraterie et de Violence Publique," fo. 3.

72 ARA, nr. 645, document 52, fo. 12.

as they were neither acknowledged as competent to do so by the Mughal authorities, nor by the Dutch or British East-India companies. To the Dutch accusation that de Schonamille's party therefore consisted of little more than a degenerated band of plunderers and robbers, Maccaffry replied "in which country of the world does it not happen, that the soldiers and sailors who lost their Chief disband and no longer maintain any discipline?"[73]

Maccaffry's colleague and contemporary, Grégoire Thousselier, suffered the same predicament as he was also jailed at Batavia and accused of piracy. He detailed the sequence of events concerning Schonamille's "War" in a letter to the Comte de Harrach, dated October 20, 1747. He belittled the accusation of piracy against Schonamille's marauders, as he claimed that they had only "rendered themselves master of a small Vessel laden with salt, of a rice boat, and of a small Vessel of Segapatnam belonging to two inhabitants of this town."[74] As to the cause of the conflict, Thousselier claimed that their acts of violence were justly aimed against "the Insults of the Infidels," as he stated that "Nobody ignores in India that the Europeans in Bengal have much to suffer from the Tiranny of the Government that Nabab Anaverdican has taken since some years, in the tone of absolute master, he doesn't even take hardship anymore to cover or colour into any pretext his vexations when he is in need of money."[75] Thousselier was diplomatically released and managed to return to the Austrian Netherlands in 1749. Yet due to the situation he addressed in his first letter, he urged Austrian Empress Maria-Theresa to grant him the authorisation to return to India, so that he could take revenge on "the Moors" by engaging in maritime plunder against "Moorish vessels":

> I only request Their Majesties to authorise me to return this year to India, in order to make way to Pondicherry or Madras, I will attribute my return to the Great colds of Europe which I was no longer able to support, which occasioned the return of almost all those who, after a long stay in India, had returned to Europe. I have maintained correspondence with all the nations, as soon as I judge it convenient I will have a friend buy a Vessel which I will arm for the place where the moors send their richest Vessels.[76]

73 ARA nr. 645, document 52, fo. 16.

74 ARA nr. 645, document 52—"Grégoire Thouzelier to the Comte de Harrach, 20 October 1747," fo. 2 vo.

75 ARA nr. 645, document 52—"Grégoire Thouzelier to the Comte de Harrach, 20 October 1747," fo. 1

76 ARA nr. 645, document 52—"Copie d'un Memoire par lequel Gregoire Thouzelier apres avoir exposé les vexations du Gouvernement Maure dans le Bengale envers les Europeens […]

## Conclusion

Piracy was considered a disruptive and dangerous reality at sea. To Ostend sailors, maritime predation was a practical reality rather than a legal discourse or identity. The Ostend pirate encounters and perceptions therefore take us to a point of view from which privateers and pirates were sometimes indistinguishable. Antony also notes how the legal distinctions Western governments made between piracy and privateering were often blurred in reality: Drake and Morgan plundered Spanish treasure galleons in the Pacific,[77] Dampier mentioned "blank" passes, and Bengali or Maratha authorities remained unrecognised by British or Ostend sailors. We may therefore conclude that Southern Netherlandish sailors also engaged in those acts and strategies which they condemned as piratical in others. Their attacks were perpetrated against whomever they considered to be a legitimate target: in Bengal this concerned "Moorish" or Armenian shipping, which was considered as fundamentally different from that of European companies or nations. Likewise, we have seen how Captain Naish considered the Spanish as inherently legitimate targets for plunder in the Pacific during the War of the Austrian Succession, despite having been a victim of illegitimate plunder himself during earlier years.

The basic tactics Ostenders used were largely the same as the pirates they encountered: operating from bases on islands, small creeks, sand banks, or river deltas useful for mobility and hiding. Like the Angria, they also employed a system of passes which defined ships as legitimate targets. In Bengal they organised themselves as a collective in defiance of their surrounding social context or context of authority. As Chinese river pirates preyed on staple commodities, North Sea privateers also preyed on enemy ships carrying staple goods during wartime. Economic circumstances often formed the trigger for such piracy: as famine and food prices in China caused upsurges of maritime robbers preying on Ostenders' staple goods near Canton, the conditions of intra-Asian trade drove Ostend merchants to piracy.

The individual role or identity of pirates or captains seemed of little importance in Southern Netherlandish shipboard sources: instead, piracy was mostly identified as a violent and lawless collective practice. In the case of De Vos' encounter with the "Dragon pirate," the direct and specific use

propose l'equipmene tde deux vaisseaux de 30 à 40 pieces de Canon pour se render au Bengale, dans la vue d'y recouvrer les Etablissemens de S M et de se venger des Maures," fo. 4 vo.

77 Antony, *Age of Sail*, p. 5.

of violence by the collectivity of the pirate crew and its maritime robbery were emphasised. In the Southern Netherlandish perception, pirates were always "others": heathens, moors, or outlaws. They were primarily indicted as lacking in state authority, whereas Ostenders in Bengal attempted to legitimise their own maritime predation as an act of privateering on unclear grounds. While revealing a different perspective than oft-cited narrative sources such as Johnson's *General History of the Robberies and Murders of the Most Notorious Pyrates*,[78] Southern Netherlandish sources show us both concrete perceptions and practices of piracy, and hint at the complexity of their legitimation.

## Bibliography

### Primary Sources

Algemeen Rijksarchief Brussel (ARA) Departement voor de Nederlanden van de Hof- en Staatskanselarij te Wenen, nr. 645, document 52. "Réponse faite et exhibée à Monsieur Reyier Stapel president et à messieurs du Conseil de Justice de Batavia par Bernard de Maccaffry ... prisonnier depuis le 19 Decembre 1745 jusques à ce jour, sur une accusation de Piraterie et de Violence Publique," 1747; "Grégoire Thouzelier to the Comte de Harrach, 20 October 1747"; "Copie d'un Memoire par lequel Gregoire Thouzelier apres avoir exposé les vexations du Gouvernement Maure dans le Bengale envers les Europeens [...] propose l'equipmene tde deux vaisseaux de 30 à 40 pieces de Canon pour se render au Bengale, dans la vue d'y recouvrer les Etablissemens de S M et de se venger des Maures." 1749.

British Library (BL), IOR/E/1/11 ff. 226–227v. "Letter 139 James Naish on board the House of Austria off of the coast of Dover to the Court reporting that their ship was taken by the pirates after some resistance." June 15, 1720.

Ghent University Library (UBG) HS 0929 Michael de Febure. *Logboek van het schip Sint-Pieter, kapitein Jan Frans Janssens, op zijn reis van Oostende naar Oost-Indië en terug*. 1721–1722.

The National Archives (TNA), HCA32/199/1. "Interrogation De Hope– cpt. Antoine Lux." 1756.

Rijksarchief Brugge (RB) TBO 132/8–12. "1941. Depot notaris F. J. Van Caillie te Oostende—Rijcx, Philips senior Oostende—band 2 januari 1720–29 december 1720—'declaratie Joannis de Vos aan Philips Rycx not. 28 Juny, 30 Juny, eersten

78 Johnson, *General History*.

July 1720 Joannis De Vos schipper en onderstuurman van den fregat schepe genaemt het Huys van Oestenrycke.'"
SAA GIC 5700. "Journal of Groot Boeck van de supercargos op de schepen de Concordia en de Marquis de Prié." Canton, 1727.
SAA GIC 5927. "Stukken uit de Economisch-historische tentoonstelling 1930: 7. Brief van kapitein Adriani aan Cobbé 1724, juni 18."
Stadsarchief Antwerpen (SAA) GIC 5757 (MF 002 K). "Chine. Résolutions et Transactions par Robert Hewer [A diary of transactions by Mr Robert Hewer]." 1726.
TNA HCA32/48/49. "The ship St Anthony of Dunkirk—captain John Tiller ... Letter of Rene Sinou to his wife." 1702–03.
TNA HCA32/48/69. "Interrogation L'Aurora of Dunkirk—cpt. Francis Demitter." 1703.
TNA HCA 32/199/1. De Hoop (Lux). "Translations of the papers seized on board the ship called De Hope, Antoine Lux ... Condemnation and Confiscation of the ship the Nancy of London." 1756.
TNA HCA 32/247/2. De Twee Gebroeders (Feliers). "Interrogation captain Pieter Feliers."

## Published Sources

Fryer, John. *A new account of East-India and Persia, in eight letters being nine years travel begun in 1672 and finished 1681*. London: printed for R. I. Chiswell, 1698.
Johnson, Charles. *A General History of the Robberies and Murders of the Most Notorious Pyrates*. London: printed for Ch. Rivington, J. Lay, and J. Stone, 1724.
*The Tryals of Joseph Dawson, Edward Forseith, William May, William Bishop, James Lewis, and John Sparkes for Several Piracies and Robberies by Them Committed ... At the Admiralty Sessions, Begun at the Old-Baily On the 29th of October, 1696, and Ended on the 6th of November.* London: Printed by John Everingham, bookseller, 1696.
Wilson, C. R., ed. *Indian Records Series. The Records of Old Fort William in Bengal, vol. 1 / Extract from Bengal Public Consultations, Fort William, October 13 1744, Range 1, Vol. XVII.* London: John Murray, 1906.

## Secondary Literature

Antony, Robert J. *Like Froth Floating on the Sea: The World of Pirates and Seafarers in Late Imperial South China* Berkeley: Center for Environmental Studies, 2003.
Antony, Robert J. *Pirates in the Age of Sail*. New York: W. W. Norton & Company, 2007.
Benton, Lauren. "Legal Spaces of Empire: Piracy and the Origins of Ocean Regionalism." *Comparative Studies in Society and History* 47, no. 4 (2005): 700–724.
Biddulph, John. *The Pirates of Malabar*. London: Smith, Elder & Co, 1907.

Calanca, Paola. "Piracy and Coastal Security in Southeastern China, 1600–1780." In *Elusive Pirates, Pervasive Smugglers: Violence and Clandestine Trade in the Greater China Seas*, edited by Robert Antony, 85–98. Hong Kong: Hong Kong University Press, 2010.

De Winter, Wim. "Southern-Netherlandish Observations and Knowledge Production of Naturalia on the Seas: The Writings of Michael de Febure (1721)." *Early Modern Low Countries* 3, no. 2 (2019): 265–82.

De Winter, Wim. "The Ostend Company's Worlds: Courtly Interactions and Local Life in Eighteenth-Century Bengal." In *An Earthly Paradise: Trade, Politics and Culture in Early-Modern Bengal*, edited by Raziuddin Aquil and Tilottama Mukherjee, 157–85. New Delhi: Manohar, 2020.

Elliot, Derek L. and Sebastian R. Prange. "Beyond Piracy: Maritime Violence and Colonial Encounters in Indian History." In *Beyond the Line: Cultural Narratives of the Southern Oceans*, edited by Michael Mann and Ineke Rhaf-Reinberger, 95–119. Berlin: Neofelis Verlag, 2014.

Harlaftis, Gelina. "Maritime History or the History of 'Thalassa.'" In *The New Ways of History: Developments in Historiography*, edited by Gelina Harlaftis, Nikos Karapidakis, Kostas Sbonias, and Vaios Vaiopoulos, 211–38. London: Bloomsbury, 2009.

Kang, David C. *East Asia Before the West: Five Centuries of Trade and Tribute*. New York: Columbia University Press, 2010.

Konstam, Angus. *Piracy: The Complete History*. London: Osprey Publishing, 2008.

Lincoln, Margarette. *British Pirates and Society, 1680–1730*. London: Routledge, 2016.

Murray, Dian. "Living and Working Conditions in Chinese Pirate Communities, 1750–1850." In *Pirates and Privateers: New Perspectives on the War on Trade in the Eighteenth and Nineteenth Centuries*, edited by David J. Starkey, E. S. van Eyck van Heslinga, and J. A. de Moor, 47–68. Exeter: University of Exeter Press, 1997.

Parmentier, Jan. *De holle compagnie. Smokkel en legale handel onder Zuidnederlandse vlag in Bengalen, ca. 1720–1744*. Hilversum: Verloren, 1992.

Parmentier, Jan. *Oostende & Co.* Gent: Ludion, 2002.

Rediker, Marcus. *Villains of All Nations: Atlantic Pirates in the Golden Age*. London: Verso, 2004.

Risso, Patricia. "Cross-Cultural Perceptions of Piracy: Maritime Violence in the Western Indian Ocean and Persian Gulf Region during a Long Eighteenth Century." *Journal of World History* 12, no. 2 (2001): 293–319.

Starkey, David J. "The Origins and Regulations of Eighteenth-Century British Privateering." In *Bandits at Sea: A Pirate Reader*, edited by C.R. Pennell, 69–81. London: New York University Press, 2007.

Thomson, Janice E. *Mercenaries, Pirates, and Sovereigns: State-Building and Extraterritorial Violence in Early Modern Europe*. Princeton: Princeton University Press, 1994.

Verhees-Van Meer, Johanna Theresia. *De Zeeuwse Kaapvaart tijdens de Spaanse Successieoorlog 1702–1713*. Middelburg: Het Koninklijk Zeeuwsch Genootschap der Wetenschappen deel 3, 1986.
Williams, Glyndwr. *The Great South Sea: English Voyages and Encounters 1570–1750*. London: Yale University Press, 1997.

## About the Author

**Wim De Winter** is a world- and maritime historian specialised in maritime history of the Indian and Pacific Ocean. He obtained his PhD at Ghent University with a dissertation on the Ostend Company (GIC). His research aims at "global microhistorical" approaches to interactions in early-modern Asia, colonial exoticism, and the perception of the environment. He is currently employed as post-doctoral researcher at KU Leuven, as part of the ERC AdG project TRANSPACIFIC, where he mainly studies the unseen exchanges and informal agents involved in piracy, smuggling, and maritime knowledge transfer in the transpacific world (from the sixteenth to the eighteenth century).

# Section III

Representations

# 6. "A Fellow! I think, in all Respects, worthy your Esteem and Favour": Fellowship and treachery in *A General History of the Pyrates*, 1724–1734

*Rebecca James**

**Abstract**

This chapter analyses the representation of the pirate in *A General History of the Pyrates* between 1724 and 1734. By examining a number of editions, it establishes that the co-operation of pirates of the *General History* is organised and maintained through fellowships. However, fellowship is most manifest in this text when it is broken, either by treachery or by the performance of bad fellowship. These broken fellowships demonstrate a tension between the values of fellowship and the way that the text hyper values venturesome behaviour. In reading across editions this chapter charts how iterations of the *General History* offer different profiles of fellowship in this ten-year period, thus changing the representation of the pirate imagined in this highly influential text.

**Keywords:** A General History of the Pyrates; Fellowship; Eighteenth-century book history; Pirate representations; Criminal Biography

Early eighteenth-century audiences who were more or less geographically distant from the depredations of Anglo-American pirates voraciously consumed material about individuals and crews, as they were mythologised and

* This work was supported by the AHRC's South, West and Wales Doctoral Training Partnership, University of Southampton and Cardiff University. Many thanks go to Dr. Stephanie Jones and Prof. Martin Kayman for their support, feedback and guidance as my supervisors—their feedback was invaluable to producing this chapter.

Coakley, J., C. Nathan Kwan and D. Wilson, *The Problem of Piracy in the Early Modern World: Maritime Predation, Empire, and the Construction of Authority at Sea.*
Taylor & Francis Group, 2024
DOI 10.5117/9789463720960_CH06

sensationalised in books and the periodical press. These factual and fictional figures prompted excitement and intrigue, and spawned fantastical tales about pirates thriving and failing in "exotic" locations. Our understanding of the eighteenth-century pirate is created as much by these tales as it is by the actions of thieves at sea. As Margarette Lincoln observes, "the history of piracy is as much about rhetoric as it is about actual events."[1] The so called "golden age" of piracy was also a "golden age" of pirate literature.[2]

Published in 1724, as the "golden age" of piracy was coming to an end, *A General History of the Pyrates* is the most influential rendering of these seductive figures. The text is constituted by a series of interconnected chapters that each focus on a pirate captain and his crew. It purports to present the "truth" about the "authentick Relations" of a collection of pirates active in the late seventeenth and early eighteenth centuries.[3] However, between 1724 and 1734 the text was proliferated in a dozen different versions. It was published in octavo, serialised, translated, expanded into a second volume, collected together with other criminal lives, and abridged as *The History and the Lives of the most Notorious Pirates*. These editions differ in form, content, language, and thus produce differing representations of the pirate.

This chapter discusses selected editions published in this ten-year period (See Table 2).

**Table 2. Editions of *A General History of the Pyrates*, 1724–1734**

| Edition | Date | Bookseller | Size |
|---|---|---|---|
| First | May 1724 | Charles Rivington, J Lacy, J Stone, Edward Easton | Octavo |
| Second | August 1724 | Thomas Warner | Octavo |
| First Abridgement | 1725 | Edward Midwinter | Duodecimo |
| Third Abridgement | 1729 | Edward Midwinter | Duodecimo |
| Compendium (collected with criminal biographies) | 1734 | James Janeway | Folio (as fascicles and bound) |

This chapter argues that the *General History* does not create a world where each individual acts exclusively to further his or her own interests. The pirates of this text are not anarchic and lawless. Instead, the *General History* imagines shipboard communities where a lack of co-operation and fidelity

1 Lincoln, *British Pirates*, p. 3.
2 Rennie, *Treasure Neverland*, p. 21.
3 Johnson, *A General History, 2nd Edition*, p. 3.

is harshly judged. This chapter argues that the *General History* imagines the pirates of the "golden age" as collaborative fellows. These fellows counter the prevailing narrative of individualism that is offered by both contemporary criminal biographies, and by many narratives of "golden age" pirates.

By extensively examining editions of the *General History* published between 1724 and 1734 this chapter establishes that the representation of the pirate fellow changes across these ten years. It asserts that fellowship is consistently expressed across the octavo and compendium editions, while the abridgements condense the text to omit elements that construct fellowships in the longer editions. It speculatively suggests that these differences could have emerged from the contemporary conditions in the eighteenth-century book market. The abridgements were produced for a larger and less affluent audience, who were used to the conventions of criminal biographies. By focusing more closely on the individual protagonists, and drawing away from the communality expressed in the octavo editions, the abridgements came to resemble criminal biographies more closely, and thus became more conventional texts. By contrast, the octavos construct a more unique literary figure within the hybrid form of these editions.

Highlighting the significance of communality within these editions challenges readings of the *General History* as an archive, or as a collection of individual criminal biographies. Reading across editions shows how the text evolved to meet the needs and expectations of audiences. It establishes that the *General History* is far from a stable, single text, and therefore that we should be alert to this when looking to the text as a source of information about the lived experience of pirates.

The pirates of the "golden age" are typically represented through individual protagonists. The captains in these texts are often individuated, and are used to embody the many crewmembers that enabled pirate adventures to take place. This is particularly exemplified in the representations of Captain Avery, who was the particular focus of pirate literature in the 1690s and 1700s. His infamous plunder of the Great Mogul's ship, the *Ganj-i-sawai*, was mythologised in pamphlets, plays and broadside ballads, as pirate was imagined to be the leader of a pirate community that lived in Madagascar.[4] Consequently, the pirate is often read as an individualistic figure. John Richetti, for example, argues that part of what makes the pirate unique is how "they embody the most dangerous and revolutionary forces of the

4 Joel Baer, Stephanie Jones, and Frederick Burwick and Manushag Powell discuss the pamphlets, plays, and broadside ballads about this pirate. Joel Baer, "Bold Captain Avery"; Jones, "Literature, Geography, Law"; Burwick and Powell, *British Pirates*, pp. 26–32.

age—the radical individualism which summarises the totally secular view of experience."[5]

The *General History* was also produced in conversation with the conventions of criminal biography. Criminal lives were a popular topic in this period—Lincoln Faller estimates that "some five or six thousand separate accounts of criminals' lives and/or trials survive from the period 1650–1800 in England."[6] These texts were typically cheap broadsides or pamphlets that described a single criminal's life. As Faller argues "highwaymen typically were shown acting as lone gunmen, it would seem, because they were at their most attractive and most entertaining when they stood outside all body politic."[7] The space literary highwaymen inhabit and the targets they pursue allow them to work alone, with only a horse for a quick getaway.

The pirate protagonist is only part of the narrative of piracy woven by the *General History*, however. Communality is also integral to its representation of the eighteenth-century pirate. The text's chapter titles signal the central duality of the captain and community by heading each chapter thusly: "Of Captain Avery and his Crew." This chapter focuses on the cooperative aspect, by analysing the concept of fellowship. Focusing on fellowship reshapes our understanding of a text that continues to influence how we imagine pirates almost three hundred years later. By analysing the rhetorical strategies the text uses to create pirate communality, this chapter resists readings that treat the *General History* as an archive and instead considers it as a representation of a kind of communality that is often obscured in criminal biographies.

Fellowship is consistently expressed across the octavo, and compendium editions, which is significant in and of itself within this text that varies extensively elsewhere. The abridgements, by contrast, condense the text to omit elements that constitute fellowship in the longer versions. The latter characters therefore more closely resemble the protagonists of contemporary criminal biography discussed above. This process conventionalises the pirate by crafting him as equivalent in many ways to the thieves based on land. Comparing across editions illuminates how the octavos craft a unique literary figure, and challenges readings of these editions as individual criminal biographies.

As outlaws the pirates of the *General History* cannot rely on modern systems of laws, regulation, and professional codes to discipline their relations.

5 Richetti, *Popular Fiction*, p. 75.

6 Singleton, "English Criminal Biography," p. 64; Faller, "Criminal Opportunities," pp. 121–22.

7 Faller, *Turned to Account*, p. 179.

They have no recourse to organise their communities beyond the bonds they form and the oaths they make. And yet, this model of homosociality is undermined by the fact that fellowship is most present in moments when it is broken. However, the text does not construct a world where everyone is out for themselves, but instead creates one where a lack of co-operation and commitment is judged harshly. This chapter thus challenges this idea that representations of pirates are lawless and anarchic.

## Pirates Are "Fellows" Aboard a Ship

The *General History* uses the word "fellow" repeatedly, although it does not use the word "fellowship." Nevertheless, fellowship is important in the text's construction of the pirate. *Samuel Johnson's Dictionary* offers useful definitions of the word "fellow," which is broadly contemporary to the *General History*: "A companion; one with whom we consort […] An associate; one united in the same affair […] One of the same kind […] Equal; peer […]."[8] These definitions, and their associated complements, emphasise companionship and collectiveness, gesturing to the existence of a peer group, a sense of similarity, and some notion of equality. This chapter discusses fellowship in relation to the fellow, by following the approaches of Jeffrey Singman and A. J. Pollard in delineating "fellow" and "fellowship."[9] It begins by arguing that the sustained use of the word "fellow," the specific assertion that the archetypal pirate is "A Fellow […] worthy your esteem and favour," and the consistent social position of the characters create the conditions where pirates can be characterised as a group of fellows on a ship.

Modifiers shape our understanding of how "fellow" is used across editions. Pirates are called "Fellow" or "Fellows" numerous times.[10] There are points where the word is "used as an informal term for a man or a male person," while meaning is more ambiguous in other moments.[11] Some characters are "bold and dissolute" fellows when they become pirates, while others are "able and desperate."[12] The first suggests a collective of moral degenerates, while the latter

8 Further definitions concern its use as an ironic statement of contempt, and the collegiate meaning. See "Page View, Page 787" in Beselke, *Dictionary of English Language*, https://johnsonsdictionaryonline.com/1755/fellow_ns (accessed August 31, 2022).

9 Pollard, *Imagining Robin Hood*, pp. 134–55; Singman, *Robin Hood.*

10 The second edition alone includes ninety instances of the word.

11 "fellow, n." *OED* Online. December 2020. https://www.oed.com/view/Entry/69094 (accessed February 9, 2020).

12 Johnson, *A General History, 2nd Edition*, pp. 24, 45.

echoes the preface's call for a fishery to employ exploited and unemployed seamen. In contrast, tried and executed pirates are often "poor" fellows—their lack of power and assumed damnation renders them pitiable. In all of these cases the term "fellow" can denote a man. However, other moments tie the categorisation of "fellow" to communality, rather than masculinity.

The second octavo adds detail that explicitly ties the "fellow" to pirate communality in "Of Capt Bartho Roberts and his Crew." One pirate outlines the characteristics of the ideal captain:

> it is my Advice, that, while we are sober, we pitch upon a Man of Courage, and skill'd in Navigation, one, who by his Council and Bravery seems best able to defend this Commonwealth, and ward us from the Dangers and Tempests of an instable Element, and the fatal Consequences of Anarchy; and such a one I take Roberts to be. A Fellow! I think, in all Respects, worthy your Esteem and Favour.[13]

"A fellow! In all respects, worthy your Esteem and Favour" is a character that acts in the best interest of the collective by displaying courage in all circumstances, by effectively commanding their negotiation of the sea space, and by managing relations to enable productivity. Calling the crew a "commonwealth" imagines this kind of fellow in service of a collective. Roberts is thus imagined as a particular kind of man who exhibits characteristics valued by the pirates. Courage, technical skill, and charisma enable him to effectively lead, control, and, most importantly, protect the pirate crew. The *General History* conceptualises a "worthy fellow" in this moment.

The prefaces to the octavo editions position Roberts as the archetypal pirate, implying that this concept of the pirate-as-fellow is relevant to the text's imagining of the pirate as a whole:

> being resolved not to weary the Reader, with tiresome Repetitions: When we found the Circumstances in Roberts's Live, and other Pyrates, either as to pyratical Articles, or any Thing else, to be the same, we thought it best to give them but once, and chose Roberts's Life for that Purpose.[14]

These prefatory remarks reinforce that the *General History* is a carefully constructed text, not a straightforward account of events. By presenting

13 Ibid., p. 209.

14 Johnson, *A General History, 1st Edition*, A4 recto; Johnson, *A General History, 2nd Edition*, A4 recto.

the chapter as the comprehensive pirate narrative these editions suggest we should read Roberts as an archetype that other pirates either conform to or deviate from. Examining the representation of homosociality alongside the invocations of "fellow" in this chapter demonstrates that the archetypal pirate embodies the qualities of the worthy pirate fellow. It therefore follows that pirates are fellows, judged to be worthy (or not) by the standards established through Roberts. However, other editions position Roberts as one of the collection, rather than as an archetype. The prefaces to the abridgements highlight Avery, Teach, and Gow. Furthermore, the abridgements substantially condense the Roberts chapter, to align with the other chapters, and omit this discussion of the "worthy fellow." These changes are indicative of how fellowship is diminished in these editions, and signals how the abridgements rework the pirate to make him more like the protagonists in other criminal biographies.

Though the pirates of the *General History* are by nature "fellows on a ship," that is not reason enough to justify my use of "fellowship" to encompass the varied depictions of co-operation and homosociality. The text uses a number of terms including "allies," "comrades," "consorts," "companions," "confederates" and "sworn friends" to describe associations, while collectively they are "compan[ies]," "crew[s]" or "commonwealth[s]." The use and meaning of these terms is unstable. Some, like "comrades" and "companions," denote a specific kind of homosociality—in this case, members of the same crew. However, others, like "consort," "confederate," "ally" and "company," refer to a variety of associations. Collectively, these terms construct fellowship. Pirates are "fellows" on a ship; therefore "fellowship" is a reasonable collective noun to use.[15]

"Fellowship" connotes co-operation, reciprocity, and joint enterprise. According to Elizabeth Archibald, fellowship is "derived from the Old English 'fēolaga' and the Old Norse 'fēlagi' originally 'one who lays down money or property in a joint undertaking' and is thus 'a partner or associate.'"[16] Archibald's etymology highlights the importance of a collective investment that helps to form a common bond. She demonstrates how by the Middle Ages ideas of communality became embedded in the definition of "fellowship" as: "the spirit that binds companions together—charity, amity,

15 "Brotherhood" or "Companionship" could have been used here. However, the *General History* uses "companion" as one of a collection of terms, therefore "companionship" more usefully describes specific relations, while brotherhood has familial associations that are only conveyed in "Of Captain Misson."

16 Archibald, "Malory's Ideal," pp. 314–15.

camaraderie" and "band of associates, followers, fighting men, crew."[17] A term rooted in shared investment develops associations with a collective spirit in Archibald's account.

Connotations of co-operation and endeavour are also embedded in eighteenth-century uses. *Samuel Johnson's Dictionary*'s entry defines "fellowship" as "companionship; consort; society"; "Association; confederacy; combination"; "Equality"; "Partnership; joint interest"; "Company; State of being together."[18] The terms used by the *General History* to describe relationships between pirates appear across these selected entries. Moreover, these definitions articulate similar ideas to those generated in Archibald's account. Fellowship retains an association with joint undertaking constituted as partnerships, while collective investment becomes implicit. The definitions, and their associated complements, emphasise that fellowship is more than physical proximity—ideas of equality, companionship, and social pleasure all connote communality and fellow feeling.

## Fellowship and Mutiny

"Of Captain Lowther and his Crew" suggests that fellowships are founded when a collective share a grievance, develop a sense of fellow feeling for one another, and then mutiny.[19] Fellowship is most stable in Lowther's chapter, which remains fixed across octavo and compendium editions despite significant changes elsewhere. However, abridgements diminish the profile of fellowship in this chapter, and consequently within the figure of the pirate. The octavos and compendium depict an action motivated by the circumstances of a collective and the behaviour of their captain, whereas the abridgements ironically represent the captain's grievance.

The octavos and compendium combine a collective grievance about circumstances aboard a ship with an onerous captain, which creates the conditions for a mutiny rooted in fellow feeling. They begin with the complaints from a group of soldiers aboard the ship who "grew very impatient and disatisfy'd, especially *Massey*, who was very loud in his Complaints"

17 Ibid., p. 311.

18 Further definitions in this entry refer to mathematical concepts, collegiate revenue sharing, and 'social pleasure." "Page View, Page 788" in Beselke, *Dictionary of English Language*, https://johnsonsdictionaryonline.com/1755/fellowship_ns (accessed August 31, 2022).

19 I approach mutiny as "the violent seizure of a ship from her officers, on the high seas." Rodger, *Wooden World*, p. 238.

that he and his men are insufficiently victualed.[20] These editions then establish a grievance on the part of the sailors when their captain takes "A pique" against his first Mate, Lowther.[21] However, Lowther "found Means to ingratiate himself into the good liking of the common Sailors," so "when Captain *Russel* order'd him to be punish'd, the Men took up Handspikes, and threatened to knock that Man down, that offered to lay hold of the Mate."[22] The men's reactions signal a deepening relationship between Lowther and the crew, as they directly oppose the authority of the captain to control discipline aboard the ship. This moment secures their homosocial bonds, as it "served but to widen the Differences between him and the Captain, and more firmly attach'd Lowther to the Ship's Company."[23] As a result, the sailors become a collective placed in opposition to their authoritarian captain.

Finally, fellow feeling develops between the two leaders of the soldiers and the sailors. Lowther and Massey "contracted an Intimacy" with one another that leads them to "[resolve] upon Measures to curb the Power that controul'd them, and to provide for themselves after another Manner."[24] Thus, they resolve to take control of their lives and support themselves when their commanders are either too ineffectual or too corrupt to do so. These two groups become fellows, engendered by the relationship between their respective leaders, which enables them to plan a mutiny.

The octavos and compendium then develop the mutiny founded in common grievance and fellow feeling into fellowship when the company become pirates. In the wake of the mutiny Lowther declares:

> *they had a good Ship under them, a parcel of brave Fellows in her, that it was not their Business to starve, or be made Slaves; and therefore, if they were all of his Mind, they should seek their Fortunes upon the Seas, as other Adventurers had done before them.* They one and all came into the Measures.[25]

20 Johnson, *A General History, 1st Edition*, p. 266; Johnson, *A General History, 2nd Edition*, p. 348; Johnson, *A General History, Compendium Edition*, p. 262.

21 Johnson, *A General History, 1st Edition*, p. 267; Johnson, *A General History, 2nd Edition*, p. 348; Johnson, *A General History, Compendium Edition*, p. 262.

22 Ibid.

23 Johnson, *A General History, 1st Edition*, p. 267; Johnson, *A General History, 2nd Edition*, p. 349; Johnson, *A General History, Compendium Edition*, p. 262.

24 Ibid.

25 Italics original. Johnson, *A General History, 1st Edition*, pp. 269–270; Johnson, *A General History, 2nd Edition*, p. 351; Johnson, *A General History, Compendium Edition*, p. 263.

These men are no longer sailors or soldiers, or even mutineers: they are now "a parcel of brave fellows." The modifier signals that the men have formed a collective identity, regardless of whether they were a sailor or soldier. The poor social and economic conditions of sailors established by the introduction are reiterated as their alternative path—they face starvation and slavery, according to Lowther, if they do not become pirates. Mutiny gives them the "good ship," while the fellowship formed between this "parcel of brave fellows" creates the opportunity to "seek their fortunes upon the seas." Fortune and adventure are tied together in this pirate fellowship as they replicate the actions of "other Adventurers." Thus, piracy is imagined as freedom and opportunity. Their fellowship is then secured when they

> knocked down the Cabins, made the Ship flush fore and aft, prepared black Colours, new named her, the Delivery, having about 50 Hands and 16 Guns, and the following short Articles were drawn up, signed and sworn to upon the Bible.[26]

They remove the signifiers of hierarchy by destroying the cabins and establish a set of rules, the "Articles," which lay out the expected conduct of members of the fellowship. They affirm their commitment to the fellowship by signing and swearing to abide by these rules.

In contrast, the abridgements diminish the sense of fellowship conveyed through these characters and their actions. The abridgements use interjections and amendments to reframe collective action as personal grievance. They turn the complaint about victuals into Massey's personal dissatisfaction:

> Massey finding he must be over-ruled by the Merchants, soon began to complain of their ill Treatment of his Men in their Allowance, saying he did not come to be a *Guinea* Slave; and that if they did not use him and his Men better, he should take other Measures.[27]

This quotation suggests that the complaint is a cover for Massey's personal dissent about his position in the garrison's hierarchy. It proposes an ironic reading of the calls for social justice as desire for power and command. This

26 Johnson, *A General History, 1st Edition*, p. 270; Johnson, *A General History, 2nd Edition*, p. 351; Johnson, *A General History, Compendium Edition*, p. 263.
27 *The History and Lives, 1st Edition*, pp. 114–15; *The History and Lives, 3rd Edition*, pp. 88–89.

description contrasts with the more sincere account of common grievance discussed above.

Similarly, the abridgements diminish the reciprocity between sailors. They begin with the same action, where "the sailors, who, upon the Captain's going to punish [Lowther] swore, They would knock down the first man that should offer to lay hands on him."[28] However, this does not create a reciprocal relationship of close homosocial bonds. Instead, Lowther takes this "favour" and "[improves it] to a general disaffection of the ship's crew."[29] Rather than showing fellow feeling, the abridgements present Lowther as a manipulator and beneficiary of affection. Likewise, Lowther and Massey's accord is transformed from a resolution to seek freedom into an agreement to "curb their Enemies, and provide for themselves some other way."[30] The use of "enemies" here turns focus away from the collective circumstances, and transforms the mutiny into a personal vendetta between Massey and Lowther and their respective authority figures. Consequently, the abridgements reframe fellowship to emphasise individual perspectives and responses.

The abridgements also replace Lowther's speech with a summary, which omits the ideas of freedom and social mistreatment that amplify fellowship in other editions. The summary states: "since they had a good ship under them, he proposed to seek their Fortunes upon the Seas, as others had done before them; which they all agreed too, calling the Ship, *The Delivery*, and swore to stand by one another."[31] Crucially, the abridgements omit the phrase "parcel of brave fellows" and remove the references to slavery and starvation. Piracy is therefore constituted solely as an opportunity to "seek their fortunes," not as an opportunity to escape oppression or degradation. Their action is not imagined as an adventure and these characters are not positioned as successors to a tradition of adventurers at sea. Instead, they echo the actions of others "before them." They do, however swear to stand together, showing a diminished, but not non-existent, notion of fellowship.

## Fellowship through Agreement

Across editions the fellowship formed between the mutinying companies of Low and Lowther is the most representative of fellowship formed between

28 *The History and Lives, 1st Edition*, pp. 115–16; *The History and Lives, 3rd Edition*, p. 89.

29 *The History and Lives, 1st Edition*, p. 116; *The History and Lives, 3rd Edition*, p. 89.

30 Ibid.

31 *The History and Lives, 1st Edition*, pp. 116–17; *The History and Lives, 3rd Edition*, p. 90.

separate companies. These imaginings of fellowship are equally significant, as they indicate that the act of piracy creates a sense of commonality and community between those who steal upon the high seas. Low and Lowther's fellowship is the most fully formed and consistently depicted fellowship in the text. It extends beyond the boundaries of a single chapter, constructing fellowship as a connecting thread, not a condition that exists in a single narrative. The *General History* represents the Low/Lowther fellowship as a successful example, and as a model for the narratives that surround it.

Low and Lowther agree to sail as companions when they meet in "Of Captain Lowther and his Crew":

> *Lowther* received them as Friends, and treated them with all imaginable Respect, inviting them as they were few in Number, and in no extraordinary Condition to pursue the Account, (as they called it) to join their Strength together, which on the Consideration aforesaid, was accepted of.[32]

The abridgements present a similar meeting: "Lowther received [Low's company] as friends, inviting them, as they were few in number, to join their strength together."[33] Across editions the word "friends" signals an affinity between the men founded in their shared occupation and/or identity as pirates. This friendship then develops into fellowship when the two groups agree to sail together through an invitation and acceptance. The companies combine to form a fellowship aboard one ship.

The importance of formalised agreements and oaths is amplified across editions through the constitution of Low and Lowther's fellowship. In the first octavo "Of Captain Low and his Crew" echoes the earlier chapter: "*Low* falls in Company with [Lowther], and after a Compliment, they join together, and go all aboard the same Vessel, which is already taken Notice of in the History of *George Lowther*."[34] The two companies informally develop a fellowship. The abridgements meanwhile call attention to the lack of formalities: "they met with *George Lowther*, who took him under his Protection as an Ally, without any formal Treaty; which *Lowe* readily agreed to."[35] This relationship centres protection rather than fellowship. However, the second octavo adds formality and structure by mimicking the relations between sovereign nations when Low:

32 Johnson, *A General History, 1st Edition*, pp. 276–77; Johnson, *A General History, 2nd Edition*, p. 358; Johnson, *A General History, Compendium Edition*, p. 264.

33 *The History and Lives, 1st Edition*, p. 119; *The History and Lives, 3rd Edition*, p. 92.

34 Johnson, *A General History, 1st Edition*, p. 284.

35 *The History and Lives, 1st Edition*, p. 123; *The History and Lives, 3rd Edition*, p. 95.

> falling in Company with *George Lowther*, another Pyrate there, who paying his Compliments to *Low*, as great Folks do to one another when they meet, and offering himself as an Ally; *Low* accepted of the Terms, and so the Treaty was presently sign'd without Plenipo's or any other Formalities.[36]

The term "ally" ties the idea of a "treaty" to the collection of terms that denotes fellowship across the text. The second octavo therefore renders the relationship between Low and Lowther more robust by mimicking the language used in negotiations of sovereign nations. Pirate ships are imagined as their own sovereign communities that can negotiate and enter into treaties with one another. Their outlaw community allows itself the impression of mechanisms used by legitimate forces. Pirates cannot turn to laws or regulations to solidify and regulate fellowships. Instead, they rely upon the mimicry of such mechanisms.

In these ways, the *General History* depicts the formation of fellowship. "Of Captain Lowther and his Crew," offers fellowship founded in common grievance, fellow feeling, and mutiny. The relationship between Low and Lowther, and their respective crews, depicts a fully realised fellowship from start to end, against which we can measure other relationships. Across chapters pirates also consistently affirm their mutual commitment to one another through formal mechanisms and their underlying oaths. These agreements show the significance of fidelity and loyalty in the world constructed by the *General History*.

## Treachery

In selected chapters the values of fellowship are most manifest in moments of crisis. The language used when pirates breach the terms of the fellowship exposes how fidelity is the core of pirate fellowship in the *General History*. The significance of treachery belies satiric readings of the text, because if vows are without value they cannot be betrayed. Crucially, these episodes are moments of crisis, not the normal conditions aboard the pirate ship.

The "Articles" of Roberts's company dictate the moment where a dissolution is permitted: "No Man to talk of breaking up their Way of Living, till each had shared a 1000 l."[37] Yet "Of Captain Bartho Roberts and his crew" features a number of episodes where pirate fellowship is broken by treachery.

36 Johnson, *A General History, 2nd Edition*, p. 368.

37 Ibid., p. 232.

In the first episode Roberts leaves part of his company in command of a prize ship.[38] Roberts and his companions send for help from those left behind, but learn:

> *Kennedy*, who was Lieutenant, and left in Absence of *Roberts*, to Command the Privateer and Prize, was gone off with both. This was Mortification with a Vengeance, and you may imagine, they did not depart without some hard Speeches from those that were left, and had suffered by their Treachery.[39]

The text carefully distinguishes between treachery and piracy. Treachery is not used to describe mutiny or encounters with those outside the fellowship. Instead, the *General History* uses the word "treachery" sparingly and specifically to describe the destruction of fellowships, where one party breaks one of the core principles. Kennedy and his companions are treacherous on two counts: firstly, because they fail to support their fellows in need, and secondly, because they deny the rest of the fellowship their fair share of the plunder as they sever their relationship. The language used in this episode is consistent across the octavos and compendium. It highlights how much of the pirate fellowship is built upon the value of one's word and expectations of a companion's fidelity in the *General History*. Overall, the language used in this episode suggests that betrayal is a moment of crisis, not the norm for these pirates.

The bigger a fellowship, the greater the chances of survival. Consequently, abandoning one's fellows to their capture or death is also treachery. The octavos and compendium show this implicitly in instances where pirates evade capture through cooperation and explicitly when Low abandons his companions in the midst of battle. Low's company become embroiled in a clash at sea after they attempt to take one of His Majesty's Ships: "with a brisk Fire on both Sides, [...] the *Ranger's* Main-Yard was shot down, and the *Greyhound* pressing close upon the disabled Sloop, *Low,* in the other, thought fit to bear away and leave his Consort a Sacrifice to his Enemy."[40] These editions emphasise the relationship between these two ships through the use of "consort." They show that Low is aware of the consequences of his selfish actions by calling the abandoned sloop "a Sacrifice to his Enemy." Those Low abandon "(seeing the Cowardice and Treachery of his Commadore

38 Johnson, *A General History, 1st Edition*, p. 165; Johnson, *A General History, 2nd Edition*, pp. 223–24; Johnson, A *General History, Compendium Edition*, p. 242.

39 Johnson, *A General History, 1st Edition*, p. 166; Johnson, *A General History, 2nd Edition*, p. 224; Johnson, *A General History, Compendium Edition*, p. 242.

40 Johnson, *A General History, 1st Edition*, p. 295; Johnson, *A General History, 2nd Edition*, p. 381; Johnson, *A General History, Compendium Edition*, p. 270.

and Leader, having ten or twelve Men kill'd and wounded, and that there was no possibility of escaping,) called out for *Quarters*."[41] The rarity of the word treachery in this text makes the use of it here striking. The text does not construct a world where pirates are only interested in furthering their own interests. Instead, it creates narratives where selfishness and acting against the interests of the collective are robustly condemned. These editions continue with a key sentiment placed in parenthesis: "had *Low's* Sloop fought half so briskly as *Harris's* did, and had stood as close to the Business, (as they were under a solemn Oath to do,) the Man of War could never have hurted them."[42] This is a moment where the text reiterates that the pirates have made an oath to defend one another with all of their capabilities. The display of martial prowess in defence of the community is central to successful fellowship. It reminds us that oaths are the foundation of pirate community, because they have no other mechanism to ensure fidelity.

However, the abridgements alter these episodes. They omit Kennedy's treachery from the Roberts chapter and change key sections of the Low chapter. Low's behaviour is transformed into a moment of cowardice and personal failing, not a betrayal of fellowship. The abridgements omit the idea that Low leaves his companions as a "Sacrifice to his Enemy"; instead "the fight was renewed with brisk fire on both sides, till the Ranger's main-yard was shot down; upon which, the Greyhound pressing close, Lowe, in the other, bore away, and left his consort."[43] These editions then eliminate the word "treachery": "seeing the cowardice of his Commadore, and that there was no possibility of escaping, called out for quarters."[44] Finally, they remove the parenthesis noting the oaths between the two ships: "for had he fought half so briskly as Harris, the Man of War could never have taken either of them."[45] The abridgements thus minimise the collective, thereby amplifying the implications for individual character and the construction of the pirates as an individualist figure.

## Bad Fellows

The octavo and compendium editions amplify the significance of fellowship through the depiction of bad fellows. "Of Captain Avery and his Crew," in

41 Ibid.
42 Ibid.
43 *The History and Lives, 1st Edition*, p. 127; *The History and Lives, 3rd Edition*, p. 98.
44 *The History and Lives, 1st Edition*, p. 127; *The History and Lives, 3rd Edition*, p. 98.
45 *The History and Lives, 1st Edition*, p. 128; *The History and Lives, 3rd Edition*, p. 99.

particular, consistently centres bad fellowship across editions, save for the abridgements. This chapter is self-consciously constructed as a corrective to the myth that flourished after Avery's capture of the *Ganj-i-sawai*, "in which Avery becomes not only a pirate, but the governor of his own kingdom."[46] The narrator specifically positions the *General History*'s account as the "true" story, claiming that other stories "were no more than false Rumours."[47] Thus, the narrative begins by refuting prior accounts, which suggests that we should read Avery's chapter differently to others in the text. Burwick and Powell place this chapter within a broader "Avery backlash that began in the 1720s," which "may have been a response to the aggrandizement" of him in the 1700s and 1710s.[48] Meanwhile, Stephanie Jones argues that this chapter is presented as both a corrective of the prevailing myth of Avery and as a counter to the narratives that follow.[49] In presenting Avery as "a greedy egomaniac, the founder of a draconian monarchy doomed to failure" she argues that the *General History* "structures Every's story as an authenticating counterpoint to the story of the entirely fictional Misson."[50] I build upon Jones's analysis to develop a new approach to Avery. Fellowship is key to the pirate from this opening chapter in the octavos and compendium, and this narrative counters those that follow because it comprehensively depicts a bad fellow.

However, the abridgements mediate between the octavos' bad fellow and the Avery myth. The narrator's initial intervention is omitted. Instead, the chapter follows the conventions of criminal biography by describing Avery's parents, his childhood, and a series of escalating misadventures that end with him aboard the Duke.[51] While the Avery of the octavo and compendium editions never makes it to Madagascar, the abridgements' Avery follows the myth by landing and building a community.[52] In this new settlement Avery "began to form a Society, and made several Degrees of Officers under him, as if he had been a Sovereign."[53] However, unlike the myth, he does not remain ruler. Instead, like the octavo and compendium version, he dies penniless in Devon, cheated by a merchant he swears "was

46 Burwick and Powell, *British Pirates*, p. 29.

47 Johnson, *A General History, 1st Edition*, p. 26; Johnson, *A General History, 2nd Edition*, p. 46; Johnson, *A General History, Compendium Edition*, p. 197.

48 Burwick and Powell, *British Pirates*, p. 29.

49 Jones, "Literature, Geography, Law," p. 75.

50 Ibid., p. 75.

51 *The History and Lives, 1st Edition*, pp. 1–5; *The History and Lives, 3rd Edition*, pp. 1–4.

52 *The History and Lives, 1st Edition*, p. 14; *The History and Lives, 3rd Edition*, p. 10.

53 Ibid.

a greater Pirate than himself."[54] The abridgements embed signals of bad fellowship within a wider narrative that minimizes communality.

Fellowship is a key quality from the beginning of the Avery chapter in the octavos and compendium. The narrative begins when Avery "insinuated himself into the good Will of several of the boldest Fellows" then "at length proposed to them, to run away with the Ship, telling them what great Wealth was to be had upon the Coasts of *India*."[55] This crew is not mistreated. Their captain's only fault is that he is a drunk.[56] Instead, making a fortune precipitates mutinous action. Furthermore, the abridgements remove the language of fellowship and the idea of making the fortunes of a collective. Instead, this Avery joins the Ship: "where he had not been long before he observed the Captain was much addicted to drunkenness. He endeavoured to spirit up, not only his own ship's crew, but" also the crew aboard another ship.[57] This action is framed as an impulsive response to an opportunity afforded by poor leadership, where a charismatic leader exploits a weakness in a hierarchical system. The octavos and compendium privilege pre-meditation and personal acquisition, while the abridgements focus on impulsive opportunism.

Furthermore, the octavo and compendium editions imply a tension between Avery's individualism and the language of fellowship he deploys. In a monologue to his captain Avery declares:

> You must know, that I am Captain of this Ship now, and this is my Cabin, therefore you must walk out; I am bound to Madagascar, with a Design of making my own Fortune, and that of all the brave Fellows joined with me.[58]

Avery separates himself from his crew by taking the captain's cabin, whereas other narratives break down this separation by taking down cabins.[59] Crucially, the pronouns used by Avery contradict the performance of fellowship expressed in the use of "brave fellows." Avery's language elevates him above his companions: "*I* am Captain," "this is *my* cabin," "*I* am bound."[60] There is

54 *The History and Lives, 1st Edition*, p. 19; *The History and Lives, 3rd Edition*, p. 14.

55 Johnson, *A General History, 1st Edition*, p. 27; Johnson, *A General History, 2nd Edition*, p. 47; Johnson, *A General History, Compendium Edition*, p. 198.

56 Ibid.

57 *The History and Lives 1st Edition*, p. 5; *The History and Lives 3rd Edition*, p. 5.

58 Johnson, *A General History, 1st Edition*, p. 28; *A General History, 2nd Edition*, p. 48; *A General History, Compendium Edition*, p. 198.

59 Johnson, *A General History, 1st Edition*, p. 270.

60 Emphasis mine.

no "we" in Avery's mutiny, merely Avery and the men who follow him, even as they are called his fellows. In this moment the "I" is privileged in a way that it is not in other chapters, which do not offer direct speech between mutineers and those they mutiny against. Thus, Avery reveals his priorities by not using collective pronouns, even as he gestures towards fellowship.

The tension between ideals of fellowship and Avery's performance is amplified when Avery's company allies with two other ships. While "the Sloops Men were rejoiced at the new Ally," Avery's perspective eschews the language of fellowship, instead calling them: "this Reinforcement."[61] Avery's insincerity is amplified as the narrator foreshadows that "tho' the Booty must be lessened to each, by being divided into so many Shares, yet he found out an Expedient not to suffer by it himself."[62] Avery's bad fellowship is signalled in both his mutiny and in this formal fellowship. He deploys the signifiers of fellowship, yet concurrent language signals an undercurrent of individualism realised across the rest of the chapter.

In contrast, the abridgements omit the language of fellowship:

> Be not frightened, saith Avery, and I'll tell you. You must note, I am now Captain of this ship; nay you must turn out for this is my Cabin and I am bound to Madagascar to make my own Fortune as well as my Companions.[63]

The personal pronouns signal Avery's self-positioning, while his companions are placed as an afterthought. This interaction finishes with a declaration of purpose across chapters—making Avery's fortune is the priority, the benefit of the others coming second. However, the less penetrating account of fellowship constructed by the abridgements makes the tension between the individualist Avery and ideas of fellowship less cohesive. Similarly, when Avery agrees to sail with the two additional ships, he "promis'd them Protection."[64] These editions cast the relationship as an offer of safety, not fellowship. Avery's narrative initiates a shift in the profile of fellowship in the abridgements of the *General History*, which sees the pirate not as a fellow but as an individual.

The octavos and compendium further deploy ideas of treachery to amplify Avery's bad fellowship. The three ships successfully attack and plunder "rich Habits and Jewels, with Vessel of Gold and Silver, and great Sums of

61 Johnson, *A General History, 1st Edition*, p. 30; *A General History, 2nd Edition*, p. 50; *A General History, Compendium Edition*, p. 198.
62 Johnson, *A General History, 1st Edition*, p. 30; *A General History, 2nd Edition*, p. 50; *A General History, Compendium Edition*, p. 198.
63 *The History and Lives 1st Edition*, p. 6; *The History and Lives 3rd Edition*, p. 5.
64 *The History and Lives 1st Edition*, p. 9; *The History and Lives 3rd Edition*, p. 7.

Money" from the *Ganj-i-sawai*.[65] Avery then convinces his crew to cheat their companions:

> telling them they now had sufficient to make them all easy: and what should hinder them from going to some Country, where they were not known, and living on Shore all the rest of their Days in Plenty; they understood what he meant: in short, they all agreed to bilk their new Allies.[66]

The text is clear that this is a knowing attempt to betray allies by stealing the proceeds of their collective action. Avery further demonstrates his failings as a fellow by tricking his allies out of their share of the plunder. The narrator then amplifies the sense of bad fellowship: "nor do I find that any of them felt any Qualms of Honour, rising in his Stomach, to hinder them from consenting to this Piece of Treachery."[67] His interjection deploys the language of betrayal used across the text to emphasise Avery's lack of honour. Therefore, this chapter depicts bad fellowship as a counter to the good fellowship expressed elsewhere.

The abridgements reframe Avery as a mock king, rather than a bad fellow. They omit references to treachery in their description of Avery's manipulations:

> by telling them, that they had nothing to fear, the Treasure now in their won Hands; and that if they would stand by him, he did not doubt but to order it so as to cheat the Sloops crew of their part, but also to make them subservient to them in their new intended settlement [...] for having secured the Men and Sloops, he strengthned his Alliance, and was hereby prepared against any Thing the others could offer.[68]

This moment allows Avery to strengthen his relationship with those aboard his ship, which in turn reinforces his position in the new settlement. The ability to successfully cheat their companions enables Avery and his crew to establish dominance over their allies. The abridgements therefore create an Avery concerned with fealty and personal relations as they mediate the

65 Johnson, *A General History, 1st Edition*, p. 31; *A General History, 2nd Edition*, p. 51; *A General History, Compendium Edition*, pp. 198–99.

66 Johnson, *A General History, 1st Edition*, pp. 32–33; *A General History, 2nd Edition*, pp. 52–53; *A General History, Compendium Edition*, p. 199.

67 Johnson, *A General History, 1st Edition*, p. 33; *A General History, 2nd Edition*, p. 53; *A General History, Compendium Edition*, p. 199.

68 *The History and Lives, 1st Edition*, p. 14; *The History and Lives, 3rd Edition*, p. 10.

Avery myth and the octavo iterations. He continues the deception as they establish a settlement: "As fast as they got ashore Avery put them to work upon his new Fortifications; and if any of them began to talk of a Division of the Treasure, he told them they had not Time at present."[69] Avery's influence upon his crew is so pervasive "that when two of the Mates began only to murmur, they were both hung up at the mast to frighten others from such attempts."[70] The pirate company then develops a hierarchy that solidifies Avery's charismatic leadership of the group as he "made several degrees of officer under him, as if he had been a Sovereign."[71] Their hierarchy is entrenched in the establishment of "degrees of officer," a middle layer of social organisation that distances Avery from "the common seamen" under him.[72] The whole society is structured around one character's leadership and personal desires. Advancement is based on Avery's preference of "those that he judged most in his interest" not on the good of the collective or any kind of homosociality.[73] Avery models himself as a king, not the leader of a fellowship.

The *General History* follows Avery's bad fellow with similar moments across other chapters. However, these are imagined as moments of crisis, not as indicative of a norm. The first octavo offers selected episodes where Roberts fails to live up to the standards of a "worthy fellow." Firstly, it describes how "the Brigantine who had hitherto lived with them, in all amicable Correspondence, and of whom they had no Mistrust, thought fit to take the Opportunity of a dark Night and leave the Commadore."[74] This moment echoes Kennedy's betrayal, discussed above. Both splinter groups run away surreptitiously, taking a ship. However, Kennedy's behaviour is a betrayal, while this action is neutrally called "the Separation of Bradley."[75]

Roberts' treatment of his fellows initiates this shift in language. The separation is

> occasioned it's thought from the Haughty and Magisterial Behaviour of *Roberts* towards them, who regarded the Brigantine only as a Tender, and as such left them no more than the Refuse of their Plunder; however, he put on a Face of Unconcern, at this his ill Conduct and Mismanagement.[76]

69 Ibid.
70 Ibid.
71 Ibid.
72 Ibid.
73 Ibid.
74 Johnson, *A General History, 1st Edition*, p. 187.
75 Ibid., p. 193.
76 Ibid., pp. 187–88.

His "Haughty and Magisterial Behaviour" directly contrasts with the perception of him as a "worthy fellow." Roberts demonstrates bad fellowship by denying those aboard the brigantine equal status. By treating them as a "tender" he affords them the status of a minor ship in service to the main ship, rather than a pirate ship in its own right. Crucially, he denies them their fair share of the plunder; an act the text calls "ill Conduct and Mismanagement." Distribution of plunder is a key concern across moments of broken fellowship. Roberts is a bad fellow here, as he fails to afford them the benefits and respect the "Articles" entitle them to. They are therefore under no obligation to reciprocate with good fellowship. Bradley and his companions do not betray the company because Roberts has already broken the fellowship.

The second octavo rewrites this episode, by replacing Bradley with Captain Anstis and adding extra detail. This change shifts the status of the captain, but for present purposes what is crucial is that the heart of this episode is consistent even as the detail changes. Anstis's complaints are identical to Bradley's: "*Roberts*, who carried himself with a haughty and magisterial Air, to [Anstis] and his Crew, he regarding the Brigantine only as a Tender, and, as such, left them no more than the Refuse of their Plunder."[77] Roberts' treatment of the Brigantine ultimately causes the split. Both the first and second octavos emphasise that covertly splitting a company is not a betrayal when good fellowship is not practised by all members.

In contrast, the abridgements focus on the individuals. They condense the second octavo's iteration, calling it "an Accident that happened at the West Indies, which had like to have ruined all, and was the occasion of the Brigantine's going off."[78] It describes an altercation between Roberts and a crewmember named Brag. As a consequence, Brag and his companions "conspired with *Anstis*, Captain of the Brigantine, who also a Malecontent on account of *Roberts*'s Haughty Carriage towards him."[79] The phrasing of this episode transforms a bad fellowship into a conflict of personalities. In pointing to "Robert's Haughty Carriage towards" Anstis the abridgement ties the two grievances together to make the separation personal. It suggests that Roberts' behaviour to the captain of the subordinate sloop causes the split, rather than his treatment of those on the secondary ship. In the absence of an overarching account of fellowship and without Kennedy's betrayal this moment becomes a warning about the excess of the captain, rather than a contrasting episode that demonstrates the importance of good fellowship performed by all members.

77 Johnson, *A General History, 2nd Edition*, p. 249.

78 *The History and Lives, 1st Edition*, p. 96; *The History and Lives, 3rd Edition*, p. 74.

79 *The History and Lives, 1st Edition*, p. 97; *The History and Lives, 3rd Edition*, p. 74.

## Conclusion

While "fellowship" is not specifically used to describe pirate homosociality in the *General History*, the sustained depiction of pirates as fellows and the imagining of the archetypal pirate as a "worthy" fellow, combined with the use of terms associated with fellowship in broadly contemporary definitions, make fellowship a critically resonant collective noun to use. The *General History* imagines the pirate as a figure fully embedded in a homosocial community structured around fidelity and reliant on the value of one's word.

By bringing ideas of fellowship to the surface of the text in moments of betrayal and bad fellowship, the *General History* strains the central premise of fellowship—that oaths rely on the value of one's word. However, these are moments of crisis in the text, not a norm. The text does not create pirate communities that are rife with insincerity. The octavo editions particularly deploy a very specific language to demonstrate how fidelity is the norm aboard the pirate ship.

The profile of fellowship varies across the ten years explored in this chapter, however. Ideas of fellowship and being a worthy fellow are more firmly articulated in the second and subsequent octavos. In contrast, the abridgements diminish the profile of fellowship by amending and condensing parts of the narrative. In doing so, the abridgements remake the pirates of the *General History* to conform to the individualisation which is at the core of criminal biography in this period. Comparing across editions demonstrates how the pirates in the octavo editions of the *General History* engage with older models of masculine relations, by creating and maintaining fellowships.

## Bibliography

Archibald, Elizabeth. "Malory's Ideal of Fellowship." *The Review of English Studies* 43, no. 171 (1992): 311–28.

Baer, Joel. "Bold Captain Avery in the Privy Council: Early Variants of a Broadside Ballad from the Pepys Collection." *Folk Music Journal* 7, no. 1 (1995): 4–26.

Beselke, Brandi, ed. *A Dictionary of English Language: A Digital Edition of the 1755 Classic by Samuel Johnson.* Last modified December 6, 2012. https://johnsonsdictionaryonline.com.

Burwick, Frederick and Manushag N. Powell. *British Pirates in Print and Performance*. Basingstoke: Palgrave Macmillan, 2015.

Faller, Lincoln B. "Criminal Opportunities in the Eighteenth Century: The 'Ready-Made' Contexts of the Popular Literature of Crime." *Comparative Literature Studies* 24, no. 2 (1987): 120–45.

Faller, Lincoln B. *Turned to Account: The Forms and Functions of Criminal Biography in Late Seventeenth- and Early Eighteenth-Century England*. Cambridge: Cambridge University Press, 1987.

*The History and Lives Of All the Most Notorious Pirates, and Their Crews*, 1st ed. London: Edward Midwinter, 1725.

*The History and Lives Of All the Most Notorious Pirates, and Their Crews*, 3rd ed. London: Edward Midwinter, 1729.

Johnson, Captain Charles. *A General History of the Pyrates from Their First Rise and Settlement, to the Present Time*, 2nd ed. London: Thomas Warner, 1724.

Johnson, Captain Charles. *A General History of the Robberies and Murders of the Most Notorious Pyrates, and Also Their Policies, Discipline and Government, From Their First Rise and Settlement in the Island of Providence, in 1717, to the Present Year 1724*, 1st ed. London: Charles Rivington, J. Stone, J. Lacey, 1724.

Johnson, Captain Charles. *A General History of the Lives and Adventures of the Most Famous Highwaymen, Murderers, Street-Robbers, &c. To Which Is Added, A Genuine Account of the Voyages and Plunders of the Most Notorious Pirates*. London: J. Janeway, 1734.

Jones, Stephanie. "Literature, Geography, Law: The Life and Adventures of Capt. John Avery (circa 1709)." *Cultural Geographies* 19, no. 1 (2012): 71–86.

Lincoln, Margarette. *British Pirates and Society, 1680–1730*. Farnham: Ashgate Publishing Limited, 2014.

Pollard, A J. *Imagining Robin Hood: The Late-Medieval Stories in Historical Context*. Abingdon: Routledge, 2004.

Rennie, Neil. *Treasure Neverland: Real and Imaginary Pirates*. Oxford: Oxford University Press, 2013.

Richetti, John J. *Popular Fiction Before Richardson: Narrative Patterns: 1700–1739*, 2nd ed. Oxford: Oxford University Press, 1992.

Rodger, Nicholas A. M. *The Wooden World: An Anatomy of the Georgian Navy*. London: Fontana Press, 1988.

Singleton, Robert R. "English Criminal Biography, 1651–1722." *Harvard Library Bulletin* 18, no. 1 (1970): 63–83.

Singman, Jeffrey. *Robin Hood: The Shaping of the Legend*. Westport: Greenwood Press, 1998.

## About the Author

**Rebecca James** completed her PhD in English Literature at the University of Southampton and Cardiff University in 2021. Her thesis, titled 'The Captain and his Fellows: Reading Editions of *A General History of the Pyrates*, 1724–1734' examines the representation of the pirate across early editions of *The General History of the Pyrates*. This research was generously supported by the AHRC's South, West and Wales Doctoral Training Partnership, and is the source of this chapter.

# 7. Henry Glasby: Atypical Pirate or a Typical Pirate?

*James Rankine*

**Abstract**

The pirates of the early eighteenth century fascinated their contemporaries and continue to enthral generations. Historical studies of pirates have prioritised the same influential rogues' gallery, but we rarely examine the lived experience of the thousands of pirates who avoided notoriety. This chapter seeks to reassess the "Golden Age" of piracy (1716–1726) through the career of Henry Glasby. Glasby's tenure as a pirate saw him progress from captive prisoner, to elected officer, to turncoat who served King's evidence. Although men like his commander, Bartholomew Roberts, are seen as emblematic of pirate life, Glasby's trajectory offers a contrasting portrait. Glasby's time aboard reveals a crew riven by division, ambivalence, and hierarchy. His dramatic exit from piracy illustrates that his experience was common.

**Keywords:** Atlantic; Authority; Desertion; Eighteenth Century; Golden Age; Piracy; Social dynamics

In the summer of 1720, Henry Glasby was the chief mate of the *Samuel*, a merchant vessel bound from London to Boston.[1] On this particular voyage, however, Glasby's path crossed with that of one of the most prolific pirates of the early eighteenth century, Bartholomew Roberts. Forced aboard the pirates' vessel, Glasby spent the next eighteen months with Roberts' crew.

1 I have used Glasby's name as spelled in the trial transcripts. However, Glasby's surname was rendered a few different ways in contemporary sources, and it is likely that Glasby was a misspelled or misheard version of Gillespie. Glasby's commanding officer identified him as "Gilespy" in an advert published soon after his capture in the *Boston Gazette*. His description of Glasby as a "North Britains" suggests he was likely Scottish. *Boston Gazette*, August 22, 1720, p. 3.

Coakley, J., C. Nathan Kwan and D. Wilson, *The Problem of Piracy in the Early Modern World: Maritime Predation, Empire, and the Construction of Authority at Sea*.
Taylor & Francis Group, 2024
DOI 10.5117/9789463720960_CH07

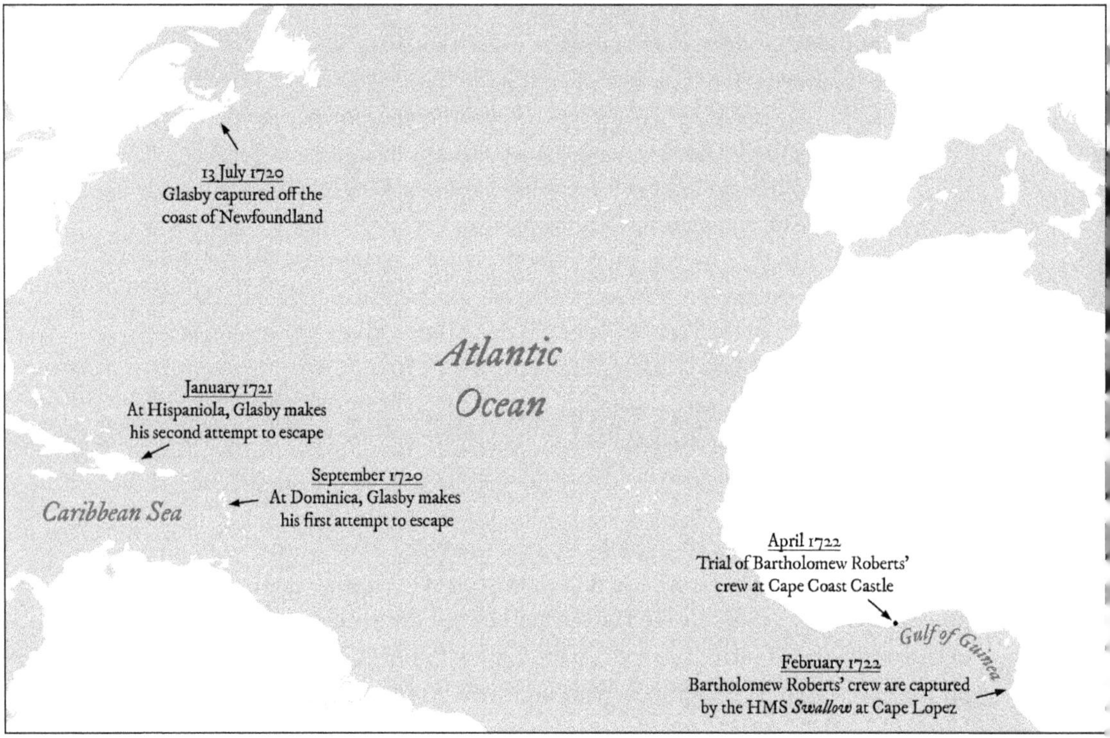
13 July 1720
Glasby captured off the coast of Newfoundland
Atlantic Ocean
January 1721
At Hispaniola, Glasby makes his second attempt to escape
September 1720
At Dominica, Glasby makes his first attempt to escape
Caribbean Sea
April 1722
Trial of Bartholomew Roberts' crew at Cape Coast Castle
Gulf of Guinea
February 1722
Bartholomew Roberts' crew are captured by the HMS *Swallow* at Cape Lopez

Glasby's experience of piracy was fraught with ambivalence, conflict, and deception. After his forcible integration into the crew, Glasby eventually rose to become the master of Roberts' flagship the *Royal Fortune*. During that same time, however, he made several concerted efforts to desert, plotted with prisoners to escape, and harboured deep misgivings about his involvement in piracy. In the critical moments after Roberts' death in battle, Glasby emerged as a crucial leader within the crew, orchestrating a coup to surrender the vessel peacefully. After being acquitted at trial, Glasby went on to serve as the key witness in the trial of scores of his former ship mates, proceedings that revealed the hidden inner workings of the crew he had lived and worked in for well over a year.

Henry Glasby was a relatively unknown and unimportant figure in comparison to his commander. Bartholomew Roberts' open embrace of criminality, his spectacular run of successful attacks, and his violent, defiant death have made him an emblematic figure of early eighteenth-century piracy.[2] Indeed, his seminal work, *The General History of the Pyrates*, Charles Johnson credited Roberts with the apocryphal and macabre motto of "a merry life and a short one" that has long served as a distillation of early eighteenth-century pirates' ethos.[3] However, Glasby's pirating career provides a stark contrast to Roberts' more well-known experience, and illustrates that rank and file crew were more diverse and very different from their notorious commanders.

Since Johnson, however, pirate crews have often been viewed as monolithic extensions of their commanders. Marcus Rediker has argued that pirate crews represented a radical challenge to Britain's emerging commercial empire, and that crews were maritime "social bandits" organised through

2 Charles Johnson suggested that Roberts managed to seize over 400 prizes in his two short years of pirating. This was almost undoubtedly an exaggeration and a comprehensive review of contemporary newspaper reports suggest the real number was probably less than a hundred. Even if Roberts managed a hundred ships, this was still an astoundingly successful run, but it is also an important reminder of the way contemporary accounts of Roberts and pirates like him often casually exaggerated their exploits. Nevertheless, Johnson's figure has often found its way into modern scholarship on early eighteenth-century piracy. Johnson, *General History*, pp. 5, 704; Rediker, *Villains of All Nations*, p. 33; Baer, *Pirates*, p. 203; Cordingly, *Under the Black Flag*, p. 109. For sampling of reporting on Roberts' career in the press see *Boston Gazette*, August 29, 1720, p. 4; *Daily Post*, December 3, 1720, p. 1; *Applebee's Original Weekly Journal*, February 4, 1721, p. 2; *Post Boy*, July 13, 1721, p. 2; *London Journal*, December 30, 1721, p. 4.

3 Despite its popularity, this quote ought to be treated with healthy skepticism. Charles Johnson attributed the utterance to Roberts, but like many details and dialogue in the *General History*, there is no way of determining whether Roberts said or even believed this sentiment, or if the words were placed in his mouth posthumously by Johnson for rhetorical effect. Johnson, *General History*, p. 244.

collective solidarity and egalitarianism.[4] More recently, Peter Leeson has taken the opposite view, arguing that pirates were shrewd capitalists organising to optimise profitable plunder in an extra-legal environment.[5] Although diametrically opposed, both perspectives approach pirate crews as composed of enthusiastic volunteers or hapless coerced prisoners. Glasby, however, does not fit neatly into either of these categories, and nor did many of his shipmates. Likewise, scholars such as Peter Earle and Mark Hanna have noted that as suppression efforts intensified, pirates relied more heavily on forced labour and impressment to compensate for their increasingly desperate situation.[6] However, we know little about how those aboard pirate ships engaged with these changing conditions and how it affected the dynamics of crew organisation and relationships. Glasby's experiences, and his extensive testimony at trial, however, offer a unique opportunity to examine more closely the inner workings of a pirate crew during this crucial period, and to offer a more complex and nuanced account of the rank-and-file pirates who represented the vast majority of early eighteenth-century pirates.

## Recruitment

Henry Glasby seemed an unlikely candidate for a life of piracy. Pirates were reputed to target young, unmarried men, seeking volunteers from among those frustrated and disaffected by their poor working conditions and marginalisation.[7] Glasby was none of these things. On the contrary he was an officer with a wife, a steady wage, and gainful employment. Yet the pirating life came to Glasby nonetheless, in the form of Bartholomew Roberts and his crew. Roberts' company came across the *Samuel* on the July 13, at the tail end of a lengthy spree of attacks off the coast of Newfoundland.[8]

4 Rediker, "'Under the Banner'"; Rediker, *Deep Blue Sea*; Rediker, "Liberty"; Rediker, "Hydrarchy and Libertalia"; Linebaugh and Rediker, *Many-Headed Hydra*; Rediker, "Seaman as Pirate"; Rediker, *Villains of All Nations*; Rediker, *Outlaws*. See also Hill, "Radical Pirates?"

5 Leeson developed this interpretive framework as early as 2007, and has continued to develop a theory of pirates as anarcho-capitalist forerunners. Leeson, "An-*arrgh*-chy"; Leeson, *Invisible Hook*; Leeson, *Anarchy Unbound*, pp. 105–31. A significant forerunner was Cyrus Karraker's work in the 1950s. Karraker, *Piracy*. Leeson takes the economic interpretation to its most extreme, but historians have made this argument before. Lane, *Pillaging Empire*, pp. 7–8; Starkey, "Pirates and Markets."

6 Earle, *Pirate Wars*, pp. 173–74, 207–8; Hanna, *Pirate Nests*, pp. 385–66.

7 Rediker, *Villains of All Nations*, pp. 42–50; Earle, *Pirate Wars*, pp. 166–69.

8 Captain Samuel Carey's reported position, admittedly limited by contemporary navigational instruments, put the *Samuel* about thirty or forty leagues east of the coast at latitude of 43°49'

After firing a warning shot, the pirates housed their macabre colours with a "Deaths Head and Cutlass in it." The *Samuel*, with a ten-man crew and six mounted guns, stood no chance against the Roberts gang, who outnumbered them ten to one in manpower, and six to one in firepower across two vessels.[9] Samuel Carey, the commander of the merchant vessel had no choice but to surrender.

The pirates spent the next forty-eight hours looting the *Samuel* of anything of value. Immediately after boarding, the pirates first robbed the crew, stripping them of their valuables and clothing at gunpoint. They then turned their attention to the *Samuel*'s holds, which they entered "like a parcel of Furies" and proceeded to thoroughly ransack.[10] Armed with axes and cutlasses, the pirates smashed their way into the boxes and barrels of goods, allegedly even opening chests by blowing the locks open with pistols. Anything of value they ferried to their own vessels, and whatever did not interest them, they simply cast into the sea.[11] The pirates even rooted out hampers of fine wines, possibly a gift sent to a wealthy Bostonian, and proceeded to use their cutlasses to cut the tops off the bottles as they "made themselves very merry."[12]

The *Boston News-Letter* account of the attack also dwelt upon the pirates' gleeful breaches of social norms. During the two-day plundering the *Samuel*, the pirates routinely cursed and swore, uttered blasphemous remarks, and openly mocked the King of Britain.[13] Perhaps most ominous of all was their alleged insistence that they would never be captured alive. If they ever faced the prospect of capture by a "Superiour power," the pirates allegedly remarked to prisoners, they would not wait to be hanged, but rather "put fire with one of their Pistolls to their Powder, and go all merrily to Hell together."[14] These anecdotes paint a familiar picture of early

north. This would place the vessel somewhere between Newfoundland and Nova Scotia. *Boston News-Letter*, August 22, 1720, p. 4; *Boston Gazette*, August 22, 1720, p. 3.

9 Captain Carey described Roberts' flagship at the time as being roughly 220 tons with twenty-six guns, accompanied by a more nimble eighty ton sloop with ten guns. *Boston Gazette*, August 22, 1720, p. 3.

10 *Boston News-Letter*, August 22, 1720, p. 4.

11 Carey alleged that the full value of what was stolen could only be estimated at around nine or ten thousand pounds, and noted that in addition to much of the cargo, Roberts' crew also stripped the *Samuel* of all her spare rigging and stores, cables, two mounted guns, as well as the ship's boat. *Boston News-Letter*, August 22, 1720, p. 4; *Boston Gazette*, August 22, 1720, p. 3.

12 *Boston News-Letter*, August 22, 1720, p. 4.

13 Ibid.

14 These anecdotes appeared in the *Boston News-Letter* account of the incident, but not Carey's statement taken before the city's public notary and published in the *Boston Gazette* on the same

eighteenth-century pirate crews' raucous behaviour, dark fatalism, and violent tendencies, but the accuracy of such reports must be viewed with caution. As Mark Hanna has pointed out, the emerging institution of the popular press played a central role in legitimating the suppression of pirates and constructing the imagery of pirates as aberrant and dangerous.[15] Indeed, Captain Carey's statement on the attack, published on the same day in the *Boston Gazette* contains none of these colourful details.[16]

Both accounts, however, agreed that after stripping the *Samuel* of cargo and supplies, Roberts and his men stole one last essential asset: the crew. With pistols once more pressed to their ribs, the sailors of the *Samuel* were forced onto the pirates' vessels.[17] As Glasby discovered, those who resisted this process faced violent coercion. Rightly fearing that he too would be forced to join the pirates' crew, Glasby hid himself away below deck. The pirates, undeterred, searched the ship until they found Glasby, dragged him from his hiding placed, gave him a sound beating and then threw him overboard for his impertinence. Despite Glasby's obvious reluctance, however, they soon fished him out of the water and impressed him into Roberts' crew anyway.[18] For Glasby, at least, his induction into pirating had not been that of a volunteer, and certainly not an embrace of the criminal status that some of his captors had displayed during the attack. Rather, it was an unwelcome, emergent situation into which he had been thrown, figuratively and literally, and that he was no obligated to survive however best he could.

By 1720, Glasby's experience was by no means out of the ordinary. The tacit support in colonial communities that had sustained pirates in the late seventeenth century had all but evaporated, while efforts to rein in unsanctioned maritime violence in both London and the colonies had intensified significantly. Pirate crews had been pushed to the margins of the Atlantic world, the risks had grown and the rewards had dwindled.[19] The Roberts crew, and pirates like them had increasingly begun to supplement the dwindling pool of willing volunteers with "forced men." Even Roberts

day. It is possible that they may have been embellishments by the publisher, but the sentiments were certainly not out of keeping with the crew's established reputation by that point. Ibid.

15 Hanna, *Pirate Nests*, pp. 377–83.

16 *Boston Gazette*, August 22, 1720, p. 3.

17 Both reports, perhaps indulging in hyperbole, specifically mentioned that the prisoners were led away at gunpoint. Ibid.; *Boston News-Letter*, August 22, 1720, p. 4.

18 The National Archives, UK (TNA), High Court Admiralty Papers (HCA) 1/99/3, f. 52. *The Tryals of the Pirates Lately Taken By his Majesty's Ship the Swallow*, March 28–April 20, 1721.

19 Hanna, *Pirate Nests*, p. 385.

himself had initially been drawn into pirating under similar circumstances. A year prior to his encounter with Glasby, Roberts was the third officer of the *Princess* a legitimate merchant vessel. When he came face to face with Howell Davis' pirate crew off the coast of Africa in 1719, however, Roberts was forcibly detained and pressed into the pirates' service.[20] To what extent and for how long Roberts was a coerced prisoner are difficult to ascertain, but he appears to have quickly overcome any compunctions about participating in piracy.[21] Shortly after his capture, locals, enraged at the atrocities committed against them by the pirates, killed Davis and several of his crew on the island of Principe. The surviving crew chose Roberts to be their new captain. He may have already emerged as a respected and powerful figure within the group, or he may simply have been selected for his navigational expertise as a former officer.

Unlike Roberts, however, Glasby did not make the transition from prisoner to pirate quite so quickly or easily. His early days aboard Roberts' ship were characterised by a marked resistance to becoming a fully-fledged pirate. For seven days, Glasby refused to affix his signature to the crew's articles, a document that pledged the signatories to a code of conduct and initiated them as part of the crew proper rather than a mere prisoner. Glasby's reluctance was understandable. His signature on such a document would have been incriminating. Mark Hanna has pointed out that although they resembled similar documents amongst late seventeenth-century privateers,

20 Captain Abraham Plumb (aka Plummer) gave a full deposition of the attack upon his return to England. Plumb testifies that he had heard, second hand, that all the men who remained with Davis did so voluntarily, and that Davis himself had explicitly stated that no men were compelled into his service. This suggests that Roberts may well have been a volunteer from the beginning, but then again, Roberts made similar claims about his own crew to witnesses, despite the fact that this was manifestly not the case. TNA, HCA 1/17, p. 154. Information of Abraham Plumb Against John Haswell. For Roberts making similar claims of his crew being entirely volunteers, see TNA, HCA 1/99/3, p. 112.

21 Plumb testified that he had heard, second hand, that all the men who remained with Davis did so voluntarily, and that Davis himself had explicitly stated that no men were compelled into his service. A veteran member of Davis' crew, Rodger Hughs testified that he was unsure whether "they were forced or took on voluntarily," an expression that implies that the matter was not as cut and dried as Davis' statement suggested. Roberts himself apparently boasted that he had only cried "crocodile tears" when he was first taken. This suggests that he may well have been a volunteer, or at least open to joining the pirates from the beginning. At the same time, Roberts may have simply been boasting or refashioning his past to fit his piratical persona. Witnesses reported that Roberts, like Davis, also claimed that he did not take men against their will, despite the fact that this was manifestly not the case. TNA, HCA 1/17, p. 154. Information of Abraham Plumb; National Register of Archives for Scotland (NRAS) 1209-116. Affidavit of Roger Hughes, April 18, 1720. For an example of Roberts making similar claims see TNA, HCA 1/99/3 p. 112.

pirates' articles functioned differently in the early eighteenth century, granting commanders leverage over crews in a situation where discipline was difficult to enforce.[22] As a recruitment tool, however, articles also imposed mutual incrimination of new members of the crew, overcoming their reluctance and pushing them to embrace rather than resist their situation. Thus, when Glasby did resist, the pirates swiftly reacted with violent intimidation, and having been "cut and abused very much," Glasby ultimately succumbed to the pressure.[23] In signing, he became, on paper at least, a member of Roberts crew, and he spent the next eighteen months living and working as a pirate amongst pirates.

Yet neither Glasby's signing of the articles, nor his participation in the labour of pirating constituted a genuine commitment to a merry, short life of piracy. Although he complied outwardly, Glasby secretly hoped to find a way out of his new pirate identity. Months after his initial capture and integration into the crew, he made his first attempt to desert in the West Indies. Accompanied by two others slipped into the trees while the crew was ashore, intending never to return. Unfortunately for the luckless trio, the pirates quickly tracked down and recaptured them. The three were then subjected to a trial, of sorts, with the jurors casually reaching a verdict and a death sentence "over a bowl of Punch."[24] Glasby's two companions were immediately shot, and so too would Glasby have been executed, he claimed, if not for the timely intercession of Valentin Ashplant, one of the "leading Pirates" who had befriended him.[25] Having narrowly escaped execution, it would have been understandable if Glasby abandoned any thought of desertion. Yet soon afterwards, he made a second risky attempt. This time with a group of around ten, Glasby crept into the jungle on a remote stretch of Hispaniola's coast with nothing more than a compass to guide them. The attempt fell apart quickly as the group became disoriented in the island's rugged interior and returned to the beach where the pirates discovered them. This time, thankfully, they managed to convince their shipmates that they had simply become lost in the jungle by accident.[26]

22 Hanna, *Pirate Nests*, pp. 385–86.

23 TNA, HCA 1/99/3, p. 52.

24 According to the testimony of Glasby at the trial, this first escape attempt likely occurred at Dominica around September, 1720. TNA, HCA 1/99/3, p. 125.

25 Valentine Ashplant remained with the crew and was captured alongside Glasby. He was convicted and executed at Cape Coast Castle on the April 3, 1722. TNA, HCA 1/99/3, pp. 52, 125–26, 181.

26 Glasby does not reveal the precise location of this botched escape, nor the precise time. Attacks attributed to Roberts were recorded in the Caribbean throughout the latter half of

That Glasby's multiple attempts to desert were not simply the act of a lone dissenter is also significant. In both cases, Glasby acted alongside others as part of a determined group of trusted allies. Indeed, some of those same men were still amongst the crew a year later when they came to face trial. Glasby's testimony about these shipmates revealed that the escape was not simply a spontaneous or opportunistic decision, but rather a carefully planned and coordinated plot developed amongst the men in secret. George Smith, Glasby noted, had "several times spoken of some Contrivance to get away." Another William Taylor, had "even to the time of their being taken [...] had almost every Night bemoaned himself with him [Glasby]."[27] One of them, Thomas Wills, remained close to Glasby for months afterwards, and never gave up on his hopes of escaping. Glasby testified that he and Wills had hatched an ambitious plan between themselves "and a few other discontented men" to get the crew of Roberts' consort vessel, the *Ranger*, so drunk they could seize control of the ship and flee.[28] In other words, Glasby's ambivalent status was not unique within Roberts' crew, there were others just like him within the crew who maintained the appearance of pirates while actively planning to escape in an opportunity arose. Just like Glasby, they too graduated from prisoners to pirates, integrating into the crew and playing their part, but never fully committing to a life of piracy.

Glasby's initiation into Roberts' crew paints a very different picture from Roberts' tenure under Davis. Indeed, the conflict and ambivalence of Glasby's recruitment are an apt illustration of how the collapse of pirates' network of support and market access as illuminated out by the scholarship of historians such as Mark Hanna and Douglas Burgess dramatically

1720 and the early months of 1721. A likely candidate for the escape attempt is Samana Bay on the northern coast of what is now the Dominican Republic. Three Dutch sailors, captured by Roberts at Dominica in January, 1721, later testified in an affidavit before the colonial council of New York that Roberts had spent weeks in the cove, careening the *Fortune* and conducting illicit trade with various contacts. TNA, HCA 1/55/3, pp. 52, 125–26; New York State Archives, A1895-78, vol. 13, pp. 77–81. Affadavit of Claus Benning, Bastin de Keyser, and George du Ref.

27 Curiously, both Smith and Taylor were among the small group of men convicted in spite of Glasby's support. Smith, witnesses testified, had eventually abandoned his hopes of escape and embraced a more amicable relationship with the crew. Two Dutch captains identified Smith specifically as having participated in plundering their ships off the West African coast. Taylor, on the other hand, was less committed. Another member of the crew noted that Taylor often "lament about a Wife and Children he had left at Bristol" and described him as "being sober and given to reading." What appears to have doomed Taylor was the fact that one witness saw him firing a gun at the *Swallow* during the final battle, a fact Taylor confessed to, though he claimed to have been both drunk and compelled at the time. TNA, HCA 1/99/3, pp. 125, 126, 135.

28 TNA, HCA 1/99/3, p. 151.

transformed relations within pirate crews.[29] To be sure, pirates used forcible recruitment to address shortages in manpower or expertise during this period, but these practical needs offer little explanation for why they also cajoled and coerced unwilling recruits into signing their articles. Glasby's experience, however, reveals why the pirates remained so peculiarly insistent that unwilling recruits sign their articles in the same way that enthusiastic volunteers did. By coercing Glasby into signing the document, Roberts and his crew had ensured that he was now deeply implicated, by his own hand, in a criminal conspiracy. This was unnecessary in order to extract labour from Glasby or any other recruit, but enforced culpability was crucial in pressuring prisoners like him into becoming pirates. Thus, the growing pattern of forcible detainment was more than simply a pragmatic response to marginalisation, but part of a canny and novel strategy to overcome the reluctance of mariners to volunteer, one that turned the growing danger of prosecution into a powerful tool of recruitment and discipline.

Glasby's early months in Roberts' crew also underscore how the traditional dichotomy between genuine pirates and hapless prisoners obscures a more nuanced and complex set of relations aboard pirate ships. Glasby made repeated and risky attempts to desert during the same period that he also integrated into the crew, living and working alongside pirates. Undoubtedly, his desire to leave pirating behind and rejoin society was very real, but so too were the working relationships he developed aboard the *Fortune* with even hardened veterans like Valentin Ashplant. The fact that Glasby was joined in these attempts by others also demonstrates that he was by no means alone in his ambivalent status. Others within the crew found themselves in the perilous middle ground between coerced prisoners and committed pirates. Glasby and the others like him developed their own strategies to survive in a very dangerous situation. In the most immediate sense, that meant forging security and gaining respect within the crew, working hard, looking the part, avoiding conflict and developing important allies. Indeed, Glasby's success in cultivating his status within the crew saved him from execution. Yet this status exposed him to significant legal peril, and in the long term, Glasby needed to escape not just from his pirate peers, but from the very role he had created in order to survive amongst his peers. Just as the pirates' recruitment strategy was more sophisticated and successful than mere brute force, so too was the careful positioning of Glasby and his companions in the ambiguous middle ground between pirate and prisoner a resourceful and inventive response.

29 Burgess, *Politics of Piracy*, pp. 200–226; Hanna, *Pirate Nests*, pp. 366–69.

## Survival Strategies and Interpersonal Networks

Although Glasby had gained some security as part of Roberts' crew, his situation grew increasingly dangerous over time. As days stretched into weeks and then months, pressure mounted on Glasby to accept that he could no longer return to his former life. A mariner who spent a few weeks aboard a pirate vessel could plausibly claim to have merely served as a forced man, but eventually a member of the crew would come to be viewed as an "old stander" not only by their ship mates, but also by an Admiralty court.[30] After capture, several members of Roberts' crew were convicted largely on the basis of the length of their time aboard rather than on testimony regarding any specific criminal actions. No witnesses testified that Roger Scot had been directly implicated in thefts or violence, for instance, and Scot claimed to have been forced into the crew just as Glasby had. Nevertheless, the court "thought that an Offender of his Standing must certainly be Guilty."[31] By 1721, Glasby's lengthy tenure with the crew had forced him to integrate more fully into its ranks. Indeed, in the final months of his pirate career Glasby served as the master of the *Royal Fortune*.

Precisely what Glasby's role as master entailed was somewhat murky. As a former officer, he certainly had the requisite skills and knowledge to perform the master's traditional role of overseeing sailing and navigation, but testimony at trial revealed that his office did not carry the authority that a master of a legitimate merchant or naval vessel would hold.[32] Glasby's rank underscored that he was not a pariah in Roberts' company, but it did not make him a "leading man" like Valentin Ashplant either.[33] Several witnesses described how some members of the crew held little regard for

30 The court that eventually tried Roberts explicitly referred to the length of time that some members of the crew had served as they condemned them, such as David Simspon whose "Plea of Force" was batted aside because "The Court knowing him to be an old offender ... very readily found him Guilty." Similar remarks regarding longstanding members were made while convicting others including Christopher Moody and Thomas Sutton. TNA, HCA 1/55/3 fols. 34, 36, 37.

31 This is one of the many small differences between the manuscript version of the trial of Roberts' crew and the printed version published a year later. The substance of both versions is largely the same, but the latter contains a number of small edits and changes that sharpen or heighten the guilt of the convicted and the innocence of the acquitted. In Scot's case, the printed version of the same line adds in "old" reading: "The Court thought, that an *old* Offender of his standing, must certainly be Guilty." I have used the manuscript version throughout this chapter. TNA, HCA 1/55/3 f. 46; *A Full Account*, p. 19. See also the cases of James Greenham and James Harris. TNA, HCA 1/55/3, fols. 73, 76.

32 On a standard merchant vessel the master held almost absolute power over the rest of the crew. Rediker, *Deep Blue Sea*, p. 84.

33 TNA, HCA 1/55/3 f. 125.

Glasby despite the fact that he "acted as Master of the said Pirate ship," but rather "did as they wou'd themselves never observing him."[34]

Yet Glasby's position was not simply ceremonial. At several points during the final months of Roberts' cruise along the West African coast, Glasby demonstrated that while he did not hold the kind of authority of someone like Roberts, he did wield a degree of influence within the crew. When Roberts seized the *Hannibal* at Calabar, one of its crew, John Wingfield later testified that Glasby "was called the Master of the Pyrate-ship *Fortune, and acted as such.*"[35] After plundering the contents, the pirates then voted to burn the *Hannibal*, but according to Wingfield it was Glasby who intervened to prevent the ship's destruction. According to Wingfield, Glasby was "the Instrument of saving her."[36] Wingfield also described how Glasby had personally opposed and successfully prevented the forced recruitment of two surgeons into Roberts' crew. Glasby was not simply an outsider to be completely disregarded, and while he may not have exercised authority over other pirates, his word clearly carried enough weight to overturn a collective decision by the crew and to intercede on behalf of potential forced recruits.

Glasby's ambiguous status, mocked and disregarded by some, respected and heeded by others, illustrates how the reality of power aboard pirate vessels was quite different from the egalitarian structures outlined in the much-vaunted pirate articles. If, as Marcus Rediker has argued, votes by pirate crews were "sacrosanct," how was it that someone like Glasby was able to overturn the decision to burn the *Hannibal*? The answer lies in a closer examination of the networks of relationships and power that lay beneath the formal structures governing the crew. At trial, as pirates and witnesses described their experiences on board, a far richer portrait of piracy emerged from beneath the façade of solidarity, equality, and egalitarianism.

As Glasby and his co-conspirators at Hispaniola demonstrated, although the pirates' increasing reliance on forcible recruitment brought in more mariners, many of them remained uncommitted to the articles they had signed and the crew they had joined. Nor were Glasby and the ten others he attempted to escape with unique in this regard. At trial, many of the defendants revealed that they had secretly harboured hopes of deserting, and even formulated clandestine plans to do so while aboard Roberts' ships. Some of these plots were led by relatively new additions to the crew, men like John Watson and Benjamin Parr who along with "three others had

34 Ibid., f. 50.
35 Ibid.
36 Ibid.

concerted Measures for their Escapes both at Callabar and Annabona."[37] Others included veterans. Thomas Wills, the same man who'd conspired with Glasby, also hatched a scheme with Robert Lillburn to escape under the flimsy cover of going "buffeloe hunting" at Cape Lopez just days before they were both captured by the *Swallow*.[38] George Smithson testified that one of these conspiracies involved "12 of 14 who had privately Contracted together to run away from the *Fortune*."[39] Glasby's delicate balancing act and ambiguous status was, in fact, quite common aboard Roberts' ship, where many found themselves caught between the need to blend in and comply with expectations on board and the desire to either escape or distance themselves from piracy in the eyes of a British court.

Of course, if Roberts' crew was nothing more than a congregation of dissatisfied pirates constantly scheming to escape, one could be forgiven for asking why none of these groups never managed to carry out their plans. Certainly, the Vice Admiralty Court asked many of the defendants this question at the trial. Many answered that they were afraid that expressing such sentiments or worse making plans to abscond was simply too dangerous.[40] It was, Isaac Russell claimed, "so dangerous to trust any Body" on board.[41] Robert Haws, likewise, complained that while he wanted out, he "dare not communicate such a Design to everybody."[42] It was easy to see why this para-

37 Ibid., p. 115.

38 Ibid., p. 151.

39 Ibid., p. 149.

40 Another common fear was the inherent dangers posed by escaping to the African coast. Relations between English pirates and coastal African communities were, at best, tense. Testimony suggests that at times Roberts conducted trade with European factors in some ports, but that in general the men were unwelcome ashore. At least one captured man, William Watts, was emblematic of this strained relationship. Initially an employee of a Mr. Glynn, an English trader at the small port of Sestos [now Cestos] in what is now Liberia, Watts encountered Roberts as a messenger ferrying trade goods from Glynn to the pirate ship. He then fled his employer to join the crew, a decision he claims to have regretted, though he could not return to shore for fear of being captured and prosecuted by Glynn. One prisoner, Thomas Diggles, frankly admitted that he did not escape when he had an opportunity to do so while ashore at Sierra Leone because "they (the Whites there) had threatened to hang em." The crew were also well aware that English pirates, including themselves, had committed violent raids on coastal communities, and that escapees could expect hostility rather than hospitality. Howell Davis' death at the hands of the residents of Principe that he had openly abused was a reminder to all that escaping the *Fortune* onto shore might well have meant leaping from the frying pan into the fire. Ibid., pp. 104, 122, 144, 145.

41 Curiously, Russell also defended himself on the basis that he had been "one of those Men in the sloop that resolutely attaqued and took Blackbeard the Pirate" on a few years before. Ibid., p. 75.

42 Ibid., p. 92.

noia and reticence set in. Some of the men in Roberts' crew were genuinely committed to piracy, many of them dangerous, volatile, and violent. Men like David Simpson, who one witness remembered distinguished himself as being "particularly cruel beyond the rest of the Pirates,"[43] or like James Phillips who was known for "carrying his Pistols sometimes about him, to terrifie new comers if they offer'd to Speak saying they ought to serve their time first."[44] Sharing a ship with men like Simpson and Phillips, many men like Russell and Haws remained isolated in their misgivings, untrusted by others and themselves unable to trust.

Roberts relied on this group of hardcore pirates to manage and discipline the crew. Though he may have claimed that his crew accepted none but volunteers, Roberts' conduct revealed a commander who was well aware of the dissatisfaction amidst the ranks. The captain and his trusted circle carefully managed who could go ashore and when, allowing only "one a mess."[45] At night the pirates chained up the boats to prevent any deserters from stealing them, and set a watch to ensure nobody meddled with them.[46] When one veteran member of the crew made a failed attempt to desert at Sierra Leone, Roberts and the hardliners in the crew organised a classic ritual of collective discipline. Williams received "two Lashes from every Man in the Company," an ordeal that left him "indisposed for several Months."[47] The punishment left its mark not only on Williams, but also on the minds of the assembled pirate crew who carried it out, and some cited William's brutal fate specifically when explaining their reluctance to organise or attempt an escape.[48]

Roberts' crew included many men like Glasby, Russell, and Haws who remained secretly disillusioned as well as some like Phillips and Simpson who were deeply committed. It was the latter group, however, that dominated both the ship and the perception of Roberts' crew as a whole. The descriptions of Roberts' crew consistently fixated on the destruction, violence, and radical rhetoric of men like Simpson and Phillips, and generalised them as

43 According to Edward Crisp, Simpson even openly boasted that "he was as good a man as Roberts." A number of other witnesses confirmed that Simpson often assumed direct authority over both the plunder and destruction of property by others in the crew. Ibid. p. 36.

44 Ibid., p. 152.

45 Roberts' rule that only one man from each mess, that is a closely bonded working unit, was allowed on shore at any time prevented close, trusted friends from seizing the opportunity to slip away together. Roberts may have developed this stricture in response to earlier desertion attempts in the West Indies like Glasby's. Ibid., p. 115.

46 Ibid., p. 127, 135 164.

47 Ibid., p. 45.

48 Ibid., pp. 65, 72, 92, 109, 114,

representative of the entire crew. The *Boston Gazette* and *Boston News-Letter* did not make any distinctions amongst the pirates when describing in lurid detail the thefts, threats, wanton destruction, and open disdain for the law. Certainly, men like Bartholomew Roberts, and some of his hardened veteran crew did embody this ethos, but these were actually a minority of the crew. For every man like William Magness, who a witness described as being "exceedingly Active at Plundering," there were several like Richard Scot who publicly drank and made merry with the crew, but privately wept at night "talking pathetically of his Wife and Child."[49] For every pirate who went out of his way to personally rob a victim, like Richard Hardy who stole everything, including the silver boot buckles of the captain of the *Diligence*, there were others like Peter La Fever, a "quiet but Discontented Man" who surreptitiously returned plundered items to their original owners.[50] Far from the unified set of enthusiastic volunteers its articles cast it as, Roberts' crew was, in fact, a diverse and somewhat fragile group, in which most pirates fell somewhere in the ambiguous middle ground just like Glasby, neither fully committed to lives as criminals nor completely innocent of participation in piracy.

Far from the unified egalitarian group that its articles cast it as, the Roberts crew consisted of a small but committed core of hardened veterans with a large collection of pirates with highly varied attitudes and levels of involvement. Although they dominated the crew, Roberts and his nucleus of forward men had to manage the internal tensions of this group with care, ensuring that the discontented remained fragmented and compliant. It was in this delicate environment that someone like Glasby was able to successfully exercise a modicum of agency. His intercessions worked because those in the middle, like Glasby, could not simply be ignored like prisoners, and keeping them quiescent required concessions as much as it required control. Though they managed to find solace and support in small groups or pairs, the majority of these wavering pirates, including Glasby, remained unable to trust enough others to organise any kind of uprising. The latent threat of that possibility, however, meant that they could not simply be ignored either, and as much as that meant keeping them in line with violence, intimidation, surveillance, it also meant granting them some privileges, security, and at times even power in the daily affairs of the ship. All of these subtle interplays of power, however, were flattened by the contemporary fixation on only the most visible and outrageous behaviours of the most radical and committed pirates.

49 Ibid., p. 61.

50 Ibid., fols. 39, 149.

## Capture and Trial

Glasby maintained his awkward position as a reluctant but respected member of the crew for months. Indeed, he was still in Roberts' crew in its final dramatic hour. After months terrorising ships along the African coast, in early 1722, Roberts' recklessness finally led to his downfall. When the HMS *Swallow*, a warship that had been tracking the pirates for weeks finally caught up with them, Roberts mistook the Royal Navy vessel for a potential prize. It was only when the *Fortune* approached too close to escape that the pirates realised that they had, in the idiom of the day, "caught a tartar." In the ensuing battle, the *Swallow* quickly overpowered the *Royal Fortune*, crippling the vessel and killing a portion of the crew, including, crucially, Bartholomew Roberts himself.[51]

Following Roberts' death in battle, the situation aboard the *Royal Fortune* was fluid and volatile. The pirates' ship was no longer under the command of its formidable captain, but it was not yet in the custody of the officers of the *Swallow* either. This period lasted for several hours, as according to the *Swallow*'s logbook and the ship's lieutenant's testimony at the trial, the pirates struck their colours at around two o'clock, but it was not until seven o'clock that evening that the prisoners were secured.[52] The lengthy negotiations were necessary because, despite their utter defeat the pirates had one last piece of extremely powerful leverage: their long hinted at threat to blow up their entire vessel rather than surrender. Over the next few hours of delicate and tense negotiations, a standoff ensured, and it was in this crucial moment that Glasby made his move.

It was Glasby, several witnesses noted, who had ordered that the *Royal Fortune*'s colours be struck in the moments after Roberts' death. Glasby's order was swiftly carried out, an indication of the reservoir of influence that Glasby commanded despite his ambiguous position. Glasby's daring decision was not without its risks. The hardline veterans within the crew, undoubtedly aware that they had little chance of surviving a trial anyway,

51 Despite some minor discrepancies Captain Ogle's report of the battle and the Lieutenant's Log concur on the major strokes of the engagement. Ogle flew French colours in order to lure Roberts into approaching too close, before hauling up her British flag and engaging. The original manuscript of Ogle's letter is in poor and rather fragile condition. The details of the battle were reprinted in the *London Gazette*. TNA, ADM 1/2242. Capt. Chaloner Ogle to Lords of Admiralty, July 26, 1722; Caird Library and Archive, Navy Board, Lieutenant's Logs (ADM/L)/S/562. Lieutenant's Logbook for HMS *Swallow*, 1718–1728, February 10, 1722; *London Gazette*, September 4, 1722, pp. 1–2.

52 Caird Library and Archive, ADM/L/S/562, Lieutenant's Logbook for HMS *Swallow*, 1718–1728, February 10, 1722; TNA, HCA 1/99/3, p. 14.

were far less enthused by the idea of surrender. Indeed, no sooner had Glasby taken the initiative in ordering the surrender than he was "threatened for it."[53] Witnesses described how James Phillips, well known as one of the most aggressive and cruel members of the crew, attempted to make good on the suicidal boasts that some of them had made for years.[54] With a lit match in hand, Phillips charged towards the hatches, announcing his intent to ignite the powder magazine.[55]

Glasby, however, had anticipated this possibility.[56] During his eighteen months aboard the *Royal Fortune*, Glasby had quietly forged friendships and alliances with many of the reluctant recruits brought into Roberts' crew, and in his final hours on board, he put that network to use. When Phillips descended below decks to carry out his threat he was met by Stephen Thomas, one of several men that Glasby had tasked with keeping watch over the powder room. In a brief struggle that followed, Thomas, with Glasby's help, tore the match from Phillips' grasp and physically subdued him, putting a decisive end to his suicidal plan.[57]

This final act of Glasby's pirating career underscores both Glasby's success in cultivating an informal support network amongst the crew, but also the extent to which he represented the majority of the crew in his actions. Glasby's ability to surrender and hold onto control of the *Fortune* derived not just from his personal position as master, but also from the fact that a significant portion of the crew were willing to comply. Those most directly involved in enacting the coup and subduing Phillips had close personal relationships with Glasby. One of them, Hugh Menzies, was a former shipmate from the *Samuel* who had been in the crew as long as Glasby had.[58] Another, Robert Lillburn, was a more recent addition to the

53 TNA, HCA 1/99/3, p. 51.

54 In May 1721, a pirate ship was blown up in a battle off the coast of Barbados. In the *Boston News-Letter* an addendum to the report speculated that the ship in question may have belonged to Roberts and wryly pointed "it is very Remarkable how their own Predictions and unlawful Imprecations are come to pass and accomplished." The writer directed readers to the report published a year earlier of the attack on the *Samuel* and the reported threats of blowing up the ship. *Boston News-Letter*, May 15, 1721, p. 2.

55 TNA, HCA 1/99/3, p. 152.

56 A similar scene had played out the day before when the *Swallow* overwhelmed Roberts' consort vessel, the *Ranger* under the command of James Skyrm. When they realised they were about to be captured, a small group of the most hardened veterans ignited a powder store in the steerage. The resulting explosion killed at least one of them, and horrendously injured several others, but failed to blow up the entire ship. Ibid., p. 150.

57 Ibid., pp. 50–52, 54, 117, 152.

58 Ibid., p. 117.

crew, who had bonded with Glasby while secretly confiding his hopes of escaping.[59] What was more significant than the small group of allies that carried out Glasby's orders, however, was that the crew made no concerted effort to stop them. For years, men like Bartholomew Roberts and James Phillips had dominated the pirate ship and its reputation at sea, but in the moment of crisis, the majority of the crew stood by, neither stopping Glasby and his friends nor leaping to Phillips' aid. In other words, it was Glasby and his pragmatic gamble on surrender not Phillips' radical desire to perish in flames that resonated with most of the crew when the moment of truth arrived. Indeed, when Isaac Sun, the lieutenant of the *Swallow*, came aboard to officially take custody of the *Royal Fortune*, the pirate he asked for, by name, was none other than Henry Glasby.

The *Swallow* captured over two hundred men from on board Roberts' ships, and most survived transportation to Cape Coast Castle, in what is now Ghana, to face a Vice Admiralty Court.[60] A significant portion of the prisoners, around seventy, were Black. They were not afforded a trial, and were simply consigned to enslavement after arriving at Cape Coast. Although little is known about the identity of disposition of these men, their fate was perhaps the most unjust of all, as Roberts and his crew appeared to have treated them with the callous inhumanity as any English slaver. Although some historians, most notably Kenneth Kinkor, have argued that pirates offered rare opportunities for liberation to black men in the Atlantic, the Roberts crew was not amongst them.[61] Several others perished of wounds or illness prior to the beginning of the trial. Ultimately, a 165 men were tried at Cape Coast.

As pirate trials go, the proceedings at Cape Coast were somewhat unusual. Faced not only with an extraordinarily large number of defendants, but also the difficulties presented by the diversity of their attitudes and levels of involvement, the court decided to try each prisoner on a case-by-case basis. This was a rare effort to more fully explore the nature, depth, and circumstances of each individual's complicity. In order to do so, the court needed a witness who was familiar with most of the crew, including its longest

59 Ibid., p. 54.

60 The men were split between the *Royal Fortune*, Roberts flagship, and the *Ranger*, Roberts' consort commanded by James Skyrm which was captured a few days prior in a very similar fashion. TNA, ADM 1/2242. Capt. Chaloner Ogle to Lords of Admiralty, July 26, 1722.

61 Kinkor, "Black Flag." See also Rediker, *Villains of All Nations*, pp. 53–56. Jeffrey Bolster amongst others has argued that pirates were much more likely to be abusers and enslavers than liberators. Bolster, *Black Jacks*, pp. 15–16. See also Ritchie, *Captain Kidd*, pp. 110, 118; Jarvis, *Eye of All Trade*, pp. 147–51, 155–56; Earle, *Pirate Wars*, pp. 171–72.

serving members, and its inner workings. Having already garnered significant goodwill with his dramatic coup aboard the *Fortune* and identified by former victims as "civil" and "of good character," Glasby was an ideal candidate to serve as King's evidence.[62] In this capacity, he was able to exercise significant influence over the proceedings in the weeks to follow. Nobody contributed more testimony, or more decisive testimony in the trial of the Roberts crew, which stretched for weeks from March 28 to April 20, than Henry Glasby. Glasby testified in eighty-six of the cases, just over half of the total of 165. The next most prolific witness, Robert Lillburn, testified in just thirty-five cases.[63]

Not only did Glasby testify more than any other witness, his testimony was also often decisive in individual cases, and closely correlated with the verdicts. Of those he testified decisively against, thirty-six in all, thirty-two were ultimately convicted, and only two were acquitted.[64] Conversely, in cases where Glasby's testimony was in favour of the defendant, twenty-seven cases in total, only two were convicted by the court, and twenty-one were ultimately acquitted.[65] This was especially telling considering that when Glasby's testimony was ambivalent, the court tended to err on the side of conviction.[66] For instance, Glasby's evidence was sometimes mixed. Of Robert Haws, the pirate who had complained of his inability to trust anyone with his hopes of escaping, Glasby noted some exculpatory factors, especially that Haws had been forced into the crew and that he had not seen Haws going aboard prizes to plunder them. At the same time, however, he also dismissed Haws' defence as a "Pretence" of being unable to escape, and that he might easily have done so.[67] In other cases, Glasby was simply unfamiliar with the details, as with Joshua Lee, of whom he could say only that he was amongst a group recruited from the *Martha* of Liverpool but that he did not know "how far with his or their Consent."[68]

62 TNA, HCA 1/99/3, pp. 50–52.

63 Figures compiled from a comprehensive analysis of the records of the trial. TNA, HCA 1/99/3.

64 The court ruled that the other two men, John Rimer and Henry Graves were to be sent to London and held at Marshalsea prison pending sentencing after further investigation. Ibid., fold. 163–64.

65 The other three men, like Rimer and Graves, were sent back to England to be held at Marshalsea. Ibid., pp. 78, 120, 136, 162.

66 This accounted for twenty-three other cases, with the bulk, eighteen, found guilty. An additional six were remanded to Marshalsea pending further investigation, and just four were acquitted outright.

67 Glasby's claim here is puzzling considering that he himself never managed an escape, and he also defended other members of the crew by noting that escape was extremely risky and difficult. It is unclear why he singled out Haws for such a harsh assessment. TNA, HCA 1/99/3, f. 91.

68 TNA, HCA 1/99/3, f. 142.

Ultimately, the trial reflected the inherent messiness and ambiguity of the many men in Roberts' crew. Close to half of the defendants were acquitted outright, a strikingly large number considering the harsh reputation of British courts.[69] Another portion of the defendants, seventeen in total, were deemed to be too difficult to assess at Cape Coast, and sent back to London to be held in Marshals prison pending further evidence. Those who survived the long, arduous voyage back to England were ultimately granted pardons and released.[70] Indeed, even amongst the seventy-four pirates convicted and sentenced to death the court recognised extenuating circumstances, commuting the sentences of twenty to seven years of hard labour.[71] Even the Vice Admiralty Court, for all its harshness, was forced to acknowledge the broad spectrum of culpability amongst the Roberts crew.

## Conclusion

Though he remains a relatively obscure figure, Henry Glasby's time as a pirate reveals much more about the history of piracy in the early eighteenth century than Roberts' far better known pirating career does. Perhaps most telling of all, shining a light on complicated tribulations of Henry Glasby reveals the extent to which Roberts' fame and stature obscure our understanding of the hundreds of men in his shadow. Of course, it is not surprising that pirate crews would have encompassed a range of motives, attitudes, deceptions, conspiracies and conflicts, but acknowledging this messy social arrangement is not the same as engaging with its historical implications.

69 Joseph Gibbs has conducted a thorough survey of the outcomes of piracy trials from the late seventeenth to the early eighteenth century, and noted that the Roberts trial was something of an outlier. Overall, the acquittal rate was closer to 25 per cent between 1670 and 1730. Gibbs, "Brevity and Severity."

70 The pirates were put aboard the *Weymouth* for transportation to London, along with Thomas Athelone, the one convicted pirate granted a respite of execution. Three of the prisoners died before the *Weymouth* weighed anchor to leave Cape Coast, likely of illness. Several others died at sea. One, John Wilden, was released at Barbado. Only seven of the original eighteen survived to disembark at Woolwich. TNA, ADM 1/1880. Captain Mungo Heardman to the Lords of Admiralty, *Weymouth* at Woolwich, April 26, 1723; TNA, ADM 1/1880. Captain Mungo Heardman to Lords of Admiralty, A List of Pirates Referred to Marshalsea, May 1, 1723; TNA, SP 42/17, f. 597. Attorney Gen's Report about Several Pyrates, August 8, 1723; TNA, HCA 1/17, f. 190. Warrant for the Pardon of Thomas Ourthelony, July 30, 1723; TNA, HCA 1/17 f. 191. Pardon for William Mead and Others, August 15, 1723.

71 TNA, HCA 1/99/3, pp. 179–80.

Too often, captains like Roberts and his witty motto of a merry and short life have come to function as ready and recognisable stand-ins for pirates. Yet in his fame, his unapologetic criminality, and in his powerful position, Roberts was only a single, and in many ways singular member of a much larger group. While it is true that Roberts and his small cadre of radical, violent "leading men" dominated that group, Glasby's experience shows that they were hardly a representative sample of the crew.

It is only when we examine more closely and seriously the broader spectrum of experiences of pirates that we can more fully apprehend the ways that they lived and worked together. The hardened veterans that beat and abducted Glasby at gunpoint boasted openly of their contempt for the law and their intentions to die rather than be captured. When the law finally arrived, however, it was Henry Glasby's sensibilities of pragmatism and survival that prevailed amongst the crew. Glasby had carefully cultivated and grown an informal interpersonal network during his time aboard the *Fortune*, allowing him to gain a degree of influence over the affairs of the ship, and more importantly to launch his daring coup during the final hours of his pirating career. It was not just Glasby's immediate circle that held this more moderate and pragmatic view, however. The ease with which a few men managed to stop James Phillips was accomplished only thanks to appeal of Glasby's approach amongst a much broader segment of the crew than those directly involved. At trial, the Roberts crew received a surprising degree of clemency from an early eighteenth-century Vice Admiralty Court. Indeed, the trial itself, and Glasby's key role within it, exposed the fact that even the court recognised the ambiguous status of many of the crew members in unusual procedures and mixed verdicts.

Yet Glasby's strange tenure aboard the *Fortune* points towards historical considerations beyond the specifics of his case. In the years following the Peace of Utrecht, the emergence of increasingly effective resistance to pirates, and the collapse of their informal networks of support created qualitatively new conditions. What Glasby's complicated pirating life offers is a glimpse into the many ways that these conditions impacted pirate crews, how they adapted and responded to pressures in novel ways. Indeed, even Roberts, when viewed against Glasby's account, emerges as a more complex figure than his larger-than-life reputation implies. Faced with the difficult conundrum of maintaining discipline in a group where true believers were often outnumbered, he was a shrewd manipulator, a careful manager, and a successful marketeer. For years, he managed to maintain his primacy over a volatile and dangerous congregation of men engaged in one of the most risky and dangerous professions of the Atlantic world.

Henry Glasby's story, of course, is just one case, within a single crew, one that serves here as an entry point into sketching out a clearer and deeper portrait of the Roberts gang. What that sketch reveals about the contours of power and loyalty within that group offers an illustration of the value of engaging with the rank and file of pirate crews on a larger scale. In Roberts' gang, the pressures of the growing backlash against pirates in the 1720s ramified through interpersonal relationships, power dynamics, and workaday life in real and concrete ways. Both Glasby and Roberts manoeuvred with clear awareness of the intensifying legal peril that hung over the crew, leveraging it as best they could in a volatile and shifting social terrain.

A fuller picture of the composition and, more importantly, the disposition of rank-and-file pirates offers us a richer understanding of piracy during this critical period. For instance, if other crews also contained significant numbers of ambivalent members as Roberts' did, then it suggests that social and economic alienation played less of a role in driving men into piracy in the 1720s than has previously been suspected. Likewise, taking into consideration the internal tensions and dynamics at play aboard the *Fortune* opens the way to new explanations for pirate behaviours from the period. For instance, given the testimony of many at trial, it is clear that Roberts' move from the western Atlantic to the African coast was likely driven in part, or even primarily, because it created a powerful deterrent to troublemakers or deserters. Finally, Glasby's story demonstrates that the egalitarian language of the pirate articles on Roberts' ship masked a much more complex web of informal power, cliques, and performance. These carefully crafted, public-facing documents ought not, and indeed cannot function as an adequate window into the ideas and attitudes of most pirates. Rather we must turn our attention towards the men like Henry Glasby, whose experiences offer rare glimpses into the anatomy and internal workings of pirate crews, and perhaps a fuller understanding of the alloyed historical realities hiding beneath the lustre of legendary figures who have dominated pirate history for three centuries.

## Bibliography

### Abbreviations

A1895-78 New York Colonial Council Minutes, New York State Archives. Albany, NY.

ADM Admiralty Office Papers, The National Archives, UK. Kew.

| | |
|---|---|
| ADM/L | Navy Board, Lieutenant's Logs, Caird Library and Archive, National Maritime Museum. Greenwich. |
| HCA | High Court Admiralty Papers, The National Archives, UK. Kew. |
| NRAS | National Register of Archives for Scotland. |
| SP | State Papers, The National Archives, UK. Kew. |

### Printed Primary Sources

*Applebee's Original Weekly Journal.* London.

*Boston Gazette.* Boston.

*Boston News-Letter.* Boston.

*Daily Post.* London.

*A Full and Exact Account, of the Trial of all the Pyrates, Lately Taken by Captain Ogle, on Board the Swallow Man of War, on the Coast of Guinea.* London: J. Roberts, 1723.

Johnson, Charles. *A General History of the Pirates.* Mineola, New York: Dover Publishing, 1999.

*London Journal.* London.

*Post Boy.* London.

### Secondary Sources

Baer, Joel. *Pirates of the British Isles.* Stroud, Gloucestershire: Tempus, 2005.

Bolster, Jeffrey. *Black Jacks: African American Seamen in the Age of Sail.* Cambridge, Massachusetts: Harvard University Press, 1997.

Burgess Jr., Douglass R. *The Politics of Piracy: Crime and Civil Disobedience in Colonial America.* Lebanon, New Hampshire: ForeEdge, 2014.

Cordingly, David. *Under the Black Flag: The Romance and Reality of Life among the Pirates.* New York: Random House, 1995.

Earle, Peter. *The Pirate Wars.* New York: Thomas Dunne Books, 2003.

Gibbs, Joseph. "The Brevity and Severity of 'Golden Age' Piracy Trials." *The International Journal of Maritime History* 31, no. 4 (2019): 729–86.

Hanna, Mark G. *Pirate Nests and the Rise of the British Empire, 1570–1740.* Chapel Hill, NC: University of North Carolina Press, 2015.

Hill, Christopher. "Radical Pirates?" In *The Origins of Anglo-American Radicalism,* edited by Margaret Jacob and James Jacob, 17–32. London: George Allen and Uniwin, 1984.

Jarvis, Michael J. *In the Eye of All Trade: Bermuda, Bermudians, and the Maritime Atlantic World, 1680–1783.* Chapel Hill, NC: North Carolina University Press, 2010.

Karraker, Cyrus H. *Piracy Was a Business*. Rindge, NH: Richard R. Smith, 1953.
Lane, Kris. *Pillaging the Empire: Piracy in the Americas, 1500–1750*. London: M. E. Sharpe, 1998.
Leeson, Peter T. "An-*arrgh*-chy: The Law and Economics of Pirate Organisation." *The Journal of Political Economy* 115, no. 6 (2007): 1049–94.
Leeson, Peter T. *The Invisible Hook: The Hidden Economics of Pirates*. Princeton: Princeton University Press, 2009.
Leeson, Peter T. *Anarchy Unbound: Why Self-Governance Works Better Than You Think*. Cambridge: Cambridge University Press, 2014.
Rediker, Marcus. "'Under the Banner of King Death': The Social World of Anglo-American Pirates, 1716–1726." *William and Mary Quarterly* 38, no. 2 (April 1981): 203–27.
Rediker, Marcus. *Between the Devil and Deep Blue Sea: Merchant Seamen, Pirates and the Anglo-American Maritime World, 1700–1750*. Cambridge: Cambridge University Press, 1987.
Rediker, Marcus. "Liberty Beneath the Jolly Roger." In *Iron Men and Wooden Women: Gender and Seafaring in the Atlantic World, 1700–1920*, edited by Margaret S. Creighton and Lisa Norling, 1–33. Baltimore: The Johns Hopkins University Press, 1996.
Rediker, Marcus. "Hydrarchy and Libertalia: The Utopian Dimensions of Atlantic Piracy in the Early Eighteenth Century." In *Pirates and Privateers: New Perspectives on the War on Trade in the Eighteenth and Nineteenth Centuries*, edited by David J. Starkey, E. S. Van Eyck van Heslinga, and J. A. Moor, 29–46. Exeter: University of Exeter Press, 1997.
Rediker, Marcus. "The Seaman as Pirate: Plunder and Social Banditry at Sea." In *Bandits at Sea: A Pirates Reader*, edited by C. R. Pennell, 139–68. New York: New York University Press, 2001.
Rediker, Marcus. *Villains of All Nations: Atlantic Pirates in the Golden Age*. Boston: Beacon Press, 2004.
Rediker, Marcus. *The Slave Ship: A Human History*. New York: Viking, 2007.
Rediker, Marcus. *Outlaws of the Atlantic: Sailors, Pirates, and Motley Crews in the Age of Sail*. Boston: Beacon Press, 2014.
Ritchie, Robert C. *Captain Kidd and the War against the Pirates*. Cambridge, Massachusetts: Harvard University Press, 1986.
Starkey, David J. "Pirates and Markets." In *Bandits at Sea: A Pirates Reader*, edited by C. R. Pennell, 107–24. New York: New York University Press, 2001.

## About the Author

**James Rankine** is originally from Brisbane, Australia, and earned his PhD in History from the University of Rochester in 2021. He is currently a Visiting Lecturer in History at the Rochester Institute of Technology. His research examines the piracy in the Atlantic world in the late-seventeenth and early-eighteenth centuries, particularly the roles of power and violence aboard pirate vessels. He is a former Andrew W. Mellon Fellow in the Digital Humanities and served as a site supervisor and teaching assistant on the Smith's Island Archeology Project.

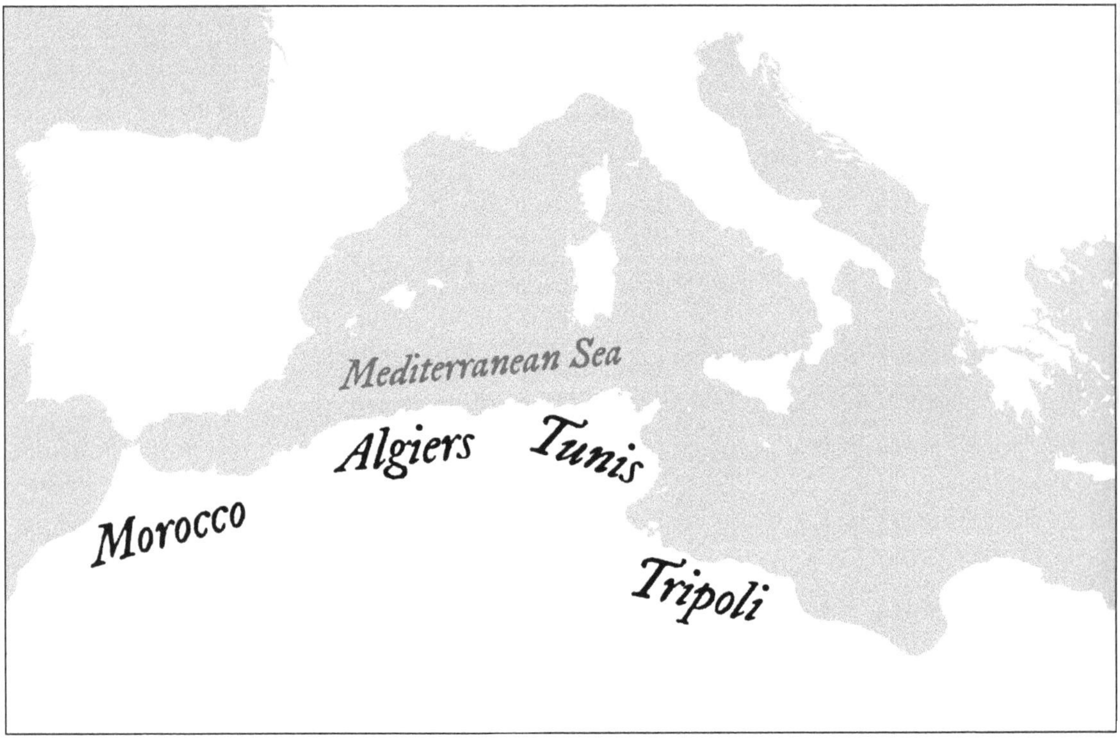
Mediterranean Sea
Algiers
Tunis
Morocco
Tripoli

# 8. "Our Affairs with the Pyratical States": The United States and the Barbary Crisis, 1784–1797

*Anna Diamantouli*

**Abstract**

The Barbary crisis (1784–1797) saw increased tensions between the United States and the North African nations of Algiers, Tunis, Tripoli, and Morocco, leading to the capture of American citizens. This chapter explores how the United States perceived the complex crisis that exposed American weaknesses, its fragile political state and lack of institutions, and which situated the new nation's citizens as unfree and threatened. Examining newspapers and diplomatic correspondence of the period, this chapter reveals how use of the word "piracy" spoke to specific anxieties permeating the United States and was politically charged. Looking at contemporary use of "Barbary piracy" it also makes clear that the term continues to reveal more about political agendas than about the "pyratical states" themselves.

**Keywords:** Piracy; Corsairs; Early Republic; Treaties; Thomas Jefferson; Islam; War on Terror

The two last decades of the eighteenth century were marked by increased tensions between the United States and the Ottoman Regencies of Algiers, Tunis, and Tripoli, and the Kingdom of Morocco. This period of hostile relations between the nations, commonly referred to from the American perspective as the Barbary crisis, saw approximately 137 American citizens held in foreign captivity following the Moroccan capture of the *Betsey* in 1784, and the Algerine captures of the *Maria* and *Dauphin* in 1785 and eleven

Coakley, J., C. Nathan Kwan and D. Wilson, *The Problem of Piracy in the Early Modern World: Maritime Predation, Empire, and the Construction of Authority at Sea.*
Taylor & Francis Group, 2024
DOI 10.5117/9789463720960_CH08

more ships in quick succession in October and November 1793. The capture of the American citizens and vessels came about as a result of the corsairing policies of the North African nations, which authorised attacks on the vessels of enemy nations with whom they were at war, or the vessels of nations who did not pay tribute. The vessels would be seized and all on board would be taken captive, held in North African bagnios, or prison-houses, and put to work. Peace and the redemption of the captives would then be demanded through the payment of ransoms, or through the negotiation and conclusion of annual tributes stipulated by treaties of peace and amity. While treaties between the North African nations and Great Britain had protected the North American colonies, following independence the US found itself alone in facing the corsairing threat without a treaty to protect its interests.

This chapter looks closely at the response in the new nation to this crisis, and more particularly at representations of Barbary piracy in diplomatic correspondence and American newspapers published during the crisis. Through such a focus, it becomes clear that the use of the term piracy in contemporaneous accounts is politically and culturally charged, taking on specific ideological significance. Indeed, American representations of Barbary piracy throughout the final two decades of the eighteenth century speak to a larger set of anxieties felt within the United States pertaining to its fragile political state and lack of institutions with which to effectively respond to the diplomatic crisis. Reference to Barbary piracy thus stresses the heightened public, diplomatic and political concern regarding the Barbary crisis but also the condition of the new nation itself during a time of great domestic uncertainty more generally. This chapter subsequently emphasises how contemporary unquestioning use of the term in media and historical scholarship to refer to North African maritime predations, reveals a similar tendency to serve political agendas and in so doing make generalised claims about the birth of a nation.

Before looking at the American response to the Barbary crisis and more closely at American representations of Barbary piracy, it is important to underscore that North African maritime predations within the late eighteenth-century context cannot be defined explicitly as piracy but are rather more closely aligned with corsairing. While the distinction is a difficult one because of the interchangeable use of the terms and the blurred boundaries in the acts of privateering and piracy themselves, the distinction can, to a certain extent, be clarified in reference to the legal framework of the period.

A corsair working within the Mediterranean context had a commission from a recognised authority to act against a designated enemy while the pirate was not commissioned and did not discriminate. This distinction, as Walter Rech puts forward, harks back to the early thirteenth century when Italian jurists "attempted to regulate the conflict between maritime republics" by distinguishing between maritime predations that took place with the approval or on behalf of an established political authority and those that lacked such authority.[1] In these terms, the maritime predations of the North African nations during the Barbary crisis were all state-endorsed and regulated corsairing activities that involved legitimate corsairs as opposed to piracy. Historians of North Africa including Daniel Panzac and Nabil Matar have emphasised this distinction. The opening lines of Daniel Panzac's *Barbary Corsairs* for example, distinguishes between the terms: "It is appropriate to stress the fundamental difference between the pirate [as] lawless [...] known in the Mediterranean as a bandit, and the corsair [...] who, in the Maghreb [another term used to describe the North African region consisting of the Barbary States], only existed [...] as one who respected the law."[2] Indeed, corsairs saw themselves as professionals and this conception was qualified in law, as Nabil Matar states, "Technically, there was no piracy among North Africans because the sailors and their captains always had the official support and cooperation of the [rulers]."[3]

Focusing on this official support, the Ottoman regency of Algiers, which captured the largest number of American captives during the Barbary crisis, corsairing was a state monopoly supervised by a minister (wakil al-kharj) who was "responsible for the provisions of the [corsairing] fleet as well as for relations with the European nations."[4] The state was in turn the principal financier of the corsairing expeditions and all income, as Martin M. Murphy highlights, was shared in adherence to strict regulations "after the dey had received his share and the port authorities and customs officials had imposed their charges."[5] This profit, which albeit by the late eighteenth century was inconsistent and uncertain, was reinvested into the wide-ranging network of actors including carpenters and armorers needed to maintain the fleet.

Like Algiers, as Priscilla Roberts and James Tull have shown, the Moroccan Sultan Sidi Muhammad Ibn Abdallah who had been in power since 1757, had "brought the independent corsairs under his royal umbrella" and attempted to "organize a state-controlled merchant marine."[6] This provided leverage for

1 Rech, *Enemies of Mankind*, p. 39.
2 Panzac, *Barbary Corsairs*, p. 9.
3 Matar, "Introduction: England and Mediterranean Captivity," p. 7.
4 Panzac, *Barbary Corsairs*, p. 320.
5 Murphy, "Barbary Pirates," p. 28.
6 Roberts and Tull, "Diplomatic Initiatives," p. 235.

the Sultan to negotiate and conclude treaties of amity and commerce with European nations including Denmark (1767), Great Britain (1760), Sweden (1763), the Republic of Venice (1765), France (1765, 1767), Spain (1767), and Portugal (1773). In Tripoli the corsairs acted on behalf of the Pasha who shared in the profits from their activities, while in Tunis corsairing was not a monopoly of the state, but as Bookin-Weiner underscores, the relationship between the ruling elite and the corsairs was symbiotic.[7] Abun-Nasr claims "important personalities of the state took part in it," expanding on the symbiotic nature of this relationship within the Tunisian context.[8]

It was not only the North African nations that sanctioned maritime predations. European and North African states alike had authorised such predations for several centuries. For example, Fuad Sha'ban asserts that "[a] s early as the sixteenth century, Queen Elizabeth I gave her royal sanction to Sir Francis Drake, who plundered Spanish ships of their valuable cargoes brought back from the American continent."[9] Indeed, as it was often more financially viable for governments to rely on "private initiative and enterprise to fight their wars" as opposed to maintaining an expensive public navy, such predations funded and facilitated the era of exploration and expansion alongside the later Revolutionary wars. The Americans alone employed approximately seven hundred commissioned ships in their Revolution.[10] Further to this, Christian corsairs, including Spaniards, Italians and the Knights of Malta also raided ships well into the seventeenth century in order to take slaves.[11]

Looking more closely at corsairing from a legal standpoint, beyond the state-sanctioned nature of North African corsairing, European actions in response also provided, Rech argues, an "'accommodating' legal framework" as evident in the treaties of amity and commerce concluded between the North African nations and their European counterparts and the system of tributes paid in exchange for "immunity from attack." As previously highlighted, European powers including France, Great Britain, Sweden, and the Netherlands, alongside Denmark and the Republic of Venice maintained such agreements, tacitly accepting corsairing as legal. However, even those governments that were not bound by treaties, having either refused or being unable to appease the North African states through payment of tributes,

7 Bookin-Weiner, "Corsairing," p. 26.
8 Abun-Nasr, *History of the Maghrib*, p. 177.
9 Sha'ban, *Islam and Arabs*, p. 68.
10 Tabarrok, "Privateers," pp. 566–77.
11 For more on Muslim slaves in France and Italy see Earle, *Corsairs*.

regarded, as Murphy emphasises, "Barbary regencies as legitimate polities, not as piratical communities, and accordingly stipulated official agreements and truces with them" recognising the commercial benefit of maintaining such relationships.[12]

Despite North African maritime predations being thus more closely affiliated within the legal framework with corsairing, American representations of Barbary piracy during the Barbary crisis proliferated. As such, it becomes evident that the term piracy is loaded and signifies on multiple levels beyond a supposedly neutral descriptor of North African activities. This becomes even more pertinent considering that in North African formal and legal correspondences as well as Royal decrees these maritime predators were referred to not as pirates (*liss albahr*, or "sea thief") but as corsairs (*qursan*).[13] While not the focus of the present chapter, this contradiction highlights that the term piracy is flexible, multivalent and assumes meaning and function in relation to the historical dynamics of the period in which it is used and the perspectives and intentions of those who use it. In this vein, "piracy" on one level spoke to the United States' efforts to negotiate its position both domestically and internationally while grappling with a complex crisis that exposed its fragile political state and which situated the new nation and its citizens as unfree and threatened. On much the same level, the term operates in contemporary conservative political discourse in response to present-day concerns.

Within the North American colonial context, the problem posed by North African corsairs to North American vessels was not particularly problematic as they remained colonies under European protection. This was because a 1682 treaty signed between Great Britain and Algiers, the largest and most militarily powerful of the North African states safeguarded both British and American shipping. This protection was strengthened with the signing of a 1703 amendment that ensured that "all ships and vessels built and fitted out in any of her Majesty's plantations in America, that have not been in England," were to be provided with passes that would prevent them from being "molested" by the Algerines.[14] As a result of this British attention, North American commerce and trade flourished in the Mediterranean and the colonies' economies became commercially dependent on British protection. Indeed, Thomas Jefferson in his role as Secretary

12 Rech, "Ambivalences of Recognition," pp. 76, 78.

13 In putting forward this point I thank Lama Elsharif for kindly and generously sharing her expertise on this topic. See also Cosman and Jones, *Medieval World*, p. 216.

14 "Articles of Peace and Commerce," p. 389.

of State would later conclude that by 1776 "about one-sixth of the wheat and flour exported from the United States, and about one-fourth in value of their dried and pickled fish, and some rice, found their best markets in the Mediterranean ports." These "articles constituted the principle part of what [was] sent into that sea" and "that [this] commerce loaded outwards, from eighty to one hundred ships, annually, of twenty thousand tons, navigated by about twelve hundred seamen."[15] In 1776 however, the year the thirteen colonies declared their independence, British protection of American shipping ended.

This key moment exposed the US Congress's inexperience in matters of international diplomacy as they were desirous of merely replacing British protection of American shipping from North African corsairing with the protection of another European power: France. A plan of the Treaty of France, drafted by John Adams, as delegate to the Second Continental Congress, expressed this desire in what the historian Gregg L. Lint argues, constituted "the first major state paper dealing with the conduct of the United States toward other sovereign powers."[16] One of the articles of the plan, which essentially constituted a guide for the future of American foreign policy, related to the protection of American shipping from the activities of the Barbary States and to this end requested French assistance in a relationship of convenience. The plan of the article required that the French king:

> [P]rotect, defend and secure, as far as in his Power, the Subjects, People and Inhabitants of the said United States and every of them, and their vessels and Effects of every kind, against all Attacks, Assaults, Violences, Injuries, Depredations or Plunderings by or from the king or Emperor of Morocco, or Fez, and the states of Algiers, Tunis and Tripoli.[17]

Congress's expectations of France as presented above both acknowledged the need to acquire protection against Barbary corsairing but also exposed a want of experience in matters of international diplomacy, as they required that France treat them with the same favour that Great Britain had provided them with in its position as colonial ruler. Indeed, the United States hoped that the French monarch would "use his Interest and Influence to procure Passes" for the protection of "the Vessels and Subjects and Inhabitants of the United States upon the Mediterranean." Congress was in essence looking

15 Jefferson, "Report of the Secretary of State Relative to the Mediterranean Trade."
16 Lint, "John Adams," pp. 313–14.
17 "Plan of the Treaties with France of 1778, Tuesday, September 17, 1776."

to replace British protection of American shipping through its issuance of maritime passes with that of French protection.

Ultimately, the Commissioners were able to negotiate the inclusion of the article concerning protection from Barbary corsairing in the final treaty, but the demands of the US were tempered. Rather than a total guarantee to "protect, defend, and secure," France agreed to:

> [E]mploy [its] good Offices and Interposition with the King or Emperor of Morocco or Fez, the Regencies of Algiers, Tunis and Tripoli, or with any of them, and also with every other Prince, State, or Power of the Coast of Barbary in Africa, and the Subjects of the said King Emperor, States and Powers, and each of them; in order to provide as fully and efficaciously as possible for the Benefit, Conveniency and Safety of the said United States, and each of them, their Subjects, People and Inhabitants, and their Vessels and Effects, against all Violence, Insult, Attacks, or Depredations on the Part of the said Princes and States of Barbary, or their subjects.[18]

Not much time passed after signing the treaty that the United States called upon French protection, signalling the urgency and necessity felt in protecting American commerce. Only six months after the signing of the treaty, the American Commissioners Benjamin Franklin, Arthur Lee, and John Adams, wrote to France's Foreign Minister, the Comte de Vergennes, Charles Gravier, calling Article 8 into effect by requesting French protection from the Barbary corsairs in line with the ratified treaty. The Commissioners informed Vergennes that they had "received Information that there are already American Vessels in Italy desirous of returning home, and that there are Merchants in Italy desirous of entering into the American Trade; but that an Apprehension of Danger from the Corsairs of Barbary is a Discouragement." The Commissioners asked for "your Excellency's Attention to this Case: and such Assistance from his Masters good Offices as was intended by" "the Eighth Article of the Treaty of Commerce."[19] However, the French did not know what this "attention" or "assistance" required on their behalf and neither did the United States itself. With no central government or head of foreign affairs, there was a lack of clarity in the intentions of Congress.

18 "Treaty of Amity and Commerce between the United States and France Signed on February 6, 1778."

19 Commissioners of the United States of America, "Extract of a Letter from the Commissioners of the United States of America to M. Le Comte De Vergennes, Dated Passy, August 28, 1778," p. 570.

Without this clarity, US foreign policy remained irresolute. While the Comte de Vergennes expressed his country's willingness to assist the United States as per Article 8 of the final Treaty, the role of the French in protecting American commerce and trade required further clarification. The Comte de Vergennes wrote to Benjamin Franklin on September 27, 1778 that "the [French] King will very willingly do whatever is in his power to satisfy the United States and to render their wishes approved of by the Princes of Barbary." However, he argued that "before the orders of the King are taken in this respect, he requires several éclaircissement"; or indeed, substantial explanation regarding something obscure or inexplicable.[20]

The Comte de Vergeness passed on the communication to Antoine-Gabriel de Sartine, the Secretary of State for the French Navy, who questioned the intentions of Congress in greater detail:

> Do the United States wish to conclude treaties with the Barbary Powers, or do they wish merely that our [French] influence should be exerted to make their flag respected by those powers? In the latter case we should never succeed, or if we should obtain liberty of commerce for the United States from some of them, it would be an illusory, temporary, and precarious permission, and would infallibly expose us, without being of the least benefit to the citizens of the United States.

Advising the United States on the best course of action it should take in regard to its foreign affairs, which involved concluding their own treaties, Sartine underscored the new nation's inexperience and also stressed the legitimacy of the practice of corsairing within the international political and diplomatic scene. Sartine highlighted that it was in the United States' best interest to conclude treaties with the individual Barbary States, something that would require Congressional instructions and financial backing: "It is necessary to know beforehand, what the instructions of the Commissioners are, and whether they are authorised to negotiate with the Barbary Powers under the mediation of France."[21] Despite the advice of the French however, the US Congress did not prioritise establishing individual treaties with the Barbary states and it did not grant its commissioners with this provision. Instead, it continued to look towards European powers with which to establish contacts and alliances. In response, the American Commissioners expressed their

20 Comte de Vergennes, "The Comte De Vergennes to Benjamin Franklin: A Translation, 27 September 1778," pp. 83–84.

21 Sartine, "M. De Sartine to Count De Vergennes, Versailles, September 21, 1778," p. 30.

concerns to the president of Congress. On November 7, 1778 they wrote that: "on the Subject of Negotiation with the Barbary States [...] [w]e do not find ourselves authorized to treat with these Powers, as they are not in Europe; And indeed, we are not furnished with Funds for making them Presents."[22]

Before any further clarification was provided to the French, and before the commissioners were granted any further provisions to negotiate treaties with the North African nations, the Franco-American Treaty of Amity and Commerce ended with the ratification of the September 1783 Treaty of Paris that heralded the end of the American Revolutionary War. With the potential for French assistance in responding to the corsairing threat coming to an end, the United States once again looked to Britain hoping that it would agree to a new commercial treaty that would allow the new country access to all ports, including those of North Africa, while under British protection. The Chancellor of the Exchequer, William Pitt, presented this proposal of the American Intercourse Bill in British Parliament in March of 1783.[23] However, despite Britain engaging in a free trade debate it declined the request in 1784 as it saw no benefit in the proposal. The United States, having looked to France and Britain to protect it from Barbary corsairing, now found itself alone, without the protection of any European power.

While under English colonial rule, one of the motivations for independence had been "to trade freely, unhampered by such inconveniences as the Navigation Acts."[24] Now free from colonial rule and no longer under the control of British treaties, the US however, as evident above, had to come to terms with a complicated and often hostile international network of diplomacy, trade, and military alliances. There was suddenly a realisation amongst American diplomats that, while they may have been desirous of becoming an equal partner in the Atlantic world and the Mediterranean following independence, they were indeed not such a partner. Instead, Jefferson expressed that Mediterranean trade and American "adventures into that sea," had now been "exposed to the depredations of the pyratical States on the coast of the Barbary."[25]

Ironically, despite the US's expression of interest to establish commercial liberal treaties with all nations, the only nation that expressed a desire to

22 Franklin, Lee, and Adams, "The Commissioners to the President of the Congress, 7 November 1778," pp. 196–200.

23 Its full title was "A Bill for the provisional establishment and regulation of trade and intercourse between the subjects of Great Britain and those of the United States of North America" in Edwards, *British Colonies*, p. 491.

24 Sears, *American Slaves*, p. 31.

25 Jefferson, "Report Relative to the Mediterranean Trade, December 28, 1790," p. 520.

do so in the years following independence was Morocco.[26] However, the US Congress ignored Morocco's requests for over a decade. Whether the United States was unable to recognise the importance of establishing direct diplomatic ties with its North African counterpart, or whether they were more preoccupied with attempts to gain money and supplies for the war effort from France, Congress failed to act regarding relations with Morocco which meant that the situation escalated.[27] In October 1784, Sidi Muhammad, frustrated by the United States' lack of response, recognising the lack of protection from European powers and hoping to attract the attention of the United States and to force their hand into establishing a commercial treaty to his satisfaction, sent his corsairs to capture an American vessel.[28] The *Betsey*, a merchant vessel from Philadelphia, which had set sail for Tenerife was thus captured shortly after leaving the port of Cadiz. Pushed into action following the capture of the *Betsey*, Congress finally followed the advice of the French and vested its commissioners with powers to negotiate treaties with Morocco and Algiers directly. Following negotiations, on June 23, 1786, Morocco became the first of the Barbary States to sign a treaty with the United States. However, relations with Morocco served as a prelude to the more serious problem posed to the United States by Algiers. While the Moroccan affairs were resolved relatively easily and quickly, 1785 marked an escalation of the Barbary crisis. With the United States still busy negotiating a peace with Morocco, Algiers, captured two American merchant ships: the Philadelphian schooner *Maria* on July 24, 1785 and the Boston brig *Dauphine* six days later. Algiers took both vessels and their cargoes, as well as the combined crew of the ships, twenty-one persons in total. Jefferson, who had been tasked with negotiating commercial treaties in his position as the United States Minister to France, expressed his hope that "the Algerines may be as easily dealt with" as Morocco.[29] However, this would prove not to be the case.

The early American response to the Barbary crisis underscored and intensified weaknesses in the government under the Articles of Confederation in not only defining who had the authority to act in response to the crisis

26 "Despatch No. 176 of 8 November 1777, Un charge d'affaires au Maroc" and "Chenier's Memoire general for 1777" in Roberts and Tull, p. 239. Roberts and Tull, who have provided an extensive analysis of American Moroccan relations in the late eighteenth century, identify this as the oft-cited document used to support the view that Morocco was the first foreign nation to recognise the new United States.

27 Buhite, *Lives at Risk*, p. 4.

28 Roberts and Tull, "Diplomatic Initiatives," p. 247.

29 Ibid., p. 234.

but also in gathering funds with which to redeem the American captives. However, despite the ratification of the Constitution in 1788, these issues persisted, and the crisis escalated in 1793 when Algiers signed a treaty with Portugal granting them access to the Atlantic and substantially increasing the threat posed to American shipping. In October 1793 Algiers captured ten American ships and 105 captives.[30] In November, an eleventh American ship, the brig *Minerva* and its seven crew members were taken, increasing the total number of "distressed and faithful" American citizens now captive in Algiers to approximately 126, some of whom had already spent over eight years in Algerine captivity.[31]

In the meantime, as each year passed with the crisis unresolved and escalating the plight of the captive American citizens of the newly independent nation generated a captive audience back home. Contemporary newspapers published not only journalistic comment but also diplomatic material, including consular correspondence and instructions, addresses to Congress, letters from diplomats involved in the negotiations, and ratified treaties. This wealth of material intended to keep the public informed of diplomatic efforts, the movements of corsairs in the Mediterranean and the Atlantic, the condition of their captives, fundraising efforts to redeem the captives, as well as debates about the best course of action for the United States to take to resolve the crisis. However, spotlighting the use of the term piracy to refer to Barbary corsairing in this material reveals the sense of the pressing nature and threatening quality of the crisis as presented and perceived, and the confusion and inability of the United States to resolve it quickly as the American captives continued their foreign captivity.

In trying to respond to the crisis, both newspaper articles and diplomatic correspondence posed questions regarding whether the US would pay tribute or not and in so doing referred to the Barbary corsairs as pirates. Some viewed becoming a tributary state itself as a failure to protect the US's independence. George Washington, for example, was inclined in 1786 "to reject the system of tribute and forced gifts as inappropriate to republican states composed of virtuous freemen." He asked, "In such an enlightened, in such a liberal age,

30 "The Brother of the Swedish Consul to D. Humphreys, Esq. November 13, 1793," in *American State Papers: Foreign Affairs* 1, pp. 414–15 (p. 415).

31 "List of American vessels captured by the corsairs of Algiers, in consequence of the peace with Spain, in 1785, and the truce with Portugal, in 1793" in Goldsborough, *Naval Chronicle*, p. 53. The American vessels and the number of crew captured by Algerine corsairs in October 1793 are as follows: the ships *Thomas* (9) and *Hope* (18) and the schooner *Despatch* (7), Oct. 8; the brig *George* (8) and the brigantines *Olive Branch* (7) and *Jane* (8), Oct. 11; the schooner *Jay* (7), Oct. 12; the ship *Minerva* (17), Oct. 18; the ship *President* (12), Oct. 23; the brig *Polly* (9), Oct. 25.

how is it possible the great maritime powers of Europe should submit to pay an annual tribute to the little piratical States of Barbary?" Paying tribute, he argued, "exemplified a lack of national honour and Old-World corruption," two qualities that the US had aimed to break free from in its independence. Instead of paying tribute, Washington supported "crush[ing] [the Algerines] into nonexistence" by establishing a navy, arguing that "such a banditti [...] might for half the sum that is paid them [as tribute] be exterminated from the Earth." For Washington, not only was it ideologically important for the US not to become a tributary state to the "piratical States," or the lawless "banditti," but it was also more economically viable for a navy to be established as it would cost the US half as much to defend itself than to pay tribute.[32] Washington's use of the term piracy, with its connotations of lawlessness, thus justified his ideological need to refuse the payment of tribute and instead to develop a navy with which to protect American interests.

Since the early 1780s, Thomas Jefferson also referred to the actions of the North African nations as piracy in his attempts to gain support for the establishment of what historian Ray Irwin has labelled an "international anti-piratical confederacy."[33] Jefferson's bid however failed and because Jefferson saw American commerce "as the bedrock of its power and position in the world system, his 'anxiety' as a result of the increasing pressure imposed on American commerce" by the threat of what he deemed piracy was "extreme."[34] Fearing that the Algerines would "ask such a tribute for a forbearance of their piracies as the US would be unwilling to pay" Jefferson grappled with the best course of action for the US to take in response to Barbary "depredations."[35] Like Washington, Jefferson now called for the establishment of a navy when writing to James Monroe in 1785, as he argued that the Barbary States are "contemptibly weak," and that Algiers is a foe "within our dimensions." Despite this assertion and the confidence it espouses, his letter simultaneously exposes the lack of an American naval military capability and signals the US's diplomatic failures highlighting that "our affairs with the pyratical states are distressing."[36] The use of the term "piratical" here is loaded, evoked as a means of reducing the North African corsairs given the term's connotations including that of lawlessness. In this sense, its use thus reveals an anxiety and defensiveness in relation to the

32 Washington, "From George Washington to Lafayette, 15 August 1786," pp. 214–16.

33 Irwin, *Diplomatic Relations*, p. 449.

34 Sofka, "Jeffersonian Idea," p. 519; Jefferson, "To John Adams from Thomas Jefferson, 6 August 1785," pp. 306–8.

35 Jefferson, "Report Relative to the Mediterranean Trade, December 28, 1790," p. 520.

36 Jefferson, "From Thomas Jefferson to James Monroe, 6 February 1785."

crisis and the United States' inability to generate an effective diplomatic response and broader political goals for the development of a navy.

Newspapers also employed the same terms relating to piracy to the same effect. On Christmas day 1793, following the capture of the eleven American ships by the Algerine corsairs in the previous two months, the *Columbian Centinel* engaged in the debate regarding how best to deal with the threat posed by Algiers. It argued, for example, that: "the United States in Congress assembled, have only to order the immediate purchase of eight or ten stout frigates [...] that this force would be sufficient to keep the pirates within their own ports, or through the payment of tribute," but "[t]hat the whole Christian world should be tributary to a nest of pirates is very extraordinary and humiliating."[37] The Barbary crisis thus took on a particular resonance for the newly independent nation in newspaper articles such as this one through the use of the term pirate as a pejorative.

Newspaper reports of the escalation of the crisis following 1793 also took on a more sinister character. The crisis not only posed a threat to American shipping, commerce and seamen but also to the physical geography of the US itself, exposing American soil to the threat of sudden invasion and attack. Alongside reporting that Algiers captured the ship *Sally* on December 26, 1793, *The Baltimore Daily Intelligencer* it argued that the threat posed by Algerine predations was not just aimed at American merchants and sailors, nor to the farmers whose goods were exported to Europe and who would lose their trade, but to the US itself in a geographical sense. The article argues that American shores were liable to attack and invasion—despite the Algerines never having threatened the US in such a way. The newspaper report states:

> The question of putting our country in a posture of defence against any sudden invasion, is before our national legislature; and, we have no doubt, will be decided to the satisfaction of all good citizens. It is singular for an independent people, who enjoy more of the bounties of nature, more freedom and prosperity than any other state or kingdom now known in the world, to leave her trade, her seacoast, her rich and flourishing towns, totally defenceless: by any little piratical fleet that scours the ocean. What shall be done? If we are attacked we must defend ourselves. By what means? By a navy.

Here, the threat posed by the "scourge of Barbary piracy" incrementally escalates from trade, which is at risk, to the seacoast towns of the US. With

37 "Of the Algerine War, Boston. Dec. 25," *Columbian Centinel,* December 25, 1793.

each word, the threat to America increases or develops by successive stages until suddenly its cities are threatened. Becoming a tributary state or being "laid under contribution" is on par with her "largest cities [being] burnt." The phrase "laid under contribution" is one often used to refer to the actions of pirates as it describes laying claim to property whether in the form of seized vessels, or in this instance the nation itself. The pejorative term "little piratical fleet" operates in the context of the pursuit of a broader political goal: the development of a navy, which the article concludes is the only defence.

Despite this demand for urgency no immediate progress was made in response to the Barbary crisis and the rampant disagreement among the people in charge of negotiations about how to respond to the crisis, whether through the payment of tribute or the development of a navy, remained unresolved until March 10, 1794, when the House of Representatives passed the Bill to provide a United States' Navy by two votes. Approval from the Senate closely followed and on March 27, the Bill became law when Congress passed "An Act to Provide a Naval Armament."[38] The Act authorised President Washington to "provide, equip and employ" six ships to protect the United States and its interests against "the depredations committed by the Algerine corsairs on the commerce of the United States."[39] Even while having passed an Act to build a navy, US Congress did not rule out the option of establishing a treaty and becoming a tributary state. While actions were being taken to protect American shipping by means of a naval force, efforts to negotiate a peace with Algiers continued. This signalled that Congress remained irresolute and undecided as to the best course of action to take, remaining in a desperate defensive posture.

Finally, on September 5, 1795, following long and difficult negotiations the US signed a treaty in Algiers entitled "The Treaty of Peace and Amity." The treaty called for the establishment of a "firm and sincere peace and amity between the President and citizens of the United States of North America, and Hassan Bashaw, Dey of Algiers, his Divan and subjects; the vessels and subjects of both nations reciprocally treating each other with civility, honour and respect."[40] In order to establish such peace America had agreed to pay Algiers $642,500, and an annual tribute of twelve thousand Algerine

38 Irwin, *Diplomatic Relations*, pp. 65–66.

39 "An Act to Provide a Naval Armament (1794)," in National Archives (Washington, DC, United States), https://www.archives.gov/education/lessons/new-us-navy/act-draft.html (accessed August 31, 2022).

40 "Treaty of Peace and Amity, Signed at Algiers September 5, 1795," in The Avalon Project: Documents in Law, History and Diplomacy, http://avalon.law.yale.edu/18th_century/bar1795t.asp (accessed August 31, 2022).

sequins ($21,600) in naval stores.[41] The President to the Senate submitted the treaty which was ratified on February 15, 1796.

Nevertheless, even after signing the treaty, obtaining the funds proved difficult for the United States and Dey Hassan Pasha grew impatient. Delays in trying to negotiate with all the creditors necessary to acquire the funds put the treaty, which American diplomats had worked so long to secure, in jeopardy and the Dey, tired of waiting, gave the United States an ultimatum.[42] On April 3, 1796, the Dey put forth "his final determination," granting the US thirty days to pay the money as stipulated in the treaty. After the thirty days, the treaty would end, and his cruisers would bring in more American vessels.[43] Adding to the United States' economic burden to resolve the crisis, it was thus necessary to present the Dey with various presents of great monetary value and to placate him while America gathered the necessary funds. Ultimately, the Treaty was put in place and the surviving American captives were free to return to the United States in 1796.

Despite the thirteen years that had lapsed since the beginning of the Barbary crisis and despite the ratification of the Constitution, the Barbary crisis ended as it had begun, with the US lacking decisive action and being unable to implement a coherent foreign policy, underscoring its fragile political state and lack of institutions with which to do so. This is also true of American negotiations with Tunis and Tripoli. The US ambassador set a price for peace with Tripoli and Tunis in 1786. However, despite this agreement "nothing was done." Jefferson reasoned years later that the "demands" which the US ambassador had agreed upon "were beyond the limits of Congress, and of Reason."[44] The commissioners agreed upon these sums, not because they were under some illusion as to the ability of the Unites States to pay. Instead, as Adams argued, feeling the pressure, they feared that if they did not agree to find the necessary money, for example by borrowing the funds from Holland, "after the delay of a [y]ear or two it may cost [the United States] a [m]illion."[45] The United States' negotiations with Tunis and Tripoli

41 Irwin, *Diplomatic Relations*, p. 72.

42 Pickering and Wolcott, "Reports of the Secretary of State, and of the Secretary of the Treasury, Relative to the Present Situation of Affairs with the Dey and Regency of Algiers. Accompanying a Confidential Message from the President of the United States, Received the 19th of January, 1797," *Daily Advertiser*, March 20, 1797.

43 Irwin, *Diplomatic Relations*, p. 72.

44 Jefferson, "III. Report on American Trade in the Mediterranean, 28 December 1790" in Boyd, *Papers of Thomas Jefferson, 4 November 1790–24 January 1791* pp. 423–30, http://founders.archives.gov/documents/Jefferson/01-18-02-0139-0004 (accessed August 31, 2022).

45 John Adams, "John Adams to Elbridge Gerry. Dated March 19, 1786," in Lint et al., *Papers of John Adams, December 1785–January 1787*, p. 215.

thus further exposed failures to direct and communicate clear instructions to its diplomats. As a result, the negotiations for the treaty with Tripoli fell through and the two nations were engaged in intermittent conflicts in the following years.

What becomes clear is the way in which the term piracy is deployed in diplomatic correspondence and newspaper articles in relation to this complex foreign crisis. Essentially, the reduction of North African corsairing to Barbary piracy with its previously mentioned connotations, betrays the difficulty diplomatic decision-makers experienced in grappling with the crisis. In the absence of a clear and coherent consensus on how to respond and the ability to enact such a response, key US figures resorted to dismissing their opposition as "little pyratical states."

Turning now to the contemporary use of the term pirate and piracy in media and historical scholarship in the US reveals a similarly reductive tendency to misrepresent North African activities in the context of the Barbary crisis. In much the same way, these designations of Barbary piracy are charged and appropriated within broader political narratives, revealing persistent anxieties with regard to how the US encountered its first foreign crisis alongside present-day threats. Designations of piracy are found in a range of popular historical accounts including but not limited to *Jefferson's War: America's First War on Terror 1801–1805*, *The Pirate Coast: Thomas Jefferson, The First Marines, and the Secret Mission of 1805*, and *Thomas Jefferson and the Tripoli Pirates: The Forgotten War that Changed American History*.[46] The use of such terms also features in several western media articles including those entitled "The Return of the Barbary Pirates," "Lessons from the Barbary Pirate Wars" and "ISIS could become the pirates of the Mediterranean," which have been published in the *Washington Times*, *The New York Times* and the *Daily Mail* respectively.[47]

Inspired by modern-day events, particularly the two Iraq Wars and the preoccupation with radical Islam since September 11, 2001, conservative politicians, journalists, historians, and activists have, since the outset of the twenty-first century, drawn parallels between the actions of "Barbary pirates" and modern-day fundamentalist groups. Looking at early American encounters with North Africa, they express the view that modern-day relations between the United States and the "Islamic world" found their roots in these early conflicts between Christians and Muslims. This so-called "Barbary Analogy" positions "Mediterranean 'pirates' of [...] North Africa

46 Wheelan, *Jefferson's War*; Zacks, *Pirate Coast*; Kilmeade and Yaeger, *Thomas Jefferson*.

47 Stewart, "Barbary Pirates"; Gettleman "Barbary Pirates."

as the historical forebears of twenty-first century Islamist militants, [or] as terrorists by another name."[48] For example, Joseph Wheelan, a one-time reporter and editor for The Associated Press, refers to the Tripolitan War as "America's First War on Terror," linking it to George W. Bush's Global War on Terror. Expanding on this, Joshua London, the journalist and author of *Victory in Tripoli*, writes in *The National Review*, a leading American conservative editorial magazine, that "there is much in the history of America's wars with the Barbary pirates that is of direct relevance to the current global war on terrorism."[49]

Similarly to the use of the term piracy in the late eighteenth century, these links diminish what Linda Colley describes as the "vital and officially recognized part of the Barbary States' revenue-raising machinery," vilifying the actions of the corsairs by putting them "on a par with [...] sea-robbers."[50] Such associations also reduce the contemporaneous actions and roles of other nations' corsairs thus introducing a religious and racial element to the designation of piracy. Indeed, as Nabil Matar posits, the use of the term piracy perpetuates and deepens a "binary between evil and good, Muslim and Christian, African and European."[51] In this instance this is a racially loaded binary that glosses over the realities of the period under study and perpetuates an interpretation of difference and enmity between the United States and the monolithic racialised other of the "Barbary states" with significant implications. For example, Brian Kilmeade, a Fox News television presenter and co-author of *Thomas Jefferson and the Tripoli Pirates*, provides a triumphalist narrative of the period focusing on the early nineteenth-century American response. He suggests that present American leaders when responding to al-Qaeda, the Taliban, or ISIS learn from Thomas Jefferson's response to the "ruthless, terrorist pirates" of the Barbary states.[52]

Thus, contemporary historical accounts generalise the Barbary crisis as one not comprised of local conflicts between the US and the individual North African nations of Morocco, Algiers, Tunis and Tripoli, but rather as a "global battle," between Christianity and Islam. Islam is thus made to figure as the US's national enemy. In this vein, Rick Forcier, the executive director of the Christian Coalition of Washington State argues that "Islamic fundamentalists" have "for centuries [...] bombed, hijacked, kidnapped,

48 Leilby, "Terrorists by Another Name."
49 London, "America's Earliest Terrorists."
50 Colley, *Captives*, pp. 44–45.
51 Matar, *British Captives*, p. 3.
52 Kilmeade, *Thomas Jefferson*.

murdered and extorted for the furthering of their religion and the glory of their god Allah." For Forcier, the term "Barbary pirates" is merely a name by which to refer to Islamic terrorists in "times past."[53] Drawing parallels not only glosses over the struggles that the United States experienced in responding to the Barbary crisis, it also creates, as Farber argues, "an implicit solidarity between the Americans of the early Republic and the present day—a timeless 'us' defined by its difference from a North African 'them.'"[54]

The use of the term pirate to refer to Barbary corsairs in newspaper articles published during the Barbary crisis, as well as in diplomatic correspondence, therefore says more of the anxieties, pressing threats, and concerns that confronted US political figures in their response to the Barbary crisis, rather than representing the nature of Barbary actions. Furthermore "piracy" then and now has been appropriated to serve temporally specific political narratives. In the eighteenth century designations of piracy were deployed in the service of the development of a navy with which to protect its interests and arguments opposing tributary payment, while in the twentieth century it operates within discussions of a racialised war against terrorism and triumphalist narratives of American exceptionalism. By noting the coherence between the term's use then and now, this chapter underscores the extent to which designations of Barbary piracy speak more profoundly of the historically entrenched perspectives of those who deploy the term, rather than of the actions of the "pyratical states" themselves.

## Bibliography

Anonymous. "A Plain Citizen, to the Honourable the Convention of the State of Pennsylvania." *Independent Gazetteer*, November 22, 1787.

Anonymous. "Extract of a Letter from London, Dated October 3." *Baltimore Daily Intelligencer, Published as the Baltimore Daily Intelligencer*, December 26, 1793.

Anonymous. "Of the Algerine War, Boston. Dec. 25." *Columbian Centinel*, December 25, 1793.

Anonymous. "Plan of the Treaties with France of 1778, Tuesday, September 17, 1776." In *Journals of the Continental Congress 1774–1779 Edited from the Original Records in the Library of Congress*, edited by Worthington Chauncey Ford. Washington, DC: Government Printing Office, 1905.

53 Forcier, *Tales of Terrorism*.

54 Farber, "Millions for Credit," p. 189.

Anonymous. "Treaty of Amity and Commerce between the United States and France Signed on February 6, 1778." In *Treaties and Other International Acts of the United States of America Documents 1–40: 1776–1818*, 2 vols. Washington DC: Government Printing Office, 1931.

Anonymous. "Articles of Peace and Commerce, between the most Serene and Mighty Princess Anne, by the Grace of God, Queen of Great Britain, France, and Ireland, Defender of the Christian Faith, &c. and the most Illustrious Lord Mustapha Dey, the Bashaw, Aga, and Governors of the Famous City and Kingdom of Algiers, in Barbary; Ratified, Confirmed, and Renewed, by George Byng, Esq; Rear-Admiral of the Red Squadron of Her Majesty's Fleet, on the 28th Day of October, Old Stile, 1703." In *A Collection of Treaties between Great Britain and Other Powers*, edited by George Chalmers, 2 vols. London: John Stockdale, Piccadilly, 1970.

Abun-Nasr, Jamil M. *A History of the Maghrib in the Islamic Period.* Cambridge: Cambridge University Press, 1987.

Bookin-Weiner, Jerome. "Corsairing in the Economy and Politics of North Africa." In *North Africa: Nation, State and Region,* edited by George Joffe, 3–33. London: Routledge, 1993.

Buhite, Russel D. *Lives at Risk: Hostages and Victims in American Foreign Policy.* Wilmington, Delaware: Scholarly Resources, 1995.

Cathcart, James. *The Captives: Eleven Years a Prisoner in Algiers.* La Porte, Indiana: Herald Print, 1899.

Colley, Linda. *Captives: Britain, Empire and the World 1600–1850.* New York: Pantheon, 2002.

Comte de Vergennes. "The Comte De Vergennes to Benjamin Franklin: A Translation, 27 September 1778." In *The Adams Papers: Papers of John Adams, September 1778–February 1779*, edited by Gregg L. Lint, C. James Taylor, Sara Georgini, Hobson Woodward, Sara B. Sikes, Amanda A. Matthews, and Sara Martin, 7 vols, 83–84. Cambridge, Massachusetts: Harvard University Press, 1989.

Commissioners of the United States of America. "Extract of a Letter from the Commissioners of the United States of America to M. Le Comte De Vergennes, Dated Passy, August 28, 1778." In *The Diplomatic Correspondence of the United States of America, from the Signing of the Definitive Treaty of Peace, 10th September 1788 to the Adoption of the Constitutions, March 4, 1789*, 1 vol, p. 570. City of Washington: Blair and Rives, 1837.

Cosman, Madeleine Pelner, and Linda Gale Jones. *Handbook to Life in the Medieval World.* Facts on File, 2008.

Earle, Peter. *Corsairs of Malta and Barbary.* London: Sidgwick & Jackson, 1970.

Edwards, Bryan. *The History, Civil and Commercial, of the British Colonies in the West Indies,* 3rd ed., 2 vols. London: John Stockdale, Piccadilly, 1801.

Farber, Hannah. "Millions for Credit: Peace with Algiers and the Establishment of America's Commercial Reputation Overseas, 1795–1796." *Journal of the Early Republic* 34, no. 2 (2014): 187–217.

Forcier, Rick. "The Tales of Terrorism." *Newsletter of the Christian Coalition of Washington State* (November 2001).

Franklin, Benjamin, Arthur Lee and John Adams. "The Commissioners to the President of the Congress, 7 November 1778." In *The Adams Papers: Papers of John Adams, September 1778–February 1779*, edited by Gregg L. Lint, C. James Taylor, Sara Georgini, Hobson Woodward, Sara B. Sikes, Amanda A. Matthews, and Sara Martin, 7 vols, 196–200. Cambridge, Massachusetts: Harvard University Press, 1989.

Gettleman, Jeffrey. "Lessons from the Barbary Pirates." *The New York Times*, April 11, 2009. http://www.nytimes.com/2009/04/12/weekinreview/12gettleman.html.

Goldsborough, Charles W. "List of American Vessels Captured by the Corsairs of Algiers, in Consequence of the Peace with Spain, in 1785, and the Truce with Portugal, in 1793." In *United States' Naval Chronicle*, p. 53. Washington City: James Wilson, 1824.

Irwin, Ray W. *The Diplomatic Relations of the United States with the Barbary Powers: 1776–1816*. Chapel Hill, NC: The University of North Carolina Press, 1931.

Jefferson, Thomas. "Report of the Secretary of State Relative to the Mediterranean Trade. Communicated to the House of Representatives December 30, 1790 and to the Senate, January 3, 1791." In *American State Papers. Documents, Legislative and Executive of the Congress of the United States, from the First Session of the First to the Second Session of the Twenty-Second Congress, Inclusive: Commencing March 3, 1769 and Ending March 3, 1833. Selected and Edited, Under the Authority of Congress, by Walter Lowrie, Secretary of the Senate, and Matthew St. Clair Clarke, Clerk of the House of Representatives*, 1 vol, p. 104. Washington: Gales and Seaton, 1833.

Jefferson, Thomas. "VIII—Report Relative to the Mediterranean Trade, December 28, 1790." In *The Writings of Thomas Jefferson*, 7 vols, 519–26. Washington, DC: Taylor & Maury, 1854.

Jefferson, Thomas. "From Thomas Jefferson to James Monroe, 6 February 1785." In *The Papers of Thomas Jefferson 2 March 1784–25 February 1785*, edited by Julian P. Boyd, 7 vols, 637–41. Princeton: Princeton University Press, 1953.

Jefferson, Thomas. "From Thomas Jefferson to Nathanel Greene, January 12, 1786." In *The Papers of Thomas Jefferson, 1 November 1785–22 June 1786*, edited by Julian P. Boyd, 7 vols, 167–69. Princeton: Princeton University Press, 1954.

Jefferson, Thomas. "To John Adams from Thomas Jefferson, 6 August 1785." In *The Adams Papers, Papers of John Adams*, edited by Gregg L. Lint, C. James Taylor, Sara Georgini, Hobson Woodward, Sara B. Sikes, Amanda A. Matthews, and Sara Martin, 7 vols, 306–8. Cambridge, MA: Harvard University Press, 2014.

Kilmeade, Brian. *Thomas Jefferson and the Tripoli Pirates: The Forgotten War that Changed American History,* Talk at the Heritage Foundation, 2017. https://www.youtube.com/watch?v=Ews2UoF51pc, November 6, 2015.

Kilmeade, Brian and Don Yaeger. *Thomas Jefferson and the Tripoli Pirates: The Forgotten War That Changed American History.* New York: Sentinel, 2015.

Leilby, Richard. "Terrorists by Another Name: The Barbary Corsairs." *Washington Post,* October 15, 2001. https://www.washingtonpost.com/archive/lifestyle/2001/10/15/terrorists-by-another-name-the-barbary-pirates/af7e4c90-b56c-4e0c-95ea-f079b889a96f.

London, Joshua. "America's Earliest Terrorists: Lessons from America's First War against Islamic Terror." *National Review,* December 16, 2005. http://www.nationalreview.com/article/216283/americas-earliest-terrorists-joshua-e-london.

London, Joshua. *Victory in Tripoli: How War with the Barbary Pirates Teaches Us How to Fight the War on Terror* (New Jersey: John Wiley and Sons, 2005)

London, Joshua. Interview with Jamie Glazov. *Free Republic,* May 9, 2006. http://www.freerepublic.com/focus/f-news/1629124/posts.

Lint, Gregg L. "John Adams on the Drafting of the Treaty Plan of 1776." *Diplomatic History* 2, no. 3 (1978): 313–20.

M. de Sartine. "M. De Sartine to Count De Vergennes, Versailles, September 21, 1778." In *The Diplomatic Correspondence of the American Revolution,* p. 30. Washington: Jackson Hall, 1818.

Matar, Nabil. *British Captives from the Mediterranean to the Atlantic, 1563–1760.* Leiden: Brill, 2014.

Matar, Nabil. "Introduction: England and Mediterranean Captivity, 1577–1704." In *Piracy, Slavery, and Redemption: Barbary Captivity Narratives from Early Modern England,* edited by Daniel J. Vitkus, 1–52. New York: Columbia University Press, 2001.

Murphy, Martin N. "The Barbary Pirates," *Mediterranean Quarterly* 24, no. 4 (2013): 19–42.

Panzac, Daniel. *Barbary Corsairs: The End of a Legend, 1800–1820.* Translated by Victoria Hobson and John E. Hawkes. Leiden: Brill, 2005.

Pickering, Timothy and Wolcott, Oliver. "Reports of the Secretary of State, and of the Secretary of the Treasury, Relative to the Present Situation of Affairs with the Dey and Regency of Algiers. Accompanying a Confidential Message from the President of the United States, Received the 19th of January, 1797." *Daily Advertiser,* March 20, 1797.

Rech, Walter. *Enemies of Mankind: Vattel's Theory of Collective Security.* Leiden: Martinus Nijhoff, 2013.

Rech, Walter. "Ambivalences of Recognition: The Position of the Barbary Corsairs in Early Modern International Law and International Politics." In *Piracy and*

*Captivity in the Mediterranean*, edited by Mario Klarer, 76–98. London: Routledge, 2019.
Roberts, Priscilla H., and James N. Tull. "Moroccan Sultan Sidi Muhammad Ibn Abdallah's Diplomatic Initiatives Toward the United States, 1777–1786." *Proceedings of the American Philosophical Society* 143, no. 2 (1999): 233–65.
Sears, Christine. *American Slaves and African Masters: Algiers and the Western Sahara, 1776–1820*. New York: Palgrave Macmillan, 2012.
Sha'ban, Fuad. *Islam and Arabs in Early American Thought*. Durham, NC: Oxford University Press, 1991.
Sofka, James R. "The Jeffersonian Idea of National Security: Commerce, the Atlantic Balance of Power, and the Barbary War, 1786–1805." *Diplomatic History* 21, no. 4 (1997): 519–44.
Stewart, Thomas C. "The Return of the Barbary Pirates." *The Washington Times*, March 10, 2015. http://www.washingtontimes.com/news/2015/mar/10/thomas-stewart-isis-may-herald-return-of-the-barba.
Tabarrok, Alexander. "The Rise, Fall, and Rise Again of Privateers." *The Independent Review* 11, no. 4 (2007): 565–77.

## About the Author

**Anna Diamantouli** is an Assistant Professor of American Literature at Deree, The American College of Greece. Her research interests lie in the post-revolutionary period and embrace transatlantic writing, the literature of slavery and captivity, and the representation of North African culture and North American literature.

# Afterword

*Claire Jowitt*

**Abstract**

Claire Jowitt's Afterword focuses on the ways a poetic depiction of early modern pirates as economic radicals is both located in a particular time and space, and simultaneously can possess meanings that resonate with other temporal and spatial contexts. Samuel Rowlands' 1613 poem "The Picture of a Pirat" focuses on the relationship between commerce and social degree. It is suggested that maritime predation, real and imagined, can operate as a cultural intermediary capable of speaking to overarching social, political, economic issues across time and space.

**Keywords:** pirates; poetry; Samuel Rowlands; commerce; economic radicals; social degree

According to Marcus Rediker, arguably the most influential historian of early modern piracy of the last forty years, "[o]ne way to think about the pirate ship is that it was a model of and for self-organized workers [...] Exploited and abused wage workers came together to create a space of autonomy where they would, as one pirate put it, have 'the choice in themselves' about how to organize their lives."[1] For blue water empires within an international capitalist system, maritime outlaws represented a direct and subversive attack on the authority of the ruling social order and early modern nation states sought to brutally "pulverize" the pirates' alternate social order.[2] From this, it is easy to see how piracy came to be romanticised by both sailors and ordinary people and maritime outlaws seen as heroes within

1 Marcus Rediker, "The History of Pirates & Stealing our Futures Back," *Asia Art Tours* (December 5, 2021), https://asiaarttours.com/marcus-rediker-the-history-of-pirates-learning-to-steal-back-our-future/ (accessed January 23, 2023).

2 Ibid.

Coakley, J., C. Nathan Kwan and D. Wilson, *The Problem of Piracy in the Early Modern World: Maritime Predation, Empire, and the Construction of Authority at Sea.*
Taylor & Francis Group, 2024
DOI 10.5117/9789463720960_AFT

histories of class struggle and social protest.[3] When the success of Golden Age pirates' marauding and commercial activities either made them too conspicuously economically successful, or they broke out on a global scale, early modern state authorities decided to eliminate the piracy and pirates they were unable to control, intervening against them with the full force of state apparatus.[4] Suppressing piracy was essential to early modern states because extra-legal violence at sea threatened the trade and order they sought to impose in home waters and overseas.[5]

Rediker's model relates primarily to Atlantic world pirates from the Golden Age challenging the merchant and naval fleets of European nation states for sovereignty at sea. His argument is located in time and space and has become central to historiography about maritime predation within these historical and geographical boundaries. At the same time, Rediker's work has also been useful in revealing throughlines concerning how the regulation of maritime predation operated in other temporal and spatial contexts both in similar and in divergent ways. This Afterword focuses briefly on an early seventeenth-century cultural representation of private maritime enterprise to focalise how specific temporal and spatial contexts govern the application of state power to the problem of piracy *and* simultaneously show the ways that patterns of engagement and rhetoric concerning maritime predation may be shared across time and space.

Histories of piracy often describe the late sixteenth century as a period of state-sponsored piracy with pirates acting as "instruments of empire" and the boundary between licit and illicit activity being highly permeable both for individual actors and as a practice.[6] After the accession of James I and VI in 1603, the boundary became less porous following the ending of the Anglo-Spanish war in 1604, and the problem of piracy amongst English seamen became particularly acute.[7] In 1613 the prolific pamphlet and verse writer Samuel Rowlands published a collection of pirate poems *More Knaves yet? The Knaves of Spades and Diamonds. With new Additions*.[8] Explicitly and in keeping with Jacobean state policies, Rowlands' collection expresses

3 See Rediker, *Deep Blue Sea*; Hill, "Radical Pirates?" On the romanticisation of pirates see Rennie, *Treasure Neverland*.

4 Robert C. Ritchie argues that after c.1670 "commercial piracy" became the dominant form, organised by merchants and tolerated by communities to enable the circulation of scarce commodities. See Ritchie, *Captain Kidd*.

5 Thomson, *Mercenaries, Pirates and Sovereigns*.

6 Appleby, *Bloody Flag*; Jowitt, *Culture of Piracy*; Murphy, "English Piracy."

7 Earle, *Pirate Wars*.

8 Rowlands, *More Knaves yet?*.

politically conventional hostility to the crime of piracy and the men who commit acts of violence at sea. In the first poem in the pirate sequence, "The Picture of a Pirat," Rowlands describes the Crown's hoped-for pirate career trajectory as a life of "outragious evils" apparently inevitably leads pirates to "Anker at the Gallowes" in punishment at Wapping, the traditional site for pirate execution.[9] Comprising just eighteen short lines, the brevity of the poem seems designed to accord with the length of a pirate's career at sea, which on average extended merely to a year or two, though more early modern pirates either abandoned the activity or died at sea than were executed.[10]

Rowlands' references to pirates in his poetry collection were topical. Though repeated and increasingly draconian royal proclamations against piracy and those that aided or abetted it were issued on behalf of King James, and some pirates were brought to justice through execution, English pirates still pillaged at will in spite of the best efforts of the English navy to reduce their numbers.[11] Since the Crown was unable to rid the seas of this large number of pirate vessels and faced increasingly vociferous international condemnation because of the pirates' attacks on foreign shipping, in 1611 the king accepted a Dutch request to search the Irish and English coasts for pirates, a revealing admission of English naval weakness.[12] In addition, James began to negotiate with pirates concerning granting them pardons, though this was recognised as a policy that undermined his prestige and position: according to the Privy Council it would have been better "for the King's honour to consume them all than to accept any to mercy."[13] In 1612, James issued a General Pardon on terms highly favourable to the pirates, allowing them to keep their plunder if they surrendered.[14] As a result, the prominence of pirates in Rowland's collection of poetry reflects the interest and anxiety generated by the pirates *and* the state policies designed to deal with them. The pirates' executions in Rowlands' verses are a description of what James had struggled to achieve politically or practically, since the king's prestige had been diminished and, there was no noticeable effect in actually reducing the number of English pirates at sea.

However, in contrast to the apparent political orthodoxy of Rowlands' other pirate poems, one poem entitled "Lightly come, Lightly goe" expresses

9 Rowlands, "The Picture of a Pirate" in Rowlands, *More Knaves yet?*, A4v.

10 See Rediker, *Villains of All Nations*, p. 9.

11 For histories of piracy in this period see, for instance, Gosse, *Piracy*; Cordingly, *Black Flag*; Hebb, *Piracy*.

12 Earle, *Pirate Wars*, p. 60.

13 Quoted in Ibid., p. 61.

14 Ibid.

an entirely different perspective on piracy, showing pirates mixing seemingly freely with Londoners in order to sell their booty, and justifying verbally and loudly their trade by depicting themselves as economic radicals whose trading practices benefit the wider metropolitan population:

> Captaine, ti's we do make things cheape or deare,
> As by our peny-worths it doth appeare.
> A yard with us is just in length a pike:
> To buy silkes so, what man is't will dislike,
> Or say we use our customers amisse?
> Your London measure (friends) comes short of this,
> Bee't three pile Velvet, Sattin, Taffaty,
> A Souldiers Pike's the Ell we measure by.
> Thus much for Mercers: next for Grocers trade,
> Our weight is like unto our measure made,
> Our pound's a Cannon bullet, good downe waight,
> In Spice, or Suger, this is no deceit.
> Then for our wines (the squeaking Vintners Art)
> We can affoord them for a penny a quarte.
> Yea fill yon pintes even by the bucket full,
> But how can this be, saith some simple gull,
> That never travail'd out of Bow bell sound?
> Marry Sim-simple heare and stand thy ground.
> That which we have, we steale from friends and foes,
> It comes good cheape, and so good cheape it goes.[15]

The pirates explicitly proclaim the remarkable value their internationally sourced goods represent to their customers and, as part of their marketing pitch, celebrate how they are able to undercut merchants' prices. For example, the pirate announces to his prospective customers that pirates measure cloth in the "Souldier's Pike's," referring to the infantry weapon of between about ten and twenty feet in length, rather than the yard of thirty-six inches or the standard tailor's "Ell" of forty-five inches. The pirates' customers can enjoy similarly exceptional deals in dried grocery goods and in vintnery, since "pirate" measures are exorbitantly generous in this case as well. Pirates charging only a penny for a quart (two pints) of wine, for instance. Such modest "pirate" prices enable expensive foreign and luxury goods to be within the reach of all sectors of the metropolitan populace.

15 Rowlands, "Lightly come, Lightly goe" in Rowlands, *More Knaves yet?*, B2v.

"Velvet, Sattin, Taffaty" and "Spice, or Suger" were imported goods normally only within the purchasing power of the wealthy, or, in the case of rich fabrics, there was a further rebellious dimension in pirates' selling them to the poor. According to the sumptuary laws, these fabrics were exclusively for the use of the elite classes, since prohibitions about wearing certain types of cloth were regulated by social degree. In 1566, for instance, a decree stated that "No man under the degree of a knight or of a lord's room [...] shall wear any hat or upper cap of velvet [...] on pain to forfeit ten shillings."[16] Pirates thus provide the poor with access to textiles from which normally they were doubly debarred, due to the price of the material and their social status.

As a result, this poem represents pirates as economic radicals and, using Rediker's terminology, as "class warriors." Through their use of bargain prices, pirates break the dominance of the London Guild Companies who regulated trade by fixing prices for goods so that the majority of the population could never afford imported luxury items. Under Elizabeth I, price regulation had been further strengthened and codified by the Statute of Artificers (1563), a law that outlined specific enforcement duties of local justices of the peace, aldermen, and local administrators.[17] This poem, then, hints at the economic challenges and class-based radicalism that Jacobean pirates represented to the mercantilism of early modern trade companies and, because in this period mercantilism is profoundly connected to the operations and rise of nation states, to government itself.[18] Pirates disrupt the legitimate business of the circulation of goods at fixed prices, changing the established channels of distribution and, through selling worldly goods cheaply, foster a popular, folk identity, like Robin Hood, as champions of the poor and deserving.[19] "Black market" practices are glamorised and exculpated in this poem. As Eric Hobsbawm (one of the key influences on Rediker's thinking about piracy) writes concerning banditry, "[i]n a society in which men live by subservience, as ancillaries to machines of metal or moving parts of human machinery, the bandit lives and dies with a straight back."[20] Rowlands' noisy, disruptive hawking pirate in "Lightly come, Lightly goe" clearly has as straight a back as any of Rediker's Golden Age incarnations.

16 Quoted in Peck, *Consuming Splendor*, p. 6.

17 Ekelund Jr., and Hébert, *Economic Theory*, p. 53.

18 For discussion of Renaissance history in terms of commodities and material culture, see Jardine, *Worldly Goods*.

19 For discussion of the political meanings of early Robin Hood stories see Pollard, "Political Ideology."

20 Hobsbawm, *Bandits*, p. 114.

This case study of the rhetorical, representational, and multivalent cross-currents in the treatment of maritime predation both situates it in a specific temporal and spatial context of the marketplace of Jacobean London *and* forges connections across time and space to Golden Age piracy when pirates are viewed as agents of international commerce. As such, it amplifies the concerns of John Coakley, C. Nathan Kwan, and David Wilson's "Problem of Piracy Network." Formed in 2019, the network studies piracy and maritime predation across the globe, time periods, and fields. Working athwart legal, cultural, political, economic, imperial and/or colonial, naval, and social history, the inaugural "The Problem of Piracy: An Interdisciplinary Conference on Plunder by Sea Across the World from the Ancient to the Modern" was held in Glasgow, Scotland, in June 2019. Forty delegates presented papers identifying key themes for the study of piracy, most notably the state's role in defining, tackling, or benefiting from maritime predation. This volume thus offers a distinctive perspective and a nuanced understanding of how "piracy" was semantically constructed and addressed in practice by states in a period of European history—roughly from the late Middle Ages to the beginning of the nineteenth century—when the forces of emergent globalisation meant that oceans and seas, and the vessels and crews who traversed them, were increasingly central to state interests.

The eight essays in the collection range widely in scale and scope from, for instance, studies of noteworthy individual actors such as Golden Age pirate Henry Glasby, to regionally focused treatments of the multi-imperial political economy of the Caribbean or the commercial and geopolitical competition in the late medieval Irish Sea world, to the cultural work undertaken by succeeding editions of arguably the most important text ever written in creating popular perceptions of pirates, *A General History of the Pyrates*. The editors frame the collection as a set of detailed case studies that pay attention to local and regional contexts, and show how jurisdiction over piracy, the practices of piracy, and the ways piracy was semantically represented differ across time as much as over space. But the connections between case studies are also clear. Each author in the collection reveals, in relation to specific locations and at certain times, how states only exercised control over maritime predation and its perpetrators intermittently and for particular purposes, and simultaneously shows how governments and other authorities sought to represent these power struggles semantically to control a narrative to their own best advantage.

Since 2019, the "Problem of Piracy Network" continues to be active and successful in promoting and encouraging an ever-wider diversity of voices to address interpretative questions and issues concerning predation at sea and its representation. It explores, for instance, the role of pirates as cultural intermediaries, active in both the formation and destruction of national, colonial, and imperial identities, and as transnational and transoceanic mediators of global geo-political forces. Coakley, Kwan, and Wilson also intend that the network will expand further the range of geographies and disciplines the network engages, to bring broader and more inclusive perspectives to how we understand the relationships between maritime predators and colonial and imperial histories and to commercial, racial, religious, legal, cultural, political, social, naval, contexts.

## Bibliography


Appleby, John C. *Under the Bloody Flag: Pirates of the Tudor Age*. Stroud: The History Press, 2009.

Cordingly, David. *Under the Black Flag: The Romance and the Reality of Life among the Pirates*. London: Random House, 1995.

Dagher-margosian, Matt. "Marcus Rediker: The History of Pirates & Stealing our Future Back." *Asia Art Tours*, December 5, 2021. https://asiaarttours.com/marcus-rediker-the-history-of-pirates-learning-to-steal-back-our-future.

Earle, Peter. *The Pirate Wars*. London: Methuen, 2003.

Ekelund Jr., Robert B., and Hébert, Robert F. *A History of Economic Theory and Method*. Long Grove, IL: Waveland Press, 2007.

Gosse, Philip. *The History of Piracy*. New York: Longmans, Green & Co., 1932.

Hebb, David D. *Piracy and the English Government 1616–1642: Policy-Making Under the Early Stuarts*. Aldershot: Scolar, 1994.

Hill, Christopher. "Radical Pirates?" In *The Collected Essays of Christopher Hill, Volume III: People and Ideas in Seventeenth-Century England*, 161–87. Amherst: University of Massachusetts Press, 1987.

Hobsbawm, Eric. *Bandits*. London: Weidenfeld and Nicolson, 1969.

Jardine, Lisa. *Worldly Goods: A New History of the Renaissance*. New York: Nan A. Tales, 1996.

Jowitt, Claire. *The Culture of Piracy, 1580–1630: English Literature and Seaborne Crime*. Aldershot: Ashgate, 2010.

Murphy, Elaine. "Early Modern English Piracy and Privateering." In *The Routledge Companion to Marine and Maritime Worlds 1400–1800*, edited by Claire Jowitt, Craig Lambert, and Steve Mentz, 368–87. London and New York: Routledge, 2020.

Peck, Linda Levy. *Consuming Splendor: Society and Culture in Seventeenth-Century England*. Cambridge: Cambridge University Press, 2005.

Pollard, A. J. "Political Ideology in the Early Stories of Robin Hood." In *Outlaws in Medieval and Early Modern England: Crime, Government and Society, c. 1066–c. 1600*, edited by John C. Appleby and Paul Dalton. Farnham: Ashgate, 2009.
Rediker, Marcus. *Between the Devil and the Deep Blue Sea: Merchant Seamen, Pirates, and the Anglo-American Maritime World, 1700–1750*. Cambridge: Cambridge University Press, 1987.
Rediker, Marcus. *Villains of All Nations*. Boston: Beacon Press, 2004.
Rennie, Neil. *Treasure Neverland: Real and Imaginary Pirates*. Oxford: Oxford University Press, 2013.
Ritchie, Robert C. *Captain Kidd and the War against the Pirates*. Cambridge, MA: Harvard University Press, 1986.
Rowlands, Samuel. *More Knaves yet? The Knaves of Spades and Diamonds. With new Additions*. London: Printed by Edward Allde for John Tap, 1613.
Thomson, Janice. *Mercenaries, Pirates and Sovereigns: State-Building and Extraterritorial Violence in Early Modern Europe*. Princeton: Princeton University Press, 1994.

## About the Author

**Claire Jowitt** (@clairejowitt) is Professor of English and History at the University of East Anglia. She is author of *The Culture of Piracy 1580–1630: English Literature and Seaborne Crime* (2010) and co-editor (with Craig Lambert and Steve Mentz) of *The Routledge Companion to Marine and Maritime Worlds 1400–1800* (2020). She co-edits (with John McAleer) the book series *Maritime Humanities 1400–1800: Cultures of the Sea* for Amsterdam University Press.

# Bibilography


Abdullah bin Abdul al Kadir. “The Hikayat Abdullah. An Annotated Translation by A. H. Hill, M. A., B.Sc.” *Journal of the Malayan Branch of the Royal Asiatic Society* 28, No. 3 (June 1955): 3–354.

Abun-Nasr, Jamil M. *A History of the Maghrib in the Islamic Period.* Cambridge: Cambridge University Press, 1987.

Amirell, Stefan Eklöf. “Civilising Pirates: Nineteenth Century British Ideas about Piracy, Race and Civilization in the Malay Archipelago.” *HumaNetten* 41 (Autumn 2018): 25–45.

Amirell, Stefan Eklöf. *Pirates of Empire: Colonisation and Maritime Violence in Southeast Asia.* Cambridge: Cambridge University Press, 2019.

Amirell, Stefan Eklöf. “The Making of the ‘Malay Pirate’ in Early Modern European Thought.” *Humanities* 9, no. 3 (2020): 91.

Amirell, Stefan Eklöf and Leos Müller, eds. *Persistent Piracy: Maritime Violence and State-Formation in Global Historical Perspective.* Basingstoke: Palgrave Macmillan, 2014.

Amirell, Stefan Eklöf, Bruce Buchan, and Hans Hägerdal, eds. *Piracy in World History.* Maritime Humanities, 1400–1800. Amsterdam: Amsterdam University Press, 2021.

Anand, Ram P., *Origin and Development of the Law of the Sea.* The Hague: Martinus Nijhoff, 1983.

Andaya, Barbara W. “Recording the Past of ‘Peoples Without History’: Southeast Asia’s Sea Nomads.” *Asian Review* 32, No. 1 (2019): 5–33.

Andaya, Leonard Y. *Leaves of the Same Tree: Trade and Ethnicity in the Straits of Melaka.* Honolulu: University of Hawai’i Press, 2008.

Anderson, Benedict. *Imagined Communities: Reflections on the Origin and Spread of Nationalism*, Revised Edition. London: Verso, 1991.

Anderson, J. L. “Piracy in the Eastern Seas, 1750–1850: Some Economic Implications.” In *Pirates and Privateers: New Perspectives on the War on Trade in the Eighteenth and Nineteenth Centuries*, edited by David J. Starkey, E. S. van Eyck van Helsinga and J. A. de Moor, 87–105. Exeter: University of Exeter Press, 1997.

Anderson, John. *Considerations Relative to the Malayan Peninsula and the British Settlements in the Straits of Malacca.* Prince of Wales Island, 1824.

Andrews, Kenneth R. *Trade, Plunder, and Settlement: Maritime Enterprise and the Genesis of the British Empire, 1480–1630.* Cambridge: Cambridge University Press, 1984.

Angster, Julia. *Erdbeeren und Piraten: Die Royal Navy und die Ordnung der Welt 1770–1860.* Göttingen: Vandenhoeck & Ruprecht, 2012.

Anonymous. "A Plain Citizen, to the Honourable the Convention of the State of Pennsylvania." *Independent Gazetteer,* November 22, 1787.

Anonymous. "Extract of a Letter from London, Dated October 3." *Baltimore Daily Intelligencer, Published as the Baltimore Daily Intelligencer,* December 26, 1793.

Anonymous. "Of the Algerine War, Boston. Dec. 25." *Columbian Centinel,* December 25, 1793.

Anonymous. "The Piracy and Slave Trade of the Indian Archipelago." *Journal of the Indian Archipelago and Eastern Asia* 3 (1849): 581–88, 629–39.

Anonymous. "Rajah Brooke and Borneo." *Chambers's Papers for the People* 3 (1850): article 34.

Anonymous. "Plan of the Treaties with France of 1778, Tuesday, September 17, 1776." In *Journals of the Continental Congress 1774–1779 Edited from the Original Records in the Library of Congress,* edited by Worthington Chauncey Ford. Washington, DC: Government Printing Office, 1905.

Anonymous. "Treaty of Amity and Commerce between the United States and France Signed on February 6, 1778." In *Treaties and Other International Acts of the United States of America Documents 1–40: 1776–1818,* 2 vols. Washington, DC: Government Printing Office, 1931.

Anonymous. "Articles of Peace and Commerce, between the most Serene and Mighty Princess Anne, by the Grace of God, Queen of Great Britain, France, and Ireland, Defender of the Christian Faith, &c. and the most Illustrious Lord Mustapha Dey, the Bashaw, Aga, and Governors of the Famous City and Kingdom of Algiers, in Barbary; Ratified, Confirmed, and Renewed, by George Byng, Esq; Rear-Admiral of the Red Squadron of Her Majesty's Fleet, on the 28th Day of October, Old Stile, 1703." In *A Collection of Treaties between Great Britain and Other Powers,* edited by George Chalmers, 2 vols. London: John Stockdale, Piccadilly, 1970.

Antony, Robert J. *Like Froth Floating on the Sea: The World of Pirates and Seafarers in Late Imperial South China.* Berkeley: Center for Environmental Studies, 2003.

Antony, Robert J. *Pirates in the Age of Sail.* New York: W. W. Norton & Company, 2007.

Antony, Robert J. "Turbulent Waters: Sea Raiding in Early Modern South East Asia." *The Mariner's Mirror* 99, no. 1 (February 2013): 23–38.

Antunes, Catia A. P. and Amelia Polonia. *Beyond Empires: Global, Self-Organizing, Cross-Imperial Networks, 1500–1800.* Leiden: Brill, 2016.

Appleby, John C. "The Problem of Piracy in Ireland, 1570–1630." In *Pirates? The Politics of Plunder, 1550–1650,* edited by Claire Jowitt, 41–55. New York: Palgrave MacMillan, 2007.

Appleby, John C. *Under the Bloody Flag: Pirates of the Tudor Age.* Stroud: The History Press, 2009.

Appleby, John C. *Women and English Piracy, 1540–1720: Partners and Victims of Crime.* Woodbridge: Boydell Press, 2013.

Archibald, Elizabeth. "Malory's Ideal of Fellowship." *The Review of English Studies* 43, no. 171 (1992): 311–28.

Baer, Joel. "Bold Captain Avery in the Privy Council: Early Variants of a Broadside Ballad from the Pepys Collection." *Folk Music Journal* 7, no. 1 (1995): 4–26.

Baer, Joel. *Pirates of the British Isles*. Stroud, Gloucestershire: Tempus, 2005.

Bahar, Matthew R. "People of the Dawn, People of the Door: Indian Pirates and the Violent Theft of an Atlantic World." *Journal of American History*, 101:2 (2014): 401–26.

Bahar, Matthew R. *Storm of the Sea: Indians and Empires in the Atlantic's Age of Sail*. New York: Oxford, 2019.

Bailyn, Bernard and Lotte Bailyn. *Massachusetts Shipping 1697–1714: A Statistical Study*. Cambridge, Massachusetts: The Belknap Press of Harvard University, 1959.

Barber, Sarah. "Brayne, William." *Oxford Dictionary of National Biography*, 2004. Accessed July 16, 2022.

Barnard, John. *Ashton's Memorial: An History of the Strange Adventures and Signal Deliverances of Mr. Philip Ashton*. Boston: Samuel Gerrish, 1725.

Barnard, Timothy P. "Celates, Rayat-Laut, Pirates: The Orang Laut and Their Decline in History." *Journal of the Malaysian Branch of the Royal Asiatic Society* 80, No. 2 (December 2007): 33–49.

Barnard, Timothy P. "Siak, Piracy and Early Modern Malay Warfare." In *Piracy and Surreptitious Activities in the Malay Archipelago and Adjacent Seas, 1600–1849*, edited by Y. H. Teddy Sim, 17–34. Singapore: Springer, 2014.

Bastin, John. *Raffles and Hastings: Private Exchanges behind the Founding of Singapore*. Singapore: National Library Board, 2014.

Bastin, John. *Sir Stamford Raffles and Some of His Friends and Contemporaries: A Memoir of the Founder of Singapore*. Singapore: World Scientific Publishing Co., 2019.

Begbie, Peter James. *The Malayan Peninsula, Embracing Its History, Manners and Customs of the Inhabitants, Politics, Natural History &c. from Its Earliest Records*. Madras: Vepery Mission Press, 1834.

Bellamy, R. Reynell, ed. *Ramblin' Jack: The Journal of Captain John Cremer, 1700–1774*. London: Jonathan Cape LTD., 1936.

Bellina, Bérénice, Roger Blench and Jean-Christophe Galipaud, eds. *Sea Nomads of Southeast Asia: From Past to Present*. Singapore: NUS Press, 2021.

Bellina, Bérénice, Roger Blench and Jean-Christophe Galipaud. "Sea Nomadism from Past to the Present." In *Sea Nomads of Southeast Asia: From Past to Present*, edited by Bérénice Bellina, Roger Blench and Jean-Christophe Galipaud, 1–27. Singapore: NUS Press, 2021.

Bentley, Jerry H., Renate Bridenthal and Kären Wigen, eds. *Seascapes: Maritime Histories, Littoral Cultures, and Transoceanic Exchanges*. Honolulu: University of Hawai'i Press, 2007.

Benton, Lauren. "Legal Spaces of Empire: Piracy and the Origins of Ocean Regionalism." *Comparative Studies in Society and History* 47, no. 4 (2005): 700–724.

Benton, Lauren. *Search for Sovereignty: Law and Geography in European Empires, 1400–1900.* Cambridge: Cambridge University Press, 2010.

Benton, Lauren. "Toward a New Legal History of Piracy: Maritime Legalities and the Myth of Universal Jurisdiction." *International Journal of Maritime History* 12, no. 1 (2011).

Benton, Lauren and Lisa Ford. *Rage for Order: The British Empire and the Origins of International Law 1800–1850.* Cambridge: Harvard University Press, 2016.

Benton, Lauren and Nathan Perl-Rosenthal, eds. *A World at Sea: Maritime Practices and Global History.* Philadelphia: University of Pennsylvania Press, 2020.

Bernaldez, Emilio. *Reseña historica de la guerra al sur de Filipinas, sostenida por las armas españolas contra los piratas de aquel archipelago, desde la conquista hasta nuestros días.* Madrid: Imprenta del memorial de ingenieros, 1857.

Beselke, Brandi, ed. *A Dictionary of English Language: A Digital Edition of the 1755 Classic by Samuel Johnson.* Last modified December 6, 2012. https://johnsonsdictionaryonline.com.

Bialuschewski, Arne. "Pirates, markets and Imperial Authority: Economic Aspects of maritime Depredations in the Atlantic World, 1716–1726." *Global Crime* 9, nos. 1–2 (2008): 53–65.

Bialuschewski, Arne. *Raiders and Natives: Cross-Cultural Relations in the Age of Buccaneers.* Athens: University of Georgia Press, 2022.

Biddulph, John. *The Pirates of Malabar.* London: Smith, Elder & Co, 1907.

Boardman, Stephen. *The Early Stewart Kings: Robert II and Robert III, 1371–1406.* Edinburgh: East Linton, 1996.

Boardman, Stephen. *The Campbells, 1250–1513.* Edinburgh: East Linton, 2005.

Boardman, Stephen. "Highland Scots and Anglo-Scottish Warfare, 1300–1513." In *England and Scotland at War, c.1296–c.1513*, edited by Andy King and David Simpkin, 231–53. Leiden: Brill, 2012.

Bolland, O. Nigel. *Colonialism and Resistance in Belize: Essays in Historical Sociology.* Barbados: University of West Indies Press, 2003.

Bolster, Jeffrey. *Black Jacks: African American Seamen in the Age of Sail.* Cambridge, Massachusetts: Harvard University Press, 1997.

Bookin-Weiner, Jerome. "Corsairing in the Economy and Politics of North Africa." In *North Africa: Nation, State and Region*, edited by George Joffe, 3–33. London: Routledge, 1993.

Boomgaard, Peter. *A World of Water: Rain, Rivers and Sea in Southeast Asian Histories.* Leiden: KITLV Press, 2007.

Botella-Ordinas, Eva. "Debating Empires, Inventing Empires: British Territorial Claims Against the Spaniards in America, 1670–1714." *Journal for Early Modern Cultural Studies* 10, no. 1 (2010): 142–68.

Branch, Jordan. *The Cartographic State: Maps, Territory, and the Origins of Sovereignty*. Cambridge: Cambridge University Press, 2014.

Breen, Colin. "The Maritime Cultural Landscape in Medieval Gaelic Ireland." In *Gaelic Ireland, c1250–c.1650: Land, Lordship, and Settlement*, edited by Patrick J. Duffy, David Edwards, and Elizabeth FitzPatrick, 418–36. Dublin: Four Courts Press, 2001.

Breen, Colin. *The Gaelic Lordship of the O'Sullivan Beare: A Landscape Cultural History*. Dublin: Four Courts Press, 2005.

Brown, Daniel. *Hugh de Lacy, First Earl of Ulster: Rising and Falling in Angevin Ireland*. Woodbridge: Boydell and Brewer, 2016.

Brown, D. E. "Brunei and the Bajau." *Borneo Research Bulletin* 3, No. 2 (December 1971): 55–8.

Brown, Michael. *The Wars of Scotland, 1214–1371*. Edinburgh: Edinburgh University Press, 2004.

Brown, Michael. *Disunited Kingdoms: Peoples and Politics in the British Isles, 1280–1460*. Harlow: Pearson, 2013.

Buhite, Russel D. *Lives at Risk: Hostages and Victims in American Foreign Policy*. Wilmington, Delaware: Scholarly Resources, 1995.

Buisseret, David. "The French Invasion of Jamaica—1694." *Jamaica Journal* 16, no. 3 (1983): 31–33.

Burgess Jr., Douglass R. *The Politics of Piracy: Crime and Civil Disobedience in Colonial America*. Lebanon, New Hampshire: ForeEdge, 2014.

Burwick, Frederick and Manushag N. Powell. *British Pirates in Print and Performance*. New York: Palgrave Macmillan, 2015.

Calanca, Paola. "Piracy and Coastal Security in Southeastern China, 1600–1780." In *Elusive Pirates, Pervasive Smugglers. Violence and Clandestine Trade in the Greater China Seas*, edited by Robert Antony, 85–98. Hong Kong: Hong Kong University Press, 2010.

Campbell, Bruce M. S. *The Great Transition: Climate, Disease and Society in the Late-Medieval World*. Cambridge: Cambridge University Press, 2016.

Campbell, Mavis Christine. *Becoming Belize: A History of an Outpost of Empire Searching for Identity, 1528–1823*. Kingston, Jamaica: University of the West Indies Press, 2011.

Campling, Liam and Alejandro Colás. *Capitalism and the Sea: The Maritime Factor in the Making of the Modern World*. London: Verso, 2021.

Carr, Anthony D. *Owen of Wales: The End of the House of Gwynedd*. Cardiff: University of Wales Press, 1991.

Cathcart, Alison. "James V, King of Scotland—and Ireland?" In *The World of the Galloglass: Kings, Warlords, and Warriors in Ireland and Scotland, 1200–1600*, edited by Séan Duffy, 124–43. Dublin: Four Courts Press, 2007.

Cathcart, Alison. "'The Forgotten' 45: Donald Dubh's Rebellion in an Archipelagic Context." *Scottish Historical Review* 91, no. 2 (October 2012): 239–64.

Cathcart, Alison. "The Maritime Dimension to Plantation in Ulster, ca. 1550–ca. 1600." *Journal of the North Atlantic* 12, SP1 (2019): 95–111.

Cathcart, James. *The Captives: Eleven Years a Prisoner in Algiers.* La Porte, Indiana: Herald Print, 1899.

Capp, Bernard. "Goodsonn, William." *Oxford Dictionary of National Biography*, 2004. Accessed July 16, 2022.

Chadwick, Mark. *Piracy and the Origins of Universal Jurisdiction: On Stranger Tides?* Leiden: Brill, 2019.

Chalmers, George, ed. *A Collection of Treaties between Great Britain and Other Powers*. 2 vols. London: John Stockdale, Piccadilly, 1970.

Chambers, A. *Granuaile: The Life and Times of Grace O'Malley.* Dublin: Wolfhound Press, 1983.

Childs, W. "Ireland's Trade with England in the Later Middle Ages." *Irish Economic and Social History*, 9 (1982): 5–33.

Childs, Wendy and Timothy O'Neill. "Overseas Trade." In *A New History of Ireland II: Medieval Ireland, 1169–1534*, edited by Art Cosgrove, 492–524. Oxford: Oxford University Press, 1987.

Chou, Cynthia. "Contesting the Tenure of Territoriality; the Orang Suku Laut." *Bijdragen tot de Taal-, Land- en Volkenkunde* 153, No. 4 (1997): 605–29.

Chou, Cynthia. "Research Trends on Southeast Asian Sea Nomads." *Kyoto Review of Southeast Asia* 7 (2006). https://kyotoreview.org/book-review/research-trends-on-southeast-asian-sea-nomads.

Clark, Grover. "The English Practice with Regard to Reprisals by Private Persons." *American Journal of International Law*, 27 (1933): 694–723.

Coakley, John. "Jamaica's Private Seafarers: Politics and Violence in a Seventeenth-Century English Colony." In *The Golden Age of Piracy: The Rise, Fall, and Enduring Popularity of Pirates*, edited by David Head. Athens, GA: The University of Georgia Press, 2018.

Coakley, John. "'The Piracies of Some Little Privateers': Language, Law and Maritime Violence in the Seventeenth-Century Caribbean." *Britain and the World* 13, no. 1 (2020): 6–26.

Coakley, John, C. Nathan Kwan, and David Wilson. "Introduction: Piracy and Occasional State Power." *The International Journal of Maritime History* 32, no. 3 (2020): 656–65.

Colley, Linda. *Captives: Britain, Empire and the World 1600–1850*. New York: Pantheon, 2002.

Cordingly, David. *Under the Black Flag: The Romance and Reality of Life among the Pirates*. New York: Random House, 1995.

Cosman, Madeleine Pelner, and Linda Gale Jones. *Handbook to Life in the Medieval World*. Facts on File, 2008.

Cowan, Edward J. "Norwegian Sunset—Scottish Dawn: Hákon IV and Alexander III." In *Scotland in the Reign of Alexander III, 1249–1286*, edited Norman H. Reid, 103–31. Edinburgh: East Linto, 1990.

Craig, Alan. "Logwood as a Factor in the Settlement of British Honduras." *Caribbean Studies* 9, no. 1 (1969): 53–62.

Crane, Elaine Forman. *Ebb Tide in New England: Women, Seaports and Social Change 1630–1800*. Boston: Northeastern University Press, 1998.

Crawfurd, John. "History and Languages of the Indian Islands." *Edinburgh Review* 23, No. 45 (1814): 151–89.

Crawfurd, John. *History of the Indian Archipelago*. 3 vols. Edinburgh: Archibald Constable and Co., 1820.

Crawfurd, John. "Malay Pirates." *The Asiatic Journal and Monthly Register for British India and Its Dependencies* 29 (January–June 1825): 243–45.

Crawfurd, John. *Journal of an Embassy from the Governor-General of India to the Courts of Siam and Cochin China Exhibiting a View of the Actual State of Those Kingdoms*. London: Henry Colburn, 1828.

Crawfurd, John. "Piracy in the Eastern Archipelago." *The Examiner* 2125 (October 21, 1848).

Crawfurd, John. *A Descriptive Dictionary of the Indian Islands & Adjacent Countries*. London: Bradbury and Evans, 1856.

Cromwell, Jesse. "Life on the Margins: (Ex) Buccaneers and Spanish Subjects on the Campeche Logwood Periphery, 1660–1716." *Itinerario* 33 (2009): 43–71.

Dalrymple, Alexander. *A Plan for Extending the Commerce of This Kingdom, and of the East-India-Company*. London, 1769.

Dampier, William. *A Collection of Voyages, In Four Volumes*. London: James and John Knapton, 1729.

Dagher-margosian, Matt. "Marcus Rediker: The History of Pirates & Stealing our Future Back." *Asia Art Tours*, December 5, 2021. https://asiaarttours.com/marcus-rediker-the-history-of-pirates-learning-to-steal-back-our-future.

Davies, J. D., Alan James, and Gijs Rommelse, eds. *Ideologies of Western Naval Powers, c. 1500–1815*. New York: Routledge, 2019.

Davies, Rees R. *The Revolt of Owain Glyn Dŵr*. Oxford: Oxford University Press, 1995.

Defoe, Daniel. *A General History of the Pyrates*. Edited by Manuel Schonhorn. New York: Dover Publications, 1999.

De Winter, Wim. "Southern-Netherlandish Observations and Knowledge Production of Naturalia on the Seas: The Writings of Michael de Febure (1721)." *Early Modern Low Countries* 3, no. 2 (2019): 265–82.

De Winter, Wim. "The Ostend Company's Worlds. Courtly Interactions and Local Life in Eighteenth-Century Bengal." In *An Earthly Paradise: Trade, Politics and Culture in Early-Modern Bengal*, edited by Raziuddin Aquil and Tilottama Mukherjee, 157–85. New Delhi: Manohar, 2020.

Deleuze, Gilles and Felix Guattari. *A Thousand Plateaus: Capitalism and Schizophrenia*. Minneapolis: University of Minnesota Press, 1987 [1980].

Ditcham, B. "The Employment of Foreign Mercenary Troops in French Royal Armies, 1415–1470." PhD Diss., University of Edinburgh, 1978.

Dow, George Francis and John Henry Edmonds. *The Pirates of the New England Coast*, reprint. New York: Dover Publications, 1996 [1923].

Duffy, Séan. "The Bruce Brothers and the Irish Sea World, 1306–29." *Cambridge Medieval Celtic Studies* 21 (1991): 55–86.

Duffy, Séan. "Ireland and the Irish Sea Region, 1014–1318." PhD Diss., Trinity College Dublin, 1993.

Duffy, Séan. "The 1169 Invasion as a Turning Point in Irish-Welsh Invasions." In *Britain and Ireland, 900–1300: Insular Responses to Medieval European Change*, edited by Brendan Smith, 98–113. Cambridge: Cambridge University Press, 1999.

Duffy, Séan. "The Prehistory of the Galloglass." In *The World of the Galloglass: Kings, Warlords, and Warriors in Ireland and Scotland, 1200–1600*, edited by Séan Duffy, 1–27. Dublin: Four Courts Press, 2007.

Dummer, Jeremiah. *A Defence of the New England Charters*. London: W. Wilkins, 1721.

Dunn, Richard S. *Sugar and Slaves: The Rise of the Planter Class in the English West Indies, 1624–1713*. New York: Norton, 1972.

Dyer, Florence E. "Captain Christopher Myngs in the West Indies." *Mariner's Mirror* xviii (1932): 168–87.

Earl, George W. *The Eastern Seas, or Voyages and Adventures in the Indian Archipelago in 1832–33–34*. London: Wm. H. Allen and Co., 1837.

Earle, Peter. *Corsairs of Malta and Barbary*. London: Sidgwick & Jackson, 1970.

Earle, Peter. *The Sack of Panama: Captain Morgan and the Battle for the Caribbean*. New York: St. Martin's, 1981.

Earle, Peter. *The Pirate Wars*. New York: Thomas Dunne Books, 2003.

Edwards, Bryan. *The History, Civil and Commercial, of the British Colonies in the West Indies*, 3rd ed., 2 vols. London: John Stockdale, Piccadilly, 1801.

Edwards, David. "Tudor Ireland: Anglicization, Mass Killing, and Security." In *The Routledge History of Genocide,* edited by Cathie Carmichael and Richard C. Maguire, 23–37. New York: Routledge, 2015.

Egan, Simon. "Richard II and the Wider Gaelic World: A Re-Assessment." *Journal of British Studies* 57, no. 2 (April 2018): 221–52.

Egan, Simon. "The Early Stewart Kings, the Lordship of the Isles, and Ireland." *Northern Studies* 49 (2018): 61–78.

Egan, Simon. "By Land and By Sea: The Role of the Maritime Sphere in the Expansion of O'Donnell Power, ca.1380–1500." *Journal of the North Atlantic* 12 (2019): 16–31.

Ekelund Jr., Robert B., and Hébert, Robert F. *A History of Economic Theory and Method.* Long Grove, IL: Waveland Press, 2007.

Ekin, Des. *The Stolen Village: Baltimore and the Barbary Pirates.* Dublin: O'Brien, 2006.

Elliot, Derek L. "Pirates, Polities and Companies: Global Politics on the Konkan Littoral, c. 1690–1756." *Working Papers* 136, no. 10 (2010): 1–43.

Elliot, Derek L. "The Politics of Capture in the Eastern Arabian Sea, c. 1700–1750." *International Journal of Maritime History* 25, no. 2 (2013): 187–98.

Elliot, Derek L. and Sebastian R. Prange. "Beyond Piracy: Maritime Violence and Colonial Encounters in Indian History." In *Beyond the Line: Cultural Narratives of the Southern Oceans*, edited by Michael Mann and Ineke Rhaf-Reinberger, 95–119. Berlin: Neofelis Verlag, 2014.

Faller, Lincoln B. "Criminal Opportunities in the Eighteenth Century: The 'Ready-Made' Contexts of the Popular Literature of Crime." *Comparative Literature Studies* 24, no. 2 (1987): 120–45.

Faller, Lincoln B. *Turned to Account: The Forms and Functions of Criminal Biography in Late Seventeenth- and Early Eighteenth-Century England.* Cambridge: Cambridge University Press, 1987.

Farber, Hannah. "Millions for Credit: Peace with Algiers and the Establishment of America's Commercial Reputation Overseas, 1795–1796." *Journal of the Early Republic* 34, no. 2 (2014): 187–217.

Finamore, Daniel. "A Mariner's Utopia: Pirates and Logwood in the Bay of Honduras." In *X Marks the Spot: The Archaeology of Piracy*, edited by Russell K. Skowronek and Charles R. Ewen, 64–78. Gainesville, Florida: University Press of Florida, 2006.

Finlayson, George. *The Mission to Siam and Hue in the Years 1821–2. From the Journal of the Late George Finlayson, Esq., Assistant Surgeon of His Majesty's 8th Light Dragoons, Surgeon and Naturalist to the Mission. With a Memoir of the Author by Sir Thomas Stamford Raffles, F.R.S.* London: John Murray, 1826.

Firth, Charles H. "The Capture of Santiago, in Cuba, by Captain Myngs, 1662." *The English Historical Review* 14, no. 55 (1899): 536–40.

Fitzmaurice, Andrew. *Sovereignty, Property and Empire, 1500–2000*. Cambridge: Cambridge University Press, 2014.

Flanagan, Marie T. "After Brian Bóraime: The High-Kingship and the Kings of Connacht." In *Medieval Dublin XVI*, edited by Séan Duffy, 218–59. Dublin, 2017.

Flavin, Susan. *Consumption and Culture in Sixteenth-Century Ireland: Saffron, Stockings, and Silk*. Woodbridge: Boydell and Brewer, 2014.

Forcier, Rick. "The Tales of Terrorism." *Newsletter of the Christian Coalition of Washington State* (November 2001).

Forrest, Thomas. *A Voyage to New Guinea and the Moluccas, from Balambangan: An Account of Magindano, Sooloo, and Other Islands; and Illustrated with Thirty Copperplates Performed in the Tartar Galley, Belonging to the Honourable East India Company, During the Years 1774, 1775, and 1776*. London: G. Scott, 1779.

Forrest, Thomas. *A Voyage from Calcutta to the Mergui Archipelago, Lying on the East Side of the Bay of Bengal*. London, 1792.

Frame, Robin. *English Lordship in Ireland, 1318–1361*. Oxford: Oxford University Press, 1982.

Fryer, John. *A new account of East-India and Persia, in eight letters being nine years travel begun in 1672 and finished 1681*. London: printed for R. I. Chiswell, 1698.

Galvin, Peter. *Patterns of Pillage: A Geography of Caribbean-Based Piracy in Spanish America, 1536–1718*. New York: Peter Lang, 1999.

García de Arboleya, José. *Historia del archipelago y sultania de Joló, y noticia de la expedición española que a órdenes del marques de la Solana acaba de destruir a los piratas joloanos*. Habana: Imprenta de M. Soler y Gelada, 1851.

Gardiner, Mark, and T. E. McNeill. "Seaborne Trade and the Commercialisation of Fifteenth- and Sixteenth-Century Ulster." *Proceedings of the Royal Irish Academy* 116 (2016): 229–62.

Gaynor, Jennifer L. "Piracy in the Offing: The Law of Lands and the Limits of Sovereignty at Sea." *Anthropological Quarterly* 85, No. 3 (2012): 817–58.

Gaynor, Jennifer L. *Intertidal History in Island Southeast Asia: Submerged Genealogy & the Legacy of Coastal Capture*. Ithaca: Cornell University Press, 2016.

Gettleman, Jeffrey. "Lessons from the Barbary Pirates." *The New York Times*, April 11, 2009. http://www.nytimes.com/2009/04/12/weekinreview/12gettleman.html.

Gibbs, Joseph. "'A Certain False, Malicious, Scandalous and Famous Libel': Sir Henry Morgan's Legal Action against a London Publisher of Alexandre Exquemelin, 1685." *International Journal of Maritime History* 30, no. 1 (2018): 3–29.

Gibbs, Joseph. "The Brevity and Severity of 'Golden Age' Piracy Trials." *The International Journal of Maritime History* 31, no. 4 (2019): 729–86.

Gosse, Philip. *The History of Piracy*. New York: Longmans, Green & Co., 1932.

Gould, Eliga. "Lines of Plunder of Crucible of Modernity? The Legal Geography of the English-Speaking Atlantic, 1660–1825." In *Seascapes: Maritime Histories,*

*Littoral Cultures, and Transoceanic Exchanges*, edited by Jerry H. Bentley, Renate Bridenthal, and Kären Wigen, 105–20. Honolulu: University of Hawai'i Press, 2007.

Graham, Gerald S. *Great Britain in the Indian Ocean: A Study of Maritime Enterprise 1810–1850*. London: Oxford University Press, 1967.

Groot, Jhr. J. P. Cornets de. "Paper Relating to the Piracies Committed in the Indian Archipelago, and to the Measures Adopted by the Netherland Government, in the Year 1815 to 1845, for Their Repression." *British Parliamentary Papers. Accounts and Papers: 1851*, vol. 26, 1.

Hahn, Steven C. "The Atlantic Odyssey of Richard Tookerman: Gentleman of South Carolina, Pirate of Jamaica, and Litigant before the King's Bench." *Early American Studies: An Interdisciplinary Journal* 15, no. 3 (2017): 539–90.

Hamilton, Alexander. *A New Account of the East-Indies Being the Observations and Remarks of Capt. Hamilton Who Spent His Time There, from the Year 1688 to 1723, Trading and Travelling, by Sea and Land, to Most of the Countries and Islands of Commerce and Navigation, Between the Cape of Good-Hope and the Island of Japan*. 2 Vols. Edinburgh: John Mosman, 1727.

Hamilton, Walter. *The East-India Gazetteer; Containing Particular Descriptions of the Empires, Kingdoms, Principalities, Provinces, Cities, Towns, Districts, Fortresses, Harbours, Rivers, Lakes, &c. of Hindostan and the Adjacent Countries, India Beyond the Ganges, and the Eastern Archipelago*. 2 vols. London: Wm. H. Allen and Co., 1828.

Hanna, Mark G. *Pirate Nests and the Rise of the British Empire, 1570–1740*. Chapel Hill: University of North Carolina Press, 2015.

Harlaftis, Gelina. "Maritime History or the History of 'Thalassa.'" In *The New Ways of History: Developments in Historiography*, edited by Gelina Harlaftis, Nikos Karapidakis, Kostas Sbonias, and Vaios Vaiopoulos, 211–38. London: Bloomsbury, 2009.

Harrington, Gordon K. "Dr. Horsfield's Report on the Island of Bangka: An Imperialist Proposal for Reform in 1813." *Selected Papers in Asian Studies: Western Conference of the Association for Asian Studies*, no. 72 (2003): 1–14.

Hayes-McCoy, Gerald A. "The Early History of Guns in Ireland." *Journal of the Galway Archaeological and Historical Society* 18, no. 1–2 (1938): 43–65.

Head, David, ed. *The Golden Age of Piracy: The Rise, Fall, and Enduring Popularity of Pirates*. Athens: University of Georgia Press, 2018.

Hebb, David D. *Piracy and the English Government 1616–1642: Policy-Making Under the Early Stuarts*. Aldershot: Scolar, 1994.

Heller-Roazen, Daniel. *The Enemy of All: Piracy and the Law of Nations*. New York: Zone Books, 2009.

Heywood, Linda M. and John K. Thornton. *Central Africans, Atlantic Creoles, and the Making of the Foundation of the Americas, 1585–1660*. New York: Cambridge University Press, 2007.

Hill, Christopher. "Radical Pirates?" In *The Origins of Anglo-American Radicalism*, edited by Margaret Jacob and James Jacob, 17–32. London: George Allen and Uniwin, 1984.

Hill, Christopher. "Radical Pirates?" In *The Collected Essays of Christopher Hill, Volume III: People and Ideas in Seventeenth-Century England*, 161–87. Amherst: University of Massachusetts Press, 1987.

Hobsbawm, Eric. *Bandits*. London: Weidenfeld and Nicolson, 1969.

Hoogervorst, Tom Gunnar. "Ethnicity and Aquatic Lifestyles: Exploring Southeast Asia's Past and Present Seascapes." *Water History* 4 (2012): 245–65.

Hopkins, Benjamin D. *Ruling the Savage Periphery: Frontier Governance and the Making of the Modern State*. Cambridge: Harvard University Press, 2020.

Horsfield, Thomas. "Report on the Island of Bangka." *Journal of the Indian Archipelago and Eastern Asia* 2 (1848): 299–336, 373–427, 705–25 and 779–824.

Irwin, Ray W. *The Diplomatic Relations of the United States with the Barbary Powers: 1776–1816*. Chapel Hill, NC: The University of North Carolina Press, 1931.

Jameson, John Franklin, ed. *Privateering and Piracy in the Colonial Period: Illustrative Documents*. New York, New York: Augustus M. Kelly, 1970.

Jardine, Lisa. *Worldly Goods: A New History of the Renaissance*. New York: Nan A. Tales, 1996.

Jarvis, Michael. *In the Eye of All Trade: Bermuda, Bermudians, and the Maritime Atlantic World, 1660–1783*. Chapel Hill, NC: University of North Carolina Press, 2010.

Jefferson, Thomas. "Report of the Secretary of State Relative to the Mediterranean Trade. Communicated to the House of Representatives December 30, 1790 and to the Senate, January 3, 1791." In *American State Papers. Documents, Legislative and Executive of the Congress of the United States, from the First Session of the First to the Second Session of the Twenty-Second Congress, Inclusive: Commencing March 3, 1769 and Ending March 3, 1833. Selected and Edited, Under the Authority of Congress, by Walter Lowrie, Secretary of the Senate, and Matthew St. Clair Clarke, Clerk of the House of Representatives*, 1 vol, p. 104. Washington: Gales and Seaton, 1833.

Jefferson, Thomas. "VIII—Report Relative to the Mediterranean Trade, December 28, 1790." In *The Writings of Thomas Jefferson,* 7 vols, 519–26. Washington, DC: Taylor & Maury, 1854.

Jefferson, Thomas. "From Thomas Jefferson to James Monroe, 6 February 1785." In *The Papers of Thomas Jefferson 2 March 1784–25 February 1785*, edited by Julian P. Boyd, 7 vols, 637–41. Princeton: Princeton University Press, 1953.

Jefferson, Thomas. "From Thomas Jefferson to Nathanel Greene, 12 January 1786." In *The Papers of Thomas Jefferson, 1 November 1785–22 June 1786*, edited by Julian P. Boyd, 7 vols, 167–69. Princeton: Princeton University Press, 1954.

Jefferson, Thomas. "To John Adams from Thomas Jefferson, 6 August 1785." In *The Adams Papers, Papers of John Adams*, edited by Gregg L. Lint, C. James Taylor, Sara Georgini, Hobson Woodward, Sara B. Sikes, Amanda A. Matthews, and Sara Martin, 7 vols, 306–8. Cambridge, MA: Harvard University Press, 2014.

Johnson, Captain Charles. *A General History of the Pyrates from Their First Rise and Settlement, to the Present Time*, 2nd ed. London: Thomas Warner, 1724.

Johnson, Captain Charles. *A General History of the Robberies and Murders of the Most Notorious Pyrates, and Also Their Policies, Discipline and Government, From Their First Rise and Settlement in the Island of Providence, in 1717, to the Present Year 1724*, 1st ed. London: Charles Rivington, J. Stone, J. Lacey, 1724.

Johnson, Captain Charles. *A General History of the Lives and Adventures of the Most Famous Highwaymen, Murderers, Street-Robbers, &c. To Which Is Added, A Genuine Account of the Voyages and Plunders of the Most Notorious Pirates.* London: J. Janeway, 1734.

Johnson, Captain Charles. *A General History of the Pirates*. Mineola, New York: Dover Publishing, 1999.

Jones, Stephanie. "Literature, Geography, Law: The Life and Adventures of Capt. John Avery (circa 1709)." *Cultural Geographies* 19, no. 1 (2012): 71–86.

Joseph, Gilbert M. "British Loggers and Spanish Governors: The Logwood Trade and Its Settlements in the Yucatan Peninsula: Part I." *Caribbean Studies* 14, no. 2 (1974): 7–37.

Joseph, Gilbert M. "British Loggers and Spanish Governors: The Logwood Trade and Its Settlements in the Yucatan Peninsula: Part II." *Caribbean Studies* 15, no. 4 (1976): 43–52.

Jowitt, Claire. *The Culture of Piracy, 1580–1630: English Literature and Seaborne Crime.* Farnham: Ashgate, 2010.

Kang, David C. *East Asia Before the West, Five Centuries of Trade and Tribute.* New York: Columbia University Press, 2010.

Karraker, Cyrus H. *Piracy Was a Business*. Rindge, NH: Richard R. Smith, 1953.

Kelleher, C. "The Gaelic O'Driscoll Lords of Baltimore, Co. Cork: Settlement, Economy, and Conflict in a Maritime Cultural Landscape." In *Lordship in Medieval Ireland: Image and Reality*, edited by L. Doran and J. Lyttleton, 130–59. Dublin: Four Courts Press, 2007.

Kelleher, C. *The Alliance of Pirates: Ireland and Atlantic Piracy in the Early Seventeenth Century.* Cork: Cork University Press 2020.

Kelly, Maria. *A History of the Black Death in Ireland.* Dublin: Tempus, 2001.

Kempe, Michael. "'Even in the Remotest Corners of the World': Globalized Piracy and International Law, 1500–1900." *Journal of Global History* 5, no. 3 (2010): 353–72.

Kempe, Michael. *Fluch der Weltmeere: Piraterie, Völkerrecht und internationale Beziehungen 1500–1900*. Frankfurt: Campus Verlag, 2010.

Kent, H. S. K. "The Historical Origins of the Three-Mile Limit." *The American Journal of International Law* 48, No. 4 (October 1954): 537–53.

Keppel, Henry. *The Expedition to Borneo of H.M.S. Dido for the Suppression of Piracy: With Extracts from the Journals of James Brook Esq. of Sarawak*. 2 vols. London: Chapman and Hall, 1846.

Keppel, Henry. *A Visit to the Indian Archipelago in the H.M. Ship Mæander. With Portions of the Private Journal of Sir James Brooke, K.C.B.* 2 vols. London: Richard Bentley, 1853.

Kilmeade, Brian and Don Yaeger. *Thomas Jefferson and the Tripoli Pirates: The Forgotten War That Changed American History*. New York: Sentinel, 2015.

Kingston, Simon. "Delusions of Dál Riada: The Coordinates of Mac Domhnaill Power, 1461–1550." In *Gaelic Ireland, c.1250–c.1650: Land, Lordship, and Settlement*, edited by Patrick J. Duffy, David Edwards, and Elizabeth FitzPatrick, 98–114. Dublin: Four Courts Press, 2001.

Knapman, Gareth. *Race and British Colonialism in Southeast Asia, 1770–1870: John Crawfurd and the Politics of Equality*. New York: Routledge, 2017.

Knapman, Gareth and Martin Müller. "Protector of Aborigines or War Criminal: Two Opposing Liberal Views of James Brooke." In *Liberalism and the British Empire in Southeast Asia*, edited by Gareth Knapman, Anthony Milner and Mary Quilty, 167–91. Abingdon: Routledge, 2019.

Konstam, Angus. *Piracy: The Complete History*. London: Osprey Publishing, 2008.

Kupperman, Karen Ordahl. *Providence Island, 1630–1641: The Other Puritan Colony*. Cambridge: Cambridge University Press, 1993.

Kwan, C. Nathan. "'Barbarian Ships Sail Freely about the Seas': Qing Reactions to the British Suppression of Piracy in South China, 1841–1856." *Asian Review of World Histories* 8, no. 1 (2020): 83–102.

Kwan, C. Nathan. "'Putting Down a Common Enemy': Piracy and Occasional Inter-state Power in South China during the Mid-Nineteenth Century," *International Journal of Maritime History* 32, no. 3 (2020): 697–712.

Lane, Kris. *Pillaging the Empire: Piracy in the Americas, 1500–1750*. London: M. E. Sharpe, 1998.

Lane, Kris. *Pillaging the Empire: Global Piracy on the High Seas, 1500–1750*. New York: Routledge, 2015.

Lapian, Adrian B. "Violence and Armed Robbery in Indonesian Seas." In *Pirates, Ports, and Coasts in Asia: Historical and Contemporary Perspectives*, edited by John Kleinen and Manon Osseweijer, 131–46. Singapore: Institute of Southeast Asian Studies, 2010.

Latimer, Jon. *Buccaneers of the Caribbean: How Piracy Forged an Empire*. Cambridge, MA: Harvard University Press, 2009.

Layton, Simon. "Discourses of Piracy in an Age of Revolutions." *Itinerario* 35, no. 2 (2011): 81–97.

Layton, Simon. "Hydras and Leviathans in the Indian Ocean World." *International Journal of Maritime History* 25, no. 2 (2013): 213–25.

Layton, Simon. "The 'Moghul's Admiral': Angrian 'Piracy' and the Rise of Bombay." *Journal of Early Modern History* 17 (2013): 75–93.

Leeson, Peter T. "An-*arrgh*-chy: The Law and Economics of Pirate Organisation." *The Journal of Political Economy* 115, no. 6 (2007): 1049–94.

Leeson, Peter T. *The Invisible Hook: The Hidden Economics of Pirates*. Princeton: Princeton University Press, 2009.

Leeson, Peter T. *Anarchy Unbound: Why Self-Governance Works Better Than You Think*. Cambridge: Cambridge University Press, 2014.

Leilby, Richard. "Terrorists by Another Name: The Barbary Corsairs." *Washington Post*, October 15, 2001. https://www.washingtonpost.com/archive/lifestyle/2001/10/15/terrorists-by-another-name-the-barbary-pirates/af7e4c90-b56c-4e0c-95ea-f079b889a96f.

Leyden, John. "On the Language and Literature of the Indo-Chinese Nations." *Asiatick Researches* 10 (1808): 158–289.

Lincoln, Margarette. *British Pirates and Society, 1680–1730*. Farnham: Ashgate, 2014.

Linebaugh, Peter and Marcus Rediker. *The Many-Headed Hydra: Sailors, Slaves, Commoners, and the Hidden History of the Revolutionary Atlantic.* Boston: Beacon Press, 2000.

Lint, Gregg L. "John Adams on the Drafting of the Treaty Plan of 1776." *Diplomatic History* 2, no. 3 (1978): 313–20.

Lipman, Andrew. *The Saltwater Frontier: Indians and the Contest for the American Coast.* New Haven: Yale University Press, 2015.

Logan, John Richardson. "Malay Amok and Piracies: What can We do to Abolish Them?" *The Journal of the Indian Archipelago and Eastern Asia* 3 (1849): 463–67.

Lomas, R. *The Fall of the House of Percy, 1368–1408*. Edinburgh: East Linton, 2007.

London, Joshua. "America's Earliest Terrorists: Lessons from America's First War against Islamic Terror." *National Review,* December 16, 2005. http://www.nationalreview.com/article/216283/americas-earliest-terrorists-joshua-e-london.

London, Joshua. *Victory in Tripoli: How War with the Barbary Pirates Teaches Us How to Fight the War on Terror.* New Jersey: John Wiley and Sons, 2005.

London, Joshua. Interview with Jamie Glazov. *Free Republic*, May 9, 2006. http://www.freerepublic.com/focus/f-news/1629124/posts.

Low, Hugh. *Sarawak: Its Inhabitants and Productions: Being Notes during a Residence in That Country with H. H. the Rajah Brooke*. London: Richard Bentley, 1848.

Lydon, James F. "The City of Waterford in the Later Middle Ages." *Decies* 12 (1979): 5–15.

MacDonald, Alastair J. *Border Bloodshed: Scotland, England and France at War, 1369–1403*. Edinburgh: East Linton, 2000.

Macfarlane, C. Alasdair. "Pirates and Publicity: The Making and Unmaking of Early Modern Pirates in English and Scottish Popular Print." *Humanities* 9, no. 1:14 (2020).

MacGettigan, Darren. "The Renaissance and the Late Medieval Lordship of Tír Chonaill, 1461–1555." In *Donegal: History and Society*, edited by William Nolan, Liam Ronayne, Mairead Dunleavy, 203–28. Dublin: Geography Society, 1995.

Mac Niocaill, Gearóid. *Na Burgéisí XII–XV Aois*. 2 vols. Dublin: Cló Morainn, 1964.

Maier, Charles S. "Transformations of Territoriality, 1600–2000." In *Transnationale Geschichte. Themen, Tendenzen und Theorien*, edited by Gunilla Budde, Sebastian Conrad and Oliver Janz, 32–55. Göttingen, 2006.

Maier, Charles S. *Once Within Borders: Territories of Power, Wealth, and Belonging since 1500*. Cambridge: Harvard University Press, 2016.

Mancall, P. C. and C. Shamma. *Governing the Sea in the Early Modern Era: Essays in Honor of Robert C. Ritchie*. California: Huntington Library, 2015.

Manickam, Sandra Khor. *Taming the Wild: Aborigines and Racial Knowledge in Colonial Malaya*. Copenhagen: NIAS Press, 2015.

Manzano, David. *El Imperio Español en Oceania*. Almuzara, 2020.

Masefield, John, ed. *Captain William Dampier: Dampier's Voyages*, 2 vols. Edinburgh and New York, 1906.

Matar, Nabil. *British Captives from the Mediterranean to the Atlantic, 1563–1760*. Leiden: Brill, 2014.

Matar, Nabil. "Introduction: England and Mediterranean Captivity, 1577–1704." In *Piracy, Slavery, and Redemption: Barbary Captivity Narratives from Early Modern England*, edited by Daniel J. Vitkus, 1–52. New York: Columbia University Press, 2001.

Mather, Cotton. *The Sad Effects of Sin: A True Relation of the Murder Committed by David Wallace, on his Companion Benjamin Stolwood*. Boston, 1713.

Matthew, E. "The Governing of the Lancastrian Lordship of Ireland in the Time of James Butler, Fourth Earl of Ormond." PhD Diss., Durham University, 1994.

Mawani, Renisa and Sebastian Prange. "Unruly Oceans: Law, Violence, and Sovereignty at Sea." *TWAILR: Third World Approaches to International Law Review*, March 12, 2021. https://twailr.com/unruly-oceans-law-violence-and-sovereignty-at-sea.

McDonald, Kevin P. *Pirates, Merchants, Settlers, and Slaves: Colonial America and the Indo-Atlantic World*. Oakland: University of California Press, 2015.

McDonald, Kevin P. "'Sailors from the Woods': Logwood Cutting and the Spectrum of Piracy." In *The Golden Age of Piracy: The Rise, Fall, and Enduring Popularity of Pirates*, edited by David Head, 50–72. Athens, GA: The University of Georgia Press, 2018.

McDonald, R. Andrew. *The Kingdom of the Isles: Scotland's Western Seaboard, c.1100–c.1336*. Edinburgh: East Linton, 1997.

McDonald, R. Andrew. *Manx Kingship in Its Irish Sea Setting, 1187–1229: King Rognvaldr and the Crovan Dynasty.* Dublin: Four Courts Press, 2007.

McErlean, Thomas, Rosemary McConkey, and Wes Forsythe. *Strangford Lough: An Archaeological Survey of the Maritime Cultural Landscape.* Belfast: Blackstaff Press, 2002.

McNamee, Colm. *The Wars of the Bruces: Scotland, England and Ireland, 1306–1328.* Edinburgh: East Linton, 2006.

Middleton, Alex. "Rajah Brooke and the Victorians." *The Historical Journal* 52, No. 2 (2010): 381–400.

Montero y Vidal, José. *Historia de la piratería malayo-mahometana en Mindanao, Jolo y Borneo. Comprende desde el descubremiento de dichas islas hasta junio de 1888.* 2 vols. Madrid: Imprente y fundación de Manuel Tello, 1888.

Müller, Martin. "Civilisation, Culture and Race in John Crawfurd's Discourses on Southeast Asia: Continuities and Changes, c.1814–1868." PhD Diss., European University Institute, 2013.

Müller, Martin. "Manufacturing Malayness: British Debates on the Malay Nation, Civilisation, Race and Language in the Early Nineteenth Century." *Indonesia and the Malay World* 42, No. 123 (2014): 170–96.

Müller, Martin. "Embodying Piracy, Textualizing the Piratical Body: Defining, Combating and Punishing Southeast Asian 'Piratical Communities' in British Discourse and Practice, c.1810–1860." In *Entre Traces Mémorielles et Marques Corporelles: Regards sur l'Ennemi de l'Antiquité à Nos Jours*, edited by Jean-Claude Caron, Laurent Lamoine and Natividad Planas, 201–21. Presses Universitaires Blaise Pascal, 2014.

Murphy, Elaine. *Ireland and the War at Sea, 1641–1653*. Woodbridge: Boydell and Brewer, 2012.

Murphy, Elaine. "Early Modern English Piracy and Privateering." In *The Routledge Companion to Marine and Maritime Worlds 1400–1800*, edited by Claire Jowitt, Craig Lambert, and Steve Mentz, 368–87. London and New York: Routledge, 2020.

Murphy, Martin N. "The Barbary Pirates," *Mediterranean Quarterly* 24, no. 4 (2013): 19–42.

Murray, Dian. "Living and Working Conditions in Chinese Pirate Communities, 1750–1850." In *Pirates and Privateers: New Perspectives on the War on Trade in the Eighteenth and Nineteenth Centuries*, edited by David J. Starkey, E. S. van Eyck van Heslinga, and J. A. de Moor, 47–68. Exeter: University of Exeter Press, 1997.

Naessens, Paul. "Gaelic Lords of the Sea: The Coastal Tower Houses of South Connemara." In *Lordship in Medieval Ireland: Image and Reality*, edited by Linda Doran and James Lyttleton, 217–35. Dublin: Four Courts Press, 2007.

Newbold, Thomas John. *Political and Statistical Account of the British Settlements of the Straits of Malacca, viz. Pinang, Malacca, and Singapore; with a History of the Malayan States on the Peninsula of Malacca*. 2 vols. London: John Murray, 1839.

Newton, Ross A. "'Good and Kind Benefactors': British Logwood Merchants and Boston's Christ Church." *Early American Studies* 11, no. 1: 15–36.

Ní Loinsigh, M. "An Assessment of Castles and Landownership in Late Medieval North Donegal." *Ulster Journal of Archaeology* 57 (1994): 145–58.

Nicholls, Kenneth. "Worlds Apart? The Ellis Two-Nation Theory on Late Medieval Ireland." *History Ireland* 7, no. 2 (March/April 1999), 22–26.

Nicholls, Kenneth. *Gaelic and Gaelicized Ireland in the Middle Ages*. Dublin: Lilliput, 2003.

Nicholson, Ranald. *Scotland: The Later Middle Ages*. Edinburgh: Oliver and Boyd, 1974.

Noor, Farish A. *The Long Shadow of the 19th Century: Critical Essays on Colonial Orientalism in Southeast Asia*. Petaling Jaya: Matahari Books, 2021.

O'Brien, A. F. "Politics, Economy, and Society: The Development of Cork and the Irish South-East Region, c.1170–c.1583." In *Cork: History and Society*, edited by Patrick O'Flanagan and Cornelius G. Buttimer, 82–155. Dublin: Geography Society, 1993.

Offen, Karl H. "British Logwood Extraction from the Mosquitia: The Origin of a Myth." *Hispanic American Historical Review* 80, no. 1 (February 2000): 113–35.

Ohlmeyer, J. "The Dunkirk of Ireland: Wexford Privateers During the 1640s." *Journal of the Wexford Historical Society* 12 (1988–89): 23–49.

O'Neill, Timothy. *Merchants and Mariners in Medieval Ireland*. Dublin: Irish Academic Press, 1987.

Oram, Richard. *The Lordship of Galloway*. Edinburgh: East Linton, 2000.

Orpen, Goddard H. *Ireland Under the Normans, 1169–1333*, edited by Séan Duffy, 2nd Edition. Dublin: Four Courts Press, 2005.

Osborn, Sherard. *Quedah; Or Stray Leaves from a Journal in Malayan Waters*. London: Longman, Brown, Green, Longmans, & Roberts, 1857.

Osterhammel, Jürgen. *The Transformation of the World: A Global History of the Nineteenth Century*. Princeton: Princeton University Press, 2014.

Ota, Atsushi. "'Pirates or Entrepreneurs?' The Migration and Trade of Sea People in Southwest Kalimantan, c.1770–1820." *Indonesia* 90 (October 2010): 67–95.

Ota, Atsushi. "Introduction." In *In the Name of the Battle against Piracy: Ideas and Practices in State Monopoly of Maritime Violence in Europe and Asia in the Period of Transition*, edited by Ota Atsushi, 1–15. Leiden: Brill, 2018.

Ota, Atsushi. "Trade, Piracy, and Sovereignty: Changing Perspectives of Piracy and Dutch Colonial State-Building in Malay Waters, c.1780–1830." In *In the Name of the Battle against Piracy: Ideas and Practices in State Monopoly of Maritime Violence in Europe and Asia in the Period of Transition*, edited by Ota Atsushi, 115–42. Leiden: Brill, 2018.

Panzac, Daniel. *Barbary Corsairs: The End of a Legend, 1800–1820*. Translated by Victoria Hobson and John E. Hawkes. Leiden: Brill, 2005.

Parmentier, Jan. *De holle compagnie. Smokkel en legale handel onder Zuidnederlandse vlag in Bengalen, ca. 1720–1744*. Hilversum: Verloren, 1992.

Parmentier, Jan. *Oostende & Co.* Gent: Ludion, 2002.

Patero, Santiago. *Sistema que conviene adopter para acabar con la piratería que los mahometanos de la Sultania de Joló ejercen en el Archipélago Filipino*. Madrid: Imprenta de Miguel Ginesta, 1872.

Pearson, Michael. "Littoral Society: The Concept and the Problems." *Journal of World History* 17, No. 4 (2006): 353–72.

Pearson, Michael. "Piracy in Asian Waters: Problems of Definition." In *Pirates, Ports, and Coasts in Asia: Historical and Contemporary Perspectives*, edited by John Kleinen and Manon Osseweijer, 15–28. Singapore: Institute of Southeast Asian Studies, 2010.

Peck, Linda Levy. *Consuming Splendor: Society and Culture in Seventeenth-Century England*. Cambridge: Cambridge University Press, 2005.

Pels, Peter and Oscar Salemink, eds. *Colonial Subjects: Essays on the Practical History of Anthropology*. Ann Arbor: University of Michigan Press, 2000.

Pennell, C. R., ed. *Bandits at Sea: A Pirates Reader*. New York: New York University Press, 2001.

Pestana, Carla Gardina. "Early English Jamaica without Pirates." *The William and Mary Quarterly* 71, no. 3 (2014): 321–60.

Pestana, Carla Gardina. *The English Conquest of Jamaica: Oliver Cromwell's Bid for Empire*. Cambridge, MA: Belknap, 2017.

Peterson, Mark. *The City-State of Boston: The Rise and Fall of an Atlantic Power, 1630–1865*. Princeton and Oxford: Princeton University Press, 2019.

Petre, J. "Domhnall Balloch, the Treaty of Ardtornish-Westminster and the MacDonald Raids of 1461–3." *Historical Research* 88, no. 242 (August 2015), 599–628.

Pitt, Steven J. J. "Building and Outfitting Ships in Colonial Boston." *Early American Studies: An Interdisciplinary Journal* 13, no. 4 (Fall 2015): 900–904.

Pitt, Steven J. J. "City upon the Atlantic Tides: Merchants, Pirates, and the Seafaring Community of Boston, 1689–1748." PhD Diss., University of Pittsburgh, 2015.

Pollard, A J. *Imagining Robin Hood: The Late-Medieval Stories in Historical Context*. Abingdon: Routledge, 2004.

Pollard, A J. "Political Ideology in the Early Stories of Robin Hood." In *Outlaws in Medieval and Early Modern England: Crime, Government and Society, c. 1066–c. 1600*, edited by John C. Appleby and Paul Dalton. Farnham: Ashgate, 2009.

Policante, Amadeo. *The Pirate Myth: Genealogies of an Imperial Concept*. London: Routledge, 2015.

Prange, Sebastian. "Asian Piracy." *Oxford Research Encyclopedia of Asian History* (September 2017). https://doi.org/10.1093/acrefore/9780190277727.013.32.

Rafferty, Matthew Taylor. "'The Law Is the Lord of the Sea': Maritime Law and Global Maritime History." In *A World at Sea: Maritime Practices and Global History*, edited by Lauren Benton and Nathan Perl-Rosenthal, 53–71. Philadelphia: University of Pennsylvania Press, 2020.

Raffles, Lady Sophia. *Memoir of the Life and Public Services of Sir Thomas Stamford Raffles, F.R.S. &c. Particularly in the Government of Java, 1811–1816, and of Bencoolen and Its Dependencies, 1817–1824; with Details of the Commerce and Resources of the Eastern Archipelago, and Selections from His Correspondence.* London: John Murray, 1830.

Raffles, Sir Thomas Stamford. "A Discourse." *Transactions of the Batavian Society of Arts and Sciences*, 8 (1816): 1–95.

Raffles, Sir Thomas. "Short Account of the Islands of Timor, Rotti, Ende, or Floris, and Sumba, or Sandal Wood Island, in the Indian Archipelago." *The Investigator; Or Quarterly Magazine* 4 (January and April 1822): 34–45.

Rankin, Hugh. *The Golden Age of Piracy*. New York: Holt, Rhinehart and Winston, 1969.

Rech, Walter. *Enemies of Mankind: Vattel's Theory of Collective Security*. Leiden: Martinus Nijhoff, 2013.

Rech, Walter. "Ambivalences of Recognition: The Position of the Barbary Corsairs in Early Modern International Law and International Politics." In *Piracy and Captivity in the Mediterranean*, edited by Mario Klarer, 76–98. London: Routledge, 2019.

Rediker, Marcus. "'Under the Banner of King Death': The Social World of Anglo-American Pirates, 1716 to 1726." *William and Mary Quarterly* 38, no. 2 (1981): 203–27.

Rediker, Marcus. *Between the Devil and the Deep Blue Sea: Merchant Seamen, Pirates, and the Anglo-American Maritime World, 1700–1750*. Cambridge: Cambridge University Press, 1993.

Rediker, Marcus. "Liberty Beneath the Jolly Roger." In *Iron Men and Wooden Women: Gender and Seafaring in the Atlantic World, 1700–1920*, edited by Margaret S. Creighton and Lisa Norling, 1–33. Baltimore: The Johns Hopkins University Press, 1996.

Rediker, Marcus. "Hydrarchy and Libertalia: The Utopian Dimensions of Atlantic Piracy in the Early Eighteenth Century." In *Pirates and Privateers: New Perspectives on the War on Trade in the Eighteenth and Nineteenth Centuries*, edited by David J. Starkey, E. S. Van Eyck van Heslinga, and J. A. Moor, 29–46. Exeter: University of Exeter Press, 1997.

Rediker, Marcus. "The Seaman as Pirate: Plunder and Social Banditry at Sea." In *Bandits at Sea: A Pirates Reader*, edited by C. R. Pennell, 139–68. New York: New York University Press, 2001.

Rediker, Marcus. *Villains of All Nations: Atlantic Pirates in the Golden Age.* Boston: Beacon Press, 2004.

Rediker, Marcus. *The Slave Ship: A Human History.* New York: Viking, 2007.

Rediker, Marcus. *Outlaws of the Atlantic: Sailors, Pirates, and Motley Crews in the Age of Sail.* Boston: Beacon Press, 2014.

Rennie, Neil. *Treasure Neverland: Real and Imaginary Pirates.* New York: Oxford, 2013.

Richetti, John J. *Popular Fiction Before Richardson: Narrative Patterns: 1700–1739*, 2nd ed. Oxford: Oxford University Press, 1992.

Risso, Patricia. "Cross-Cultural Perceptions of Piracy: Maritime Violence in the Western Indian Ocean and Persian Gulf Region during a Long Eighteenth Century." *Journal of World History* 12, no. 2 (2001): 293–319.

Ritchie, Robert C. *Captain Kidd and the War against the Pirates.* Cambridge, MA: Harvard, 1986.

Roberts, Priscilla H., and James N. Tull. "Moroccan Sultan Sidi Muhammad Ibn Abdallah's Diplomatic Initiatives Toward the United States, 1777–1786." *Proceedings of the American Philosophical Society* 143, no. 2 (1999): 233–65.

Rodger, Nicholas A. M. *The Wooden World: An Anatomy of the Georgian Navy.* London: Fontana Press, 1988.

Rowlands, Samuel. *More Knaves yet? The Knaves of Spades and Diamonds. With new Additions.* London: Printed by Edward Allde for John Tap, 1613.

Rubin, Alfred P. *The Law of Piracy.* Newport, RI: Naval War College Press, 1988.

Sather, Clifford. "Sulu's Political Jurisdiction over the Bajau Laut." *Borneo Research Bulletin* 3, No. 2 (December 1971): 58–62.

Satsuma, Shinsuke. "Plunder and Free Trade: British Privateering and Its Abolition in 1856 in Global Perspective." In *In the Name of the Battle against Piracy: Ideas and Practices in State Monopoly of Maritime Violence in Europe and Asia in the Period of Transition*, edited by Ota Atsushi, 43–65. Leiden: Brill, 2018.

Schmitt, Carl. *The Theory of the Partisan: A Commentary/Remark on the Concept of the Political.* Michigan State University Press, 2004 [1963].

Schmitt, Carl. "The Concept of Piracy (1937)." *Humanity: An International Journal of Human Rights, Humanitarianism, and Development* 2, No. 1 (Spring 2011): 27–29.

Schmitt, Carl. *Land and Sea: A World Historical Meditation.* Candor: Telos Press Publishing, 2015 [1942].

Scott, David. "Colonial Governmentality." *Social Text* 43 (Autumn 1995): 191–220.

Scott, James C. *The Art of Not Being Governed: An Anarchist History of Upland Southeast Asia.* New Haven: Yale University Press, 2009.

Sears, Christine. *American Slaves and African Masters: Algiers and the Western Sahara, 1776–1820.* New York: Palgrave Macmillan, 2012.

Sha'ban, Fuad. *Islam and Arabs in Early American Thought.* Durham, NC: Oxford University Press, 1991.

Sicking, Louis. "The Pirate and the Admiral: Europeanisation and Globalisation of Maritime Conflict Managements." *Journal of the History of International Law,* 20 (2018): 429–47.

Simms, Katharine. "Niall Garbh O'Donnell, King of Tír Conaill, 1422–39." *Donegal Annual* 12 (1977): 7–21.

Simms, Katharine. "The 'King's Friend': O'Neill, the Crown, and the Earldom of Ulster." In *England and Ireland in the Later Middle Ages*, edited by James F. Lydon, 214–36. Dublin: Irish Academic Press, 1981.

Simms, Katharine. "Late Medieval Donegal." In *Donegal: History and Society*, edited by William Nolan, Liam Ronayne, Mairead Dunleavy, 183–202. Dublin: Geography Society, 1995.

Simms, Katharine. "Late Medieval Tír Eoghain: The Kingdom of the Great O'Neill." In *Tyrone: History and Society*, edited by Charles Dillon and Henry A. Jefferies, 127–62. Dublin: Geography Society, 2000.

Simms, Katharine. "The Ulster Revolt 1404—An Anti-Lancastrian Dimension?" In *Ireland and the English World in the Late Middle Ages*, edited by Brendan Smith, 141–60. Basingstoke: Palgrave MacMillan, 2009.

Singleton, Robert R. "English Criminal Biography, 1651–1722." *Harvard Library Bulletin* 18, no. 1 (1970): 63–83.

Singman, Jeffrey. *Robin Hood: The Shaping of the Legend*. Westport: Greenwood Press, 1998.

Skeat, Walter W. and Henry N. Ridley. "The Orang Laut of Singapore." *Journal of the Straits Branch of the Royal Asiatic Society* 33 (January 1900): 247–50.

Smith, Brendan. "Late Medieval Ireland and the English Connection: Waterford and Bristol, ca.1360–1460." *Journal of British Studies* 50, no. 3 (July 2011): 546–65.

Smith, J. Beverly. "The Last Phase of the Glyn Dŵr Rebellion." *Bulletin of the Board of Celtic Studies* 12 (1966–67): 250–60.

Sofka, James R. "The Jeffersonian Idea of National Security: Commerce, the Atlantic Balance of Power, and the Barbary War, 1786–1805." *Diplomatic History* 21, no. 4 (1997): 519–44.

Sopher, David, E. *The Sea Nomads: A Study of the Maritime Boatpeople of Southeast Asia*. Singapore: National Museum Singapore, 1977.

Starkey, David J. "Pirates and Markets." In *Bandits at Sea: A Pirates Reader*, edited by C. R. Pennell, 107–24. New York: New York University Press, 2001.

Starkey, David J. "The Origins and Regulations of Eighteenth-Century British Privateering." In *Bandits at Sea: A Pirate Reader*, edited by C.R. Pennell, 69–81. London: New York University Press, 2007.

Starkey, David J., E. S. Van Eyck Van Heslinga, and J. A. De Moor, eds. *Pirates and Privateers: New Perspectives on the War on Trade in the Eighteenth and Nineteenth Centuries*. Exeter: University of Exeter Press, 1997.

Starkey, David J. and Matthew McCarthy. "A Persistent Phenomenon: Private Prize-Taking in the British Atlantic World, c.1540–1856." In *Persistent Piracy: Maritime Violence and State Formation in Global Historical Perspective,* edited by Stefan Amirell and Leos Müller, 131–51. London: Palgrave Macmillan, 2014.

Stein, Stanley J. and Barbara H. Stein. *Silver, Trade, and War: Spain and America in the Making of Early Modern Europe.* Baltimore: Johns Hopkins University Press, 2000.

Steinberg, Philip E. *The Social Construction of the Ocean.* Cambridge: Cambridge University Press, 2001.

Steinberg, Philip E. "Of Other Seas: Metaphors and Materialities in Maritime Regions." *The Atlantic Studies* 10, No. 2 (2013): 156–69.

Stewart, Thomas C. "The Return of the Barbary Pirates." *The Washington Times*, March 10, 2015. http://www.washingtontimes.com/news/2015/mar/10/thomas-stewart-isis-may-herald-return-of-the-barba.

Stringer, Keith J. "Periphery and Core in Thirteenth-Century Scotland: Alan son of Roland, Lord of Galloway and Constable of Scotland." In *Medieval Scotland: Crown, Lordship and Community*, edited by Alexander Grant and Keith J. Stringer, 82–113. Edinburgh: Edinburgh University Press, 1998.

Sumption, Jonathan. *Divided Houses: The Hundred Years War.* London: Faber and Faber, 2012.

Tabarrok, Alexander. "The Rise, Fall, and Rise Again of Privateers." *The Independent Review* 11, no. 4 (2007): 565–77.

Tagliacozzo, Eric. "A Necklace of Fins: Marine Goods Trading in Maritime Southeast Asia, 1780–1860." *International Journal of Asian Studies* 1, No. 1 (2004): 23–48.

Tarling, Nicholas. *Piracy and Politics in the Malay World: A Study of British Imperialism in Nineteenth-Century South-East Asia.* Nendeln: Kraus Reprint, 1978 [1963].

Teitler, Ger, A. M. C. van Dissel and J. N. F. M. á Campo. *Zeeroof en Zeeroofbestrijding in de Indische archipel (19de eeuw).* Amsterdam: De Bataafsche Leeuw, 2005.

Templer, John C. (ed.). *The Private Letters of Sir James Brooke, K. C. B., Rajah of Sarawak: Narrating the Events from 1838 to the Present Time.* 3 vols. London: Richard Bentley, 1853.

Thomson, Janice E. *Mercenaries, Pirates, and Sovereigns: State-Building and Extraterritorial Violence in Early Modern Europe.* Princeton: Princeton University Press, 1994.

Thomson, John Turnbull. "Description of the Eastern Coast of Johore and Pahang, with Adjacent Islands." *The Journal of the Indian Archipelago and Eastern Asia* 5 (1851): 85–92 and 135–54.

Thornton, A. P. "The English at Campeachy." *The Jamaican Historical Review* 2 (1953): 27–38.

Tuck, Anthony. *Richard II and the English Nobility.* London: Edward Arnold, 1973.

Uring, Nathaniel. *A History of the Voyages and Travels of Captain Nathaniel Uring.* London, 1726.

Verhees-Van Meer, Johanna Theresia. *De Zeeuwse Kaapvaart tijdens de Spaanse Successieoorlog 1702–1713.* Middelburg: Het Koninklijk Zeeuwsch Genootschap der Wetenschappen deel 3, 1986.

Verschuer, F. H. van. "De Badjo's." *Tijdschift van het Aardrijkskundig Genootshap* Zevende deel (7th part) 1883: 1–7.

Wagner, Kim A. *Thuggee: Banditry and the British in Early Nineteenth-Century India.* Delhi: Primus Books, 2014.

Warren, James F. *Iraun and Balangini: Globalization, Maritime Raiding and the Birth of Ethnicity.* Singapore: Singapore University Press, 2002.

Warren, James F. *The Sulu Zone 1768–1898: The Dynamics of External Trade, Slavery, and Ethnicity in the Transformation of a Southeast Asian Maritime State.* Singapore: NUS Press, 2007 [1981].

Warren, James F. "The Global Economy and the Sulu Zone: Connections, Commodities and Culture." *Crossroads* 3 (2011): 53–124.

Watkins, Josh. "Spatial Imaginaries Research in Geography: Synergies, Tensions, and New Directions." *Geography Compass* 9, no. 9 (2015): 508–22.

Whatley, A. T. "Historical Sketch of the Law of Piracy." *The Law Magazine and Review* New Series 3 (1874): 536–55 and 618–39.

White, Dean G. "The Reign of Edward VI in Ireland: Some Political, Social, and Economic Aspects." *Irish Historical Studies* 14, no. 55 (1965): 197–211.

Whitson, Agnes. *The Constitutional Development of Jamaica, 1660–1729.* Manchester, 1929.

Whitworth, Sir Charles. *State of the Trade of Great Britain in its Imports and Exports, Progressively from the Year 1697.* London, 1776.

Wigen, Kären. "Introduction." In *Seascapes: Maritime Histories, Littoral Cultures, and Transoceanic Exchanges*, edited by Jerry H. Bentley, Renate Bridenthal, and Kären Wigen, 1–18. Honolulu: University of Hawai'i Press, 2007.

Williams, Glyndwr. *The Great South Sea: English Voyages and Encounters 1570–1750.* London: Yale University Press, 1997.

Wilson, Arthur M. "The Logwood Trade in the Seventeenth and Eighteenth Centuries." In *Essays in the History of Modern Europe*, edited by Douglas McKay, 1–15. New York: Harper & Brothers, 1936.

Wilson, C. R., ed. *Indian Records Series. The Records of Old Fort William in Bengal, vol. 1/ Extract from Bengal Public Consultations, Fort William, October 13 1744, Range 1, Vol. XVII.* London: John Murray, 1906.

Wilson, David. "From the Caribbean to Craignish: Imperial Authority and Piratical Voyages in the Early Eighteenth-Century Atlantic Commons." *Itinerario* 42, no. 3 (2018): 430–60.

Wilson, David. "European Colonisation, Law, and Indigenous Marine Dispossession: Historical Perspectives on the Construction and Entrenchment of Unequal Marine Governance." *Maritime Studies*, 20 (2021): 387–407.

Wilson, David. *Suppressing Piracy in the Early Eighteenth Century: Pirates, Merchants and British Imperial Authority in the Atlantic and Indian Oceans*. Woodbridge: Boydell Press, 2021.

Woodard, Colin. *The Republic of Pirates*. New York: Harcourt, 2007.

Zahedieh, Nuala. "The Merchants of Port Royal, Jamaica, and the Spanish Contraband Trade, 1655–1692." *The William and Mary Quarterly* 43, no. 4 (1986): 570–93.

Zahedieh, Nuala. "Trade, Plunder, and Economic Development in Early English Jamaica, 1655–89." *Economic History Review* 39, no. 2 (1986): 205–22.

Zahedieh, Nuala. "Modyford, Sir Thomas, First Baronet (c. 1620–1679), Planter and Colonial Governor." *Oxford Dictionary of National Biography*, 2004. Accessed February 4, 2022.

# Index